THE LOGIC OF SCIENTIFIC DISCOVERY

KARL R. POPPER

Martino Publishing
Mansfield Centre, CT
2014

Martino Publishing
P.O. Box 373,
Mansfield Centre, CT 06250 USA

ISBN 978-161427-7-439

© *2014 Martino Publishing*

Cover design by T. Matarazzo

Printed in the United States of America On 100% Acid-Free Paper

THE LOGIC OF SCIENTIFIC DISCOVERY

KARL R. POPPER

BASIC BOOKS, INC.
NEW YORK

Printed in the United States of America

Library of Congress Catalog Card Number 59-8371

TO MY WIFE
who is responsible for the revival of this book

TRANSLATORS' NOTE

The Logic of Scientific Discovery is a translation of *Logik der Forschung*, published in Vienna in the autumn of 1934 (with the imprint '1935'). The translation was prepared by the author, with the assistance of Dr. Julius Freed and Lan Freed.

The original text of 1934 has been left unchanged for the purpose of the translation. As usual, the translation is a little longer than the original. Words and phrases for which no equivalent exists had to be paraphrased. Sentences had to be broken up and rearranged—the more so as the text to be translated was highly condensed: it had been several times drastically cut, to comply with the publisher's requirements. Yet the author decided against augmenting the text, and also against restoring cut passages.

In order to bring the book up to date, the author has added new appendices and new footnotes. Some of these merely expand the text, or correct it; but others explain where the author has changed his mind, or how he would now reframe his arguments.

All new additions—new appendices and new footnotes—are marked by starred numbers; and where old footnotes have been expanded, the expansion is also marked by a star (unless it consists only of a reference to the English edition of a book originally quoted from a German edition).

In these new starred additions, references will be found to a sequel to this volume, entitled *Postscript: After Twenty Years* (not previously published). Its chapters and sections are also preceded by starred numbers. (Since it has no appendices, all references to appendices, whether starred or not, refer to the present volume.) The two volumes treat of the same problems. Though they complement each other, they are independent.

It should also be mentioned that the numbering of the chapters of the present volume has been changed. In the original, they were numbered i to ii (part i), and i to viii (part ii). They are now numbered through from i to x.

CONTENTS

PART I

Introduction to the Logic of Science

PART II

Some Structural Components of a Theory of Experience

CONTENTS

CONTENTS

APPENDICES

NEW APPENDICES

Theories are nets: only he who casts will catch.

NOVALIS

PREFACE TO THE FIRST EDITION, 1934

The hint that man has, after all, solved his most stubborn problems . . .
is small solace to the philosophic connoisseur; for what he cannot help
fearing is that philosophy will never get so far as to pose a genuine
problem.   M. SCHLICK (1930).

I for my part hold the very opposite opinion, and I assert that whenever
a dispute has raged for any length of time, especially in philosophy, there
was, at the bottom of it, never a problem about mere words, but always
a genuine problem about things.   I. KANT (1786).

A scientist engaged in a piece of research, say in physics, can
attack his problem straight away. He can go at once to the heart of
the matter: to the heart, that is, of an organized structure. For a
structure of scientific doctrines is already in existence; and with it,
a generally accepted problem-situation. This is why he may leave it
to others to fit his contribution into the framework of scientific
knowledge.

The philosopher finds himself in a different position. He does not
face an organized structure, but rather something resembling a heap
of ruins (though perhaps with treasure buried underneath). He cannot
appeal to the fact that there is a generally accepted problem-situation;
for that there is no such thing is perhaps the one fact which is generally
accepted. Indeed it has by now become a recurrent question in
philosophical circles whether philosophy will ever get so far as to
pose a genuine problem.

Nevertheless there are still some who do believe that philosophy
can pose genuine problems about things, and who therefore still hope
to get these problems discussed, and to have done with those depressing
monologues which now pass for philosophical discussions. And if by
chance they find themselves unable to accept any of the existing
creeds, all they can do is to begin afresh from the beginning.

VIENNA, *Autumn 1934.*

PREFACE TO THE ENGLISH EDITION, 1958

In my old preface of 1934 I tried to explain—too briefly, I am afraid—my attitude towards the then prevailing situation in philosophy, and especially towards linguistic philosophy and the school of language analysts of those days. In this new preface I intend to explain my attitude towards the present situation, and towards the two main schools of language analysts of today. Now as then, language analysts are important to me; not only as opponents, but also as allies, in so far as they seem to be almost the only philosophers left who keep alive some of the traditions of rational philosophy.

Language analysts believe that there are no genuine philosophical problems, or that the problems of philosophy, if any, are problems of linguistic usage, or of the meaning of words. I, however, believe that there is at least one philosophical problem in which all thinking men are interested. It is the problem of cosmology: *the problem of understanding the world—including ourselves, and our knowledge, as part of the world.* All science is cosmology, I believe, and for me the interest of philosophy as well as of science lies solely in the contributions which they have made to it. For me, at any rate, both philosophy and science would lose all their attraction if they were to give up that pursuit. Admittedly, understanding the functions of language is an important part of it; but explaining away our problems as merely linguistic 'puzzles' is not.

Language analysts regard themselves as practitioners of a method peculiar to philosophy. I think they are wrong, for I believe in the following thesis.

Philosophers are as free as others to use any method in searching for truth. *There is no method peculiar to philosophy.*

A second thesis which I should like to propound here is this.

The central problem of epistemology has always been and still is the problem of the growth of knowledge. *And the growth of knowledge can be studied best by studying the growth of scientific knowledge.*

15

I do not think that the study of the growth of knowledge can be replaced by the study of linguistic usages, or of language systems.

And yet, I am quite ready to admit that there is a method which might be described as 'the one method of philosophy'. But it is not characteristic of philosophy alone; it is, rather, the one method of all *rational discussion*, and therefore of the natural sciences as well as of philosophy. The method I have in mind is that of stating one's problem clearly and of examining its various proposed solutions *critically*.

I have italicized the words *'rational discussion'* and *'critically'* in order to stress that I equate the rational attitude and the critical attitude. The point is that whenever we try to propose a solution to a problem, we ought to try as hard as we can to overthrow our solution, rather than defend it. Few of us, unfortunately, practise this precept; but other people, fortunately, will supply the criticism for us if we fail to supply it ourselves. Yet criticism will be fruitful only if we state our problem as clearly as we can and put our solution in a sufficiently definite form—a form in which it can be critically discussed.

I do not deny that something which may be called 'logical analysis' can play a role in this process of clarifying and scrutinizing our problems and our proposed solutions; and I do not assert that the methods of 'logical analysis or 'language analysis' are necessarily useless. My thesis is, rather, that these methods are far from being the only ones which a philosopher can use with advantage, and that they are in no way characteristic of philosophy. They are no more characteristic of philosophy than of any other scientific or rational inquiry.

It may perhaps be asked what other 'methods' a philosopher might use. My answer is that though there are any number of different 'methods', I am really not interested in enumerating them. I do not care what methods a philosopher (or anybody else) may use so long as he has an interesting problem, and so long as he is sincerely trying to solve it.

Among the many methods which he may use—always depending, of course, on the problem in hand—one method seems to me worth mentioning. It is a variant of the (at present unfashionable) historical method. It consists, simply, in trying to find out what other people have thought and said about the problem in hand: why they had to face it: how they formulated it: how they tried to solve it. This seems to me important because it is part of the general method of rational discussion. If we ignore what other people are thinking, or have

16

thought in the past, then rational discussion must come to an end, though each of us may go on happily talking to himself. Some philosophers have made a virtue of talking to themselves; perhaps because they felt that there was nobody else worth talking to. I fear that the practice of philosophizing on this somewhat exalted plane may be a symptom of the decline of rational discussion. No doubt God talks mainly to Himself because He has no-one worth talking to. But philosophers should know that they are no more godlike than other men.

There are several interesting historical reasons for the widespread belief that what is called 'linguistic analysis' is the true method of philosophy.

One such reason is the correct belief that *logical paradoxes*, like that of the liar ('I am now lying') or those found by Russell, Richard, and others, need the method of linguistic analysis for their solution, with its famous distinction between meaningful (or 'well-formed') and meaningless linguistic expressions. This correct belief is then combined with the mistaken belief that the traditional problems of philosophy arise from the attempt to solve *philosophical paradoxes* whose structure is analogous to that of *logical paradoxes*, so that the distinction between meaningful and meaningless talk must be of central importance for philosophy also. That this belief is mistaken can be shown very easily. It can be shown, in fact, by logical analysis. For this reveals that a certain characteristic kind of reflexivity or self-reference which is present in all logical paradoxes is absent from all the so-called philosophical paradoxes—even from Kant's antinomies.

The main reason for exalting the method of linguistic analysis, however, seems to have been the following. It was felt that the so-called '*new way of ideas*' of Locke, Berkeley, and Hume, that is to say the psychological or rather pseudo-psychological method of analysing our ideas and their origin in our senses, should be replaced by a more 'objective' and a less genetic method. It was felt that we should analyse words and their meanings or usages rather than 'ideas' or 'conceptions' or 'notions'; that we should analyse propositions or statements or sentences rather than 'thoughts' or 'beliefs' or 'judgments'. I readily admit that this replacement of Locke's 'new way of ideas' by a 'new way of words' was an advance, and one that was urgently needed.

It is understandable that those who once saw in the 'new way of ideas' the one true method of philosophy may thus have turned to the belief that the 'new way of words' is the one true method of philosophy. From this challenging belief I strongly dissent. But I will make only two critical comments on it. First, the 'new way of ideas' should never have been taken for the main method of philosophy, let alone for its one true method. Even Locke introduced it merely as a method of dealing with certain preliminaries (preliminaries for a science of ethics); and it was used by both Berkeley and Hume chiefly as a weapon for harrying their opponents. Their own interpretation of the world—the world of things and of men—which they were anxious to impart to us was never based upon this method. Berkeley did not base his religious views on it, nor Hume his determinism, or his political theories.

But my gravest objection to the belief that either the 'new way of ideas' or the 'new way of words' is the main method of epistemology —or perhaps even of philosophy—is this.

The problem of epistemology may be approached from two sides: (1) as the problem of ordinary or *common-sense knowledge*, or (2) as the problem of *scientific knowledge*. Those philosophers who favour the first approach think, rightly, that scientific knowledge can only be an extension of common-sense knowledge, and they also think, wrongly, that common-sense knowledge is the easier of the two to analyse. In this way these philosophers come to replace the 'new way of ideas' by an analysis of *ordinary language*—the language in which common-sense knowledge is formulated. They replace the analysis of vision or perception or knowledge by the analysis of the phrases 'I see' or 'I perceive', or 'I know', 'I believe', 'I hold that it is probable'; or perhaps by that of the word 'perhaps'.

Now to those who favour this approach to the theory of knowledge I should reply as follows. Although I agree that scientific knowledge is merely a development of ordinary knowledge or common-sense knowledge, I contend that the most important and most exciting problems of epistemology must remain completely invisible to those who confine themselves to analysing ordinary or common-sense knowledge or its formulation in ordinary language.

I wish to refer here only to one example of the kind of problem I have in mind: the problem of the *growth* of our knowledge. A little reflection will show that most problems connected with the growth of our knowledge must necessarily transcend any study which is confined to common-sense knowledge as opposed to scientific knowledge.

18

For the most important way in which common-sense knowledge grows is, precisely, by turning into scientific knowledge. Moreover, it seems clear that the growth of scientific knowledge is the most important and interesting case of the growth of knowledge.

It should be remembered, in this context, that almost all the problems of traditional epistemology are connected with the problem of the growth of knowledge. I am inclined to say even more: from Plato to Descartes, Leibniz, Kant, Duhem and Poincare; and from Bacon, Hobbes, and Locke, to Hume, Mill, and Russell, the theory of knowledge was inspired by the hope that it would enable us not only to know more about knowledge, but also to contribute to the advance of knowledge—of scientific knowledge, that is. (The only possible exception to this rule among the great philosophers I can think of is Berkeley.) Most of the philosophers who believe that the characteristic method of philosophy is the analysis of ordinary language seem to have lost this admirable optimism which once inspired the rationalist tradition. Their attitude, it seems, has become one of resignation, if not despair. They not only leave the advancement of knowledge to the scientists: they even define philosophy in such a way that it becomes, by definition, incapable of making any contribution to our knowledge of the world. The self-mutilation which this so surprisingly persuasive definition requires does not appeal to me. There is no such thing as an essence of philosophy, to be distilled and condensed into a definition. A definition of the word 'philosophy' can only have the character of a convention, of an agreement; and I, at any rate, see no merit in the arbitrary proposal to define the word 'philosophy' in a way which may well prevent a student of philosophy from trying to contribute, *qua* philosopher, to the advancement of our knowledge of the world.

Also, it seems to me paradoxical that philosophers who take pride in specializing in the study of ordinary language nevertheless believe that they know enough about cosmology to be sure that it is in essence so different from philosophy that philosophy cannot make any contribution to it. And indeed they are mistaken. For it is a fact that purely metaphysical ideas—and therefore philosophical ideas—have been of the greatest importance for cosmology. From Thales to Einstein, from ancient atomism to Descartes's speculation about matter, from the speculations of Gilbert and Newton and Leibniz and Boscovic about forces to those of Faraday and Einstein about fields of forces, metaphysical ideas have shown the way.

Such are, in brief, my reasons for believing that even within the province of epistemology, the first approach mentioned above—that is to say, the analysis of knowledge by way of an analysis of ordinary language—is too narrow, and that it is bound to miss the most interesting problems.

Yet I am far from agreeing with all those philosophers who favour that other approach to epistemology—the approach by way of an analysis of scientific knowledge. In order to explain more easily where I disagree and where I agree, I am going to sub-divide the philosophers who adopt this second approach into two groups—the goats and the sheep, as it were.

The first group consists of those whose aim is to study 'the language of science', and whose chosen philosophical method is the construction of artificial model languages; that is to say, the construction of what they believe to be models of 'the language of science'.

The second group does not confine itself to the study of the language of science, or any other language, and it has no such chosen philosophical method. Its members philosophize in many different ways, because they have many different problems which they want to solve; and any method is welcome to them if they think that it may help them to see their problems more clearly, or to hit upon a solution, however tentative.

I turn first to those whose chosen method is the construction of artificial models of the language of science. Historically, they too take their departure from the 'new way of ideas'. They too replace the (pseudo-) psychological method of the old ' new way' by linguistic analysis. But perhaps owing to the spiritual consolations offered by the hope for knowledge that is 'exact' or 'precise' or 'formalized', the chosen object of their linguistic analysis is 'the language of *science*' rather than ordinary language. Yet unfortunately there seems to be no such thing as 'the language of science'. It therefore becomes necessary for them to construct one. However, the construction of a full-scale working model of a language of science—one in which we could operate a real science such as physics—turns out a little difficult in practice; and for this reason we find them engaged in the construction of intricate working models in miniature—of vast systems of minute gadgets.

In my opinion, this group of philosophers gets the worst of both

20

worlds. By their method of constructing miniature model languages they miss the most exciting problems of the theory of knowledge—those connected with its advancement. For the intricacy of the outfit bears no relation to its effectiveness, and practically no scientific theory of any interest can be expressed in these vast systems of minutiae. These model languages have no bearing on either science or common sense.

Indeed, the models of 'the language of science' which these philosophers construct have nothing to do with the language of modern science. This may be seen from the following remarks which apply to the three most widely known model languages. (They are referred to in notes 13 and 15 to appendix *vii, and in note *2 to section 38.) The first of these model languages lacks even the means of expressing identity. As a consequence, it cannot express an equation: it does not contain even the most primitive arithmetic. The second model language works only as long as we do not add to it the means of proving the usual theorems of arithmetic—for example, Euclid's theorem that there is no greatest prime number, or even the principle that every number has a successor. In the third model language—the most elaborate and famous of all—mathematics can again not be formulated; and what is still more interesting, there are no measurable properties expressible in it. For these reasons, and for many others, the three model languages are too poor to be of use to any science. They are also, of course, essentially poorer than ordinary languages, including even the most primitive ones.

The limitations mentioned were imposed upon the model languages simply because otherwise the solutions offered by the authors to their problems would not have worked. This fact can be easily proved, and it has been partly proved by the authors themselves. Nevertheless, they all seem to claim two things: (a) that their methods are, in some sense or other, capable of solving problems of the theory of scientific knowledge, or in other words, that they are applicable to science (while in fact they are applicable with any precision only to discourse of an extremely primitive kind), and (b) that their methods are 'exact' or 'precise'. Clearly these two claims cannot both be upheld.

Thus the method of constructing artificial model languages is incapable of tackling the problems of the growth of our knowledge; and it is even less able to do so than the method of analysing ordinary languages, simply because these model languages are poorer than ordinary languages. It is a result of their poverty that they yield only the most crude and the most misleading model of the growth of knowledge—the model of an accumulating heap of observation statements.

I now turn to the last group of epistemologists—those who do not pledge themselves in advance to any philosophical method, and who make use, in epistemology, of the analysis of scientific problems, theories, and procedures, and, most important, of scientific discussions. This group can claim, among its ancestors, almost all the great philosophers of the West. (It can claim even the ancestry of Berkeley despite the fact that he was, in an important sense, an enemy of the very idea of rational scientific knowledge, and that he feared its advance.) Its most important representatives during the last two hundred years were Kant, Whewell, Mill, Peirce, Duhem, Poincare, Meyerson, Russell, and—at least in some of his phases—Whitehead. Most of those who belong to this group would agree that scientific knowledge is the result of the growth of common-sense knowledge. But all of them discovered that scientific knowledge can be more easily studied than common-sense knowledge. For it is *common-sense knowledge writ large*, as it were. Its very problems are enlargements of the problems of common-sense knowledge. For example, it replaces the Humean problem of 'reasonable belief' by the problem of the reasons for accepting or rejecting scientific theories. And since we possess many detailed reports of the discussions pertaining to the problem whether a theory such as Newton's or Maxwell's or Einstein's should be accepted or rejected, we may look at these discussions as if through a microscope that allows us to study in detail, and objectively, some of the more important problems of 'reasonable belief'.

This approach to the problems of epistemology gets rid (as do the other two mentioned) of the pseudo-psychological or 'subjective' method of the new way of ideas (a method still used by Kant). But it also allows us to analyse scientific problem-situations and scientific discussions. And it can help us to understand the history of scientific thought.

I have tried to show that the most important of the traditional problems of epistemology—those connected with the *growth of knowledge*—transcend the two standard methods of linguistic analysis and require the analysis of scientific knowledge. But the last thing I wish to do is to advocate yet another dogma. Even the analysis of science—the 'philosophy of science'—is threatening to become a fashion, a specialism. Yet philosophers should not be specialists. For myself, I am interested in science and in philosophy only because I want to learn something about the riddle of the world in which we live, and the riddle of man's knowledge of that world. And I believe that only a revival of interest in these riddles can save the sciences and philosophy from narrow specialization and from an obscurantist faith in the expert's special skill and in his personal knowledge and authority; a faith that so well fits our 'post-rationalist' and 'post-critical' age, proudly dedicated to the destruction of the tradition of rational philosophy, and of rational thought itself.

PENN, BUCKINGHAMSHIRE, *Spring 1958.*

PART I

INTRODUCTION TO THE LOGIC OF SCIENCE

CHAPTER I

A SURVEY OF SOME FUNDAMENTAL PROBLEMS

A SCIENTIST, whether theorist or experimenter, puts forward statements, or systems of statements, and tests them step by step. In the field of the empirical sciences, more particularly, he constructs hypotheses, or systems of theories, and tests them against experience by observation and experiment.

I suggest that it is the task of the logic of scientific discovery, or the logic of knowledge, to give a logical analysis of this procedure; that is, to analyse the method of the empirical sciences.

But what are these 'methods of the empirical sciences'? And what do we call 'empirical science'?

1. *The Problem of Induction.*

According to a widely accepted view—to be opposed in this book —the empirical sciences can be characterized by the fact that they use *'inductive methods'*, as they are called. According to this view, the logic of scientific discovery would be identical with inductive logic, *i.e.* with the logical analysis of these inductive methods.

It is usual to call an inference 'inductive' if it passes from *singular statements* (sometimes also called 'particular' statements), such as accounts of the results of observations or experiments, to *universal statements*, such as hypotheses or theories.

Now it is far from obvious, from a logical point of view, that we are justified in inferring universal statements from singular ones, no matter how numerous; for any conclusion drawn in this way may always turn out to be false: no matter how many instances of white swans we may have observed, this does not justify the conclusion that *all* swans are white.

The question whether inductive inferences are justified, or under what conditions, is known as *the problem of induction*.

The problem of induction may also be formulated as the question of how to establish the truth of universal statements which are based on experience, such as the hypotheses and theoretical systems of the empirical sciences. For many people believe that the truth of these universal statements is '*known by experience*'; yet it is clear that an account of an experience—of an observation or the result of an experiment—can in the first place be only a singular statement and not a universal one. Accordingly, people who say of a universal statement that we know its truth from experience usually mean that the truth of this universal statement can somehow be reduced to the truth of singular ones, and that these singular ones are known by experience to be true; which amounts to saying that the universal statement is based on inductive inference. Thus to ask whether there are natural laws known to be true appears to be only another way of asking whether inductive inferences are logically justified.

Yet if we want to find a way of justifying inductive inferences, we must first of all try to establish a *principle of induction*. A principle of induction would be a statement with the help of which we could put inductive inferences into a logically acceptable form. In the eyes of the upholders of inductive logic, a principle of induction is of supreme importance for scientific method: '. . . this principle', says Reichenbach, 'determines the truth of scientific theories. To eliminate it from science would mean nothing less than to deprive science of the power to decide the truth or falsity of its theories. Without it, clearly, science would no longer have the right to distinguish its theories from the fanciful and arbitrary creations of the poet's mind.'[1]

Now this principle of induction cannot be a purely logical truth like a tautology or an analytic statement. Indeed, if there were such a thing as a purely logical principle of induction, there would be no problem of induction; for in this case, all inductive inferences would have to be regarded as purely logical or tautological transformations, just like inferences in deductive logic. Thus the principle of induction must be a synthetic statement; that is, a statement whose negation is not self-contradictory but logically possible. So the question arises why such a principle should be accepted at all, and how we can justify its acceptance on rational grounds.

[1] H. Reichenbach, *Erkenntnis* I, 1930, p. 186 (*cf.* also p. 64 f.).

Some who believe in inductive logic are anxious to point out, with Reichenbach, that 'the principle of induction is unreservedly accepted by the whole of science and that no man can seriously doubt this principle in everyday life either'.[2] Yet even supposing this were the case—for after all, 'the whole of science' might err—I should still contend that a principle of induction is superfluous, and that it must lead to logical inconsistencies.

That inconsistencies may easily arise in connection with the principle of induction should have been clear from the work of Hume;[*1] also, that they can be avoided, if at all, only with difficulty. For the principle of induction must be a universal statement in its turn. Thus if we try to regard its truth as known from experience, then the very same problems which occasioned its introduction will arise all over again. To justify it, we should have to employ inductive inferences; and to justify these we should have to assume an inductive principle of a higher order; and so on. Thus the attempt to base the principle of induction on experience breaks down, since it must lead to an infinite regress.

Kant tried to force his way out of this difficulty by taking the principle of induction (which he formulated as the 'principle of universal causation') to be 'a priori valid'. But I do not think that his ingenious attempt to provide an a priori justification for synthetic statements was successful.

My own view is that the various difficulties of inductive logic here sketched are insurmountable. So also, I fear, are those inherent in the doctrine, so widely current today, that inductive inference, although not 'strictly valid', *can attain some degree of 'reliability' or of 'probability'*. According to this doctrine, inductive inferences are 'probable inferences'.[3] 'We have described', says Reichenbach, 'the principle of induction as the means whereby science decides upon truth. To be more exact, we should say that it serves to decide upon probability. For it is not given to science to reach either truth or falsity . . . but scientific statements can only attain continuous degrees of

---

[2] Reichenbach *ibid.*, p. 67.

[*1] The decisive passages from Hume are quoted in appendix *vii, text to footnotes 4, 5, and 6; see also note 2 to section 81, below.

[3] Cf. J. M. Keynes, *A Treatise on Probability* (1921); O. Külpe, *Vorlesungen über Logic* (ed. by Selz, 1923); Reichenbach (who uses the term 'probability implications'), *Axiomatik der Wahrscheinlichkeitrechnung, Mathem. Zeitschr.* **34** (1932); and in many other places.

probability whose unattainable upper and lower limits are truth and falsity'.[4]

At this stage I can disregard the fact that the believers in inductive logic entertain an idea of probability that I shall later reject as highly unsuitable for their own purposes (see section 80, below). I can do so because the difficulties mentioned are not even touched by an appeal to probability. For if a certain degree of probability is to be assigned to statements based on inductive inference, then this will have to be justified by invoking a new principle of induction, appropriately modified. And this new principle in its turn will have to be justified, and so on. Nothing is gained, moreover, if the principle of induction, in its turn, is taken not as 'true' but only as 'probable'. In short, like every other form of inductive logic, the logic of probable inference, or 'probability logic', leads either to an infinite regress, or to the doctrine of *apriorism*.[*2]

The theory to be developed in the following pages stands directly opposed to all attempts to operate with the ideas of inductive logic. It might be described as the theory of *the deductive method of testing*, or as the view that a hypothesis can only be empirically *tested*—and only *after* it has been advanced.

Before I can elaborate this view (which might be called 'deductivism', in contrast to 'inductivism'[5]) I must first make clear the distinction between the *psychology of knowledge* which deals with empirical facts, and the *logic of knowledge* which is concerned only with logical relations. For the belief in inductive logic is largely due to a confusion of psychological problems with epistemological ones. It may be worth noticing, by the way, that this confusion spells trouble not only for the logic of knowledge but for its psychology as well.

[4] Reichenbach, *Erkenntnis* 1, 1930, p. 186.
[*2] See also chapter x, below, especially note 2 to section 81, and chapter *ii of the *Postscript* for a fuller statement of this criticism.
[5] Liebig (in *Induktion und Deduktion*, 1865) was probably the first to reject the inductive method from the standpoint of natural science; his attack is directed against Bacon. Duhem (in *La Théorie physique, son objet et sa structure*, 1906; English translation by P. P. Wiener: *The Aim and Structure of Physical Theory*, Princeton, 1954) held pronounced deductivist views. (*But there are also inductivist views to be found in Duhem's book, for example in the third chapter, Part One, where we are told that only experiment, induction, and generalization have produced Descartes's law of diffraction; *cf.* the English translation, p. 455.) See also V. Kraft, *Die Grundformen der Wissenschaftlichen Methoden*, 1925; and Carnap, *Erkenntnis* 2, 1932, p. 440.

2. *Elimination of Psychologism.*

I said above that the work of the scientist consists in putting forward and testing theories.

The initial stage, the act of conceiving or inventing a theory, seems to me neither to call for logical analysis nor to be susceptible of it. The question how it happens that a new idea occurs to a man—whether it is a musical theme, a dramatic conflict, or a scientific theory—may be of great interest to empirical psychology; but it is irrelevant to the logical analysis of scientific knowledge. This latter is concerned not with *questions of fact* (Kant's *quid facti*?), but only with questions of *justification or validity* (Kant's *quid juris*?). Its questions are of the following kind. Can a statement be justified? And if so, how? Is it testable? Is it logically dependent on certain other statements? Or does it perhaps contradict them? In order that a statement may be logically examined in this way, it must already have been presented to us. Someone must have formulated it, and submitted it to logical examination.

Accordingly I shall distinguish sharply between the process of conceiving a new idea, and the methods and results of examining it logically. As to the task of the logic of knowledge—in contradistinction to the psychology of knowledge—I shall proceed on the assumption that it consists solely in investigating the methods employed in those systematic tests to which every new idea must be subjected if it is to be seriously entertained.

Some might object that it would be more to the purpose to regard it as the business of epistemology to produce what has been called a '*rational reconstruction*' of the steps that have led the scientist to a discovery—to the finding of some new truth. But the question is: what, precisely, do we want to reconstruct? If it is the processes involved in the stimulation and release of an inspiration which are to be reconstructed, then I should refuse to take it as the task of the logic of knowledge. Such processes are the concern of empirical psychology but hardly of logic. It is another matter if we want to reconstruct rationally the *subsequent tests* whereby the inspiration may be discovered to be a discovery, or become known to be knowledge. In so far as the scientist critically judges, alters, or rejects his own inspiration we may, if we like, regard the methodological analysis undertaken here as a kind of 'rational reconstruction' of the corresponding thought-processes. But this reconstruction would not describe these processes

31

as they actually happen: it can give only a logical skeleton of the procedure of testing. Still, this is perhaps all that is meant by those who speak of a 'rational reconstruction' of the ways in which we gain knowledge.

It so happens that my arguments in this book are quite independent of this problem. However, my view of the matter, for what it is worth, is that there is no such thing as a logical method of having new ideas, or a logical reconstruction of this process. My view may be expressed by saying that every discovery contains 'an irrational element', or 'a creative intuition', in Bergson's sense. In a similar way Einstein speaks of '. . . the search for those highly universal . . . laws from which a picture of the world can be obtained by pure deduction. There is no logical path', he says, 'leading to these . . . laws. They can only be reached by intuition, based upon something like an intellectual love ('*Einfühlung*') of the objects of experience'.[1]

3. *Deductive Testing of Theories.*

According to the view that will be put forward here, the method of critically testing theories, and selecting them according to the results of tests, always proceeds on the following lines. From a new idea, put up tentatively, and not yet justified in any way—an anticipation, a hypothesis, a theoretical system, or what you will—conclusions are drawn by means of logical deduction. These conclusions are then compared with one another and with other relevant statements, so as to find what logical relations (such as equivalence, derivability, compatiblity, or incompatibility) exist between them.

We may if we like distinguish four different lines along which the testing of a theory could be carried out. First there is the logical comparison of the conclusions among themselves, by which the internal consistency of the system is tested. Secondly, there is the investigation of the logical form of the theory, with the object of determining whether it has the character of an empirical or scientific theory, or whether it is, for example, tautological. Thirdly, there is the com-

[1] Address on Max Planck's 6oth birthday. The passage quoted begins with the words, 'The supreme task of the physicist is to search for those universal laws . . .', etc. (quoted from A. Einstein, *Mein Weltbild*, 1934, p. 168; English translation by A. Harris: *The World as I see It*, 1935, p. 125). Similar ideas are found earlier in Liebig, *op. cit.*; *cf.* also Mach, *Principien der Wärmelehre* (1896), p. 443 *ff.* * The German word '*Einfühlung*' is difficult to translate. Harris translates: 'sympathetic understanding of experience'.

parison with other theories, chiefly with the aim of determining whether the theory would constitute a scientific advance should it survive our various tests. And finally, there is the testing of the theory by way of empirical applications of the conclusions which can be derived from it.

The purpose of this last kind of test is to find out how far the new consequences of the theory—whatever may be new in what it asserts —stand up to the demands of practice, whether raised by purely scientific experiments, or by practical technological applications. Here too the procedure of testing turns out to be deductive. With the help of other statements, previously accepted, certain singular statements— which we may call 'predictions'—are deduced from the theory; especially predictions that are easily testable or applicable. From among these statements, those are selected which are not derivable from the current theory, and more especially those which the current theory contradicts. Next we seek a decision as regards these (and other) derived statements by comparing them with the results of practical applications and experiments. If this decision is positive, that is, if the singular conclusions turn out to be acceptable, or *verified*, then the theory has, for the time being, passed its test: we have found no reason to discard it. But if the decision is negative, or in other words, if the conclusions have been *falsified*, then their falsification also falsifies the theory from which they were logically deduced.

It should be noticed that a positive decision can only temporarily support the theory, for subsequent negative decisions may always overthrow it. So long as a theory withstands detailed and severe tests and is not superseded by another theory in the course of scientific progress, we may say that it has 'proved its mettle' or that it is '*corroborated*'.*1

Nothing resembling inductive logic appears in the procedure here outlined. I never assume that we can argue from the truth of singular statements to the truth of theories. I never assume that by force of 'verified' conclusions, theories can be established as 'true', or even as merely 'probable'.

In this book I intend to give a more detailed analysis of the methods of deductive testing. And I shall attempt to show that, within the framework of this analysis, all the problems can be dealt

*1 For this term, see note *1 before section 79, and section *29 of my *Postscript*.

with that are usually called '*epistemological*'. Those problems, more especially, to which inductive logic gives rise, can be eliminated without creating new ones in their place.

4. *The Problem of Demarcation.*

Of the many objections which are likely to be raised against the view here advanced, the most serious is perhaps the following. In rejecting the method of induction, it may be said, I deprive empirical science of what appears to be its most important characteristic; and this means that I remove the barriers which separate science from metaphysical speculation. My reply to this objection is that my main reason for rejecting inductive logic is precisely that *it does not provide a suitable distinguishing mark* of the empirical, non-metaphysical, character of a theoretical system; or in other words, that *it does not provide a suitable 'criterion of demarcation'.*

The problem of finding a criterion which would enable us to distinguish between the empirical sciences on the one hand, and mathematics and logic as well as 'metaphysical' systems on the other, I call the *problem of demarcation.*[1]

This problem was known to Hume who attempted to solve it.[2] With Kant it became the central problem of the theory of knowledge. If, following Kant, we call the problem of induction 'Hume's problem', we might call the problem of demarcation 'Kant's problem'.

Of these two problems—the source of nearly all the other problems of the theory of knowledge—the problem of demarcation is, I think, the more fundamental. Indeed, the main reason why epistemologists with empiricist leanings tend to pin their faith to the 'method of induction' seems to be their belief that this method alone can provide a suitable criterion of demarcation. This applies especially to those empiricists who follow the flag of 'positivism'.

The older positivists wished to admit, as scientific or legitimate, only those *concepts* (or notions or ideas) which were, as they put it, 'derived from experience'; those concepts, that is, which they believed to be logically reducible to elements of sense-experience, such as sensations (or sense–data), impressions, perceptions, visual or auditory

---

[1] With this (and also with sections 1 to 6 and 13 to 24) *cf.* my note: *Erkenntnis* 3, 1933, p. 426; *It is now here reprinted, in translation, as appendix *i.

[2] *Cf.* the last sentence of his *Enquiry Concerning Human Understanding.* *With the next paragraph, compare for example the quotation from Reichenbach in the text to note 1, section 1.

memories, and so forth. Modern positivists are apt to see more clearly that science is not a system of concepts but rather a system of *statements*.*¹ Accordingly, they wish to admit, as scientific or legitimate, only those statements which are reducible to elementary (or 'atomic') statements of experience—to 'judgments of perception' or 'atomic propositions' or 'protocol-sentences' or what not.*² It is clear that the implied criterion of demarcation is identical with the demand for an inductive logic.

Since I reject inductive logic I must also reject all these attempts to solve the problem of demarcation. With this rejection, the problem of demarcation gains in importance for the present inquiry. Finding an acceptable criterion of demarcation must be a crucial task for any epistemology which does not accept inductive logic.

Positivists usually interpret the problem of demarcation in a *naturalistic* way; they interpret it as if it were a problem of natural science. Instead of taking it as their task to propose a suitable convention, they believe they have to discover a difference, existing in the nature of things, as it were, between empirical science on the one hand and metaphysics on the other. They are constantly trying to prove that metaphysics by its very nature is nothing but nonsensical twaddle—'sophistry and illusion', as Hume says, which we should 'commit to the flames'.*³

If by the words 'nonsensical' or 'meaningless' we wish to express no more, by definition, than 'not belonging to empirical science', then the characterization of metaphysics as meaningless nonsense would be trivial; for metaphysics has usually been defined as non-empirical. But of course, the positivists believe they can say much more about metaphysics than that some of its statements are non-

---

*¹ When I wrote this paragraph I overrated the 'modern positivists', as I now see. I should have remembered that *in this respect* the promising beginning of Wittgenstein's *Tractatus*—'The world is the totality of facts, not of things'—was cancelled by its end which denounced the man who 'had given no meaning to certain signs in his propositions'. See also my *Open Society and its Enemies*, chapter 11, section ii, and chapter *i of my *Postscript*, especially sections *11 (note 5), *24 (the last five paragraphs), and *25.

*² Nothing depends on names, of course. When I invented the new name 'basic statement' (or 'basic proposition'; see below, sections 7 and 28) I did so only because I needed a term *not* burdened with the connotation of a perception statement. But unfortunately it was soon adopted by others, and used to convey precisely the kind of meaning which I wished to avoid. *Cf.* also my *Postscript*, *29.

*³ Hume thus condemned his own *Enquiry* on its last page, just as later Wittgenstein condemned his own *Tractatus* on its last page. (See note 2 to section 10.)

empirical. The words 'meaningless' or 'nonsensical' convey, and are meant to convey, a derogatory evaluation; and there is no doubt that what the positivists really want to achieve is not so much a successful demarcation as the final overthrow[3] and the annihilation of metaphysics. However this may be, we find that each time the positivists tried to say more clearly what 'meaningful' meant, the attempt led to the same result—to a definition of 'meaningful sentence' (in contradistinction to 'meaningless pseudo-sentence') which simply reiterated the criterion of demarcation of their *inductive logic*.

This 'shows itself' very clearly in the case of Wittgenstein, according to whom every meaningful proposition must be *logically reducible*[4] to elementary (or atomic) propositions, which he characterizes as descriptions or 'pictures of reality'[5] (a characterization, by the way, which is to cover all meaningful propositions). We may see from this that Wittgenstein's criterion of meaningfulness coincides with the inductivists' criterion of demarcation, provided we replace their words 'scientific' or 'legitimate' by 'meaningful'. And it is precisely over the problem of induction that this attempt to solve the problem of demarcation comes to grief: positivists, in their anxiety to annihilate metaphysics, annihilate natural science along with it. For scientific laws, too, cannot be logically reduced to elementary statements of experience. If consistently applied, Wittgenstein's criterion of meaningfulness rejects as meaningless those natural laws the search for which, as Einstein says,[6] is 'the supreme task of the physicist': they can never be accepted as genuine or legitimate statements. This view, which tries to unmask the problem of induction as an empty pseudo-problem, has been expressed by Schlick[*4] in the following words:

[3] Carnap, *Erkenntnis* 2, 1932, p. 219 ff. Earlier Mill had used the word 'meaningless' in a similar way, *no doubt under the influence of Comte; cf. Comte's *Early Essays on Social Philosophy*, ed. by H. D. Hutton, 1911, p. 223. See also my *Open Society*, note 51 to chapter 11.

[4] Wittgenstein, *Tractatus Logico-Philosophicus* (1918 and 1922), Proposition 5. *As this was written in 1934, I am dealing here of course *only* with the *Tractatus*.

[5] Wittgenstein, *op. cit.*, Proposition 4.01; 4.03; 2.221.

[6] *Cf.* note 1 to section 2.

[*4] The idea of treating scientific laws as pseudo-propositions—thus solving the problem of induction—was attributed by Schlick to Wittgenstein. (*Cf.* my *Open Society*, notes 46 and 51 f. to chapter 11.) But it is really much older. It is part of the instrumentalist tradition which can be traced back to Berkeley, and further. (See for example my paper 'Three Views Concerning Human Knowledge', in *Contemporary British Philosophy* 1956; and 'A Note on Berkeley as a Precursor of Mach', in *The British Journal for the Philosophy of Science* iv, 4, 1953, pp. 26 ff., now in my *Conjectures and Refutations*, 1959. Further references in note *1 before section 12 (p. 59). The problem is also treated in my *Postscript*, sections *11 to *14, and *19 to *26.)

'The problem of induction consists in asking for a logical justification of *universal statements* about reality . . . We recognize, with Hume, that there is no such logical justification: there can be none, simply because *they are not genuine* statements.'[7]

This shows how the inductivist criterion of demarcation fails to draw a dividing line between scientific and metaphysical systems, and why it must accord them equal status; for the verdict of the positivist dogma of meaning is that both are systems of meaningless pseudo-statements. Thus instead of eradicating metaphysics from the empirical sciences, positivism leads to the invasion of metaphysics into the scientific realm.[8]

In contrast to these anti-metaphysical stratagems—anti-metaphysical in intention, that is—my business, as I see it, is not to bring about the overthrow of metaphysics. It is, rather, to formulate a suitable characterization of empirical science, or to define the concepts 'empirical science' and 'metaphysics' in such a way that we shall be able to say of a given system of statements whether or not its closer study is the concern of empirical science.

My criterion of demarcation will accordingly have to be regarded as a *proposal for an agreement or convention*. As to the suitability of any such convention opinions may differ; and a reasonable discussion of these questions is only possible between parties having some purpose in common. The choice of that purpose must, of course, be ultimately a matter of decision, going beyond rational argument.*[5]

Thus anyone who envisages a system of absolutely certain, irrevocably true statements[9] as the end and purpose of science will certainly reject the proposals I shall make here. And so will those

---

[7] Schlick, *Naturwissenschaften* 19, 1931, p. 156. (The italics are mine.) Regarding natural laws Schlick writes (p. 151), 'It has often been remarked that, strictly, we can never speak of an absolute verification of a law, since we always, so to speak, tacitly make the reservation that it may be modified in the light of further experience. If I may add, by way of parenthesis', Schlick continues, 'a few words on the logical situation, the above-mentioned fact means that a natural law, in principle, does not have the logical character of a statement, but is, rather, a prescription for the formation of statements.' *('Formation' no doubt was meant to include transformation or derivation.) Schlick attributed this theory to a personal communication of Wittgenstein's. See also section *12 of my *Postscript*.

[8] *Cf.* Section 78 (for example note 1). *See also my *Open Society*, notes 46, 51, and 52 to chapter 11, and my paper 'The Demarcation between Science and Metaphysics', contributed in January 1955 to the planned Carnap volume of the *Library of Living Philosophers*, edited by P. A. Schilpp.

*[5] I believe that a reasonable discussion is always possible between parties interested in truth, and ready to pay attention to each other. (*Cf.* my *Open Society*, chapter 24.)

[9] This is Dingler's view; *cf.* note 1 to section 19.

who see 'the essence of science . . . in its dignity', which they think resides in its 'wholeness' and its 'real truth and essentiality'.[10] They will hardly be ready to grant this dignity to modern theoretical physics in which I and others see the most complete realization to date of what I call 'empirical science'.

The aims of science which I have in mind are different. I do not try to justify them, however, by representing them as the true or the essential aims of science. This would only distort the issue, and it would mean a relapse into positivist dogmatism. There is only *one* way, as far as I can see, of arguing rationally in support of my proposals. This is to analyse their logical consequences: to point out their fertility—their power to elucidate the problems of the theory of knowledge.

Thus I freely admit that in arriving at my proposals I have been guided, in the last analysis, by value judgments and predilections. But I hope that my proposals may be acceptable to those who value not only logical rigour but also freedom from dogmatism; who seek practical applicability, but are even more attracted by the adventure of science, and by discoveries which again and again confront us with new and unexpected questions, challenging us to try out new and hitherto undreamed-of answers.

The fact that value judgments influence my proposals does not mean that I am making the mistake of which I have accused the positivists—that of trying to kill metaphysics by calling it names. I do not even go so far as to assert that metaphysics has no value for empirical science. For it cannot be denied that along with metaphysical ideas which have obstructed the advance of science there have been others—such as speculative atomism—which have aided it. And looking at the matter from the psychological angle, I am inclined to think that scientific discovery is impossible without faith in ideas which are of a purely speculative kind, and sometimes even quite hazy; a faith which is completely unwarranted from the point of view of science, and which, to that extent, is 'metaphysical'.[11]

Yet having issued all these warnings, I still take it to be the first task of the logic of knowledge to put forward a *concept of empirical science*, in order to make linguistic usage, now somewhat uncertain,

[10] This is the view of O. Spann (*Kategorienlehre*, 1924).

[11] *Cf.* also: Planck, *Positivismus und reale Aussenwelt* (1931) and Einstein, *Die Religiosität der Forschung*, in *Mein Weltbild* (1934), p. 43; English translation by A. Harris: *The World as I See It* (1935), p. 23 *ff*. *See also section 85, and my *Postscript*.

as definite as possible, and in order to draw a clear line of demarcation between science and metaphysical ideas—even though these ideas may have furthered the advance of science throughout its history.

5. *Experience as a Method.*

The task of formulating an acceptable definition of the idea of empirical science is not without its difficulties. Some of these arise from *the fact that there must be many theoretical systems* with a logical structure very similar to the one which at any particular time is the accepted system of empirical science. This situation is sometimes described by saying that there are a great many—presumably an infinite number—of 'logically possible worlds'. Yet the system called 'empirical science' is intended to represent only *one* world: the 'real world' or the 'world of our experience'.[*1]

In order to make this idea a little more precise, we may distinguish three requirements which our empirical theoretical system will have to satisfy. First, it must be *synthetic*, so that it may represent a non-contradictory, a *possible* world. Secondly, it must satisfy the criterion of demarcation (*cf.* sections 6 and 21), *i.e.* it must not be metaphysical, but must represent a world of possible *experience*. Thirdly, it must be a system distinguished in some way from other such systems as the one which represents *our* world of experience.

But how is the system that represents our world of experience to be distinguished? The answer is: by the fact that it has been submitted to tests, and has stood up to tests. This means that it is to be distinguished by applying to it that deductive method which it is my aim to analyse, and to describe.

'Experience', on this view, appears as a distinctive *method* whereby one theoretical system may be distinguished from others; so that empirical science seems to be characterized not only by its logical form but, in addition, by its distinctive *method*. (This, of course, is also the view of the inductivists, who try to characterize empirical science by its use of the inductive method.)

The theory of knowledge whose task is the analysis of the method or procedure peculiar to empirical science, may accordingly be described as a theory of the empirical method—*a theory of what is usually called experience.*

[*1] *Cf.* appendix *x.

39

6. Falsifiability as a Criterion of Demarcation.

The criterion of demarcation inherent in inductive logic—that is, the positivistic dogma of meaning—is equivalent to the requirement that all the statements of empirical science (or all 'meaningful' statements) must be capable of being finally decided, with respect to their truth *and* falsity; we shall say that they must be *'conclusively decidable'*. This means that their form must be such that *to verify them and to falsify them* must both be logically possible. Thus Schlick says: '. . . a genuine statement must be capable of *conclusive verification*'[1]; and Waismann says still more clearly: 'If there is no possible way to *determine whether a statement is true* then that statement has no meaning whatsoever. For the meaning of a statement is the method of its verification.'[2]

Now in my view there is no such thing as induction.[*1] Thus inference to theories, from singular statements which are 'verified by experience' (whatever that may mean), is logically inadmissible. Theories are, therefore, *never* empirically verifiable. If we wish to avoid the positivist's mistake of eliminating, by our criterion of demarcation, the theoretical systems of natural science,[*2] then we must choose a criterion which allows us to admit to the domain of empirical science even statements which cannot be verified.

But I shall certainly admit a system as empirical or scientific only if it is capable of being *tested* by experience. These considerations suggest that not the *verifiability* but the *falsifiability* of a system is to be taken as a criterion of demarcation.[*3] In other words: I shall not

[1] Schlick, *Naturwissenschaften* 19, 1931, 1931, p. 150.
[2] Waismann, *Erkenntnis* I, 1930, p. 229.
[*1] I am not, of course, here considering so-called 'mathematical induction'; what I am denying is that there is such a thing as induction in the so-called 'inductive sciences'; that there are either 'inductive procedures' or 'inductive inferences'.
[*2] In his *Logical Syntax* (1937, p. 321 f.) Carnap admitted that this was a mistake (with a reference to my criticism); and he did so even more fully in 'Testability and Meaning', recognizing the fact that universal laws are not only 'convenient' for science but even 'essential' (*Philosophy of Science* 4, 1937, p. 27). But in his inductivist *Logical Foundations of Probability* (1950), he returns to a position very like the one here criticized: finding that universal laws have zero probability (p. 511), he is compelled to say (p. 575) that though they need not be expelled from science, science can very well do without them.
[*3] Note that I suggest falsifiability as a criterion of demarcation, but *not of meaning*. Note, moreover, that I have already (section 4) sharply criticized the use of the idea of meaning as a criterion of demarcation, and that I attack the dogma of meaning again, even more sharply, in section 9. It is therefore a sheer myth (though any number of refutations of my theory have been based upon this myth) that I ever proposed falsifiability as a criterion of meaning. Falsifiability separates two kinds of perfectly meaningful statements: the falsifiable and the non-falsifiable. It draws a line inside meaningful language, not around it. See also Appendix *i, and chapter *i of my *Postscript*, especially sections *17 and *19.

require of a scientific system that it shall be capable of being singled out, once and for all, in a positive sense; but I shall require that its logical form shall be such that it can be singled out, by means of empirical tests, in a negative sense: *it must be possible for an empirical scientific system to be refuted by experience.*[3]

(Thus the statement, 'It will rain or not rain here tomorrow' will not be regarded as empirical, simply because it cannot be refuted; whereas the statement, 'It will rain here tomorrow' will be regarded as empirical.)

Various objections might be raised against the criterion of demarcation here proposed. In the first place, it may well seem somewhat wrong-headed to suggest that science, which is supposed to give us positive information, should be characterized as satisfying a negative requirement such as refutability. However, I shall show, in sections 31 to 46, that this objection has little weight, since the amount of positive information about the world which is conveyed by a scientific statement is the greater the more likely it is to clash, because of its logical character, with possible singular statements. (Not for nothing do we call the laws of nature 'laws': the more they prohibit the more they say.)

Again, the attempt might be made to turn against me my own criticism of the inductivist criterion of demarcation; for it might seem that objections can be raised against falsifiability as a criterion of demarcation similar to those which I myself raised against verifiability.

This attack would not disturb me. My proposal is based upon an *asymmetry* between verifiability and falsifiability; an asymmetry which results from the logical form of universal statements.*[4] For these are never derivable from singular statements, but can be contradicted by singular statements. Consequently it is possible by means of purely deductive inferences (with the help of the *modus tollens* of classical logic) to argue from the truth of singular statements to the falsity of universal statements. Such an argument to the falsity of universal statements is the only strictly deductive kind of inference that proceeds, as it were, in the 'inductive direction'; that is, from singular to universal statements.

A third objection may seem more serious. It might be said that

[3] Related ideas are to be found, for example, in Frank, *Die Kausalität und ihre Grenzen* (1931), ch. I, §10 (p. 15 f.); Dubislav, *Die Definition* (3rd edition 1931), p. 100 f. (Cf. also note 1 to section 4, above.)

*[4] This asymmetry is now more fully discussed in section *22 of my *Postscript*.

even if the asymmetry is admitted, it is still impossible, for various reasons, that any theoretical system should ever be conclusively falsified. For it is always possible to find some way of evading falsification, for example by introducing *ad hoc* an auxiliary hypothesis, or by changing *ad hoc* a definition. It is even possible without logical inconsistency to adopt the position of simply refusing to acknowledge any falsifying experience whatsoever. Admittedly, scientists do not usually proceed in this way, but logically such procedure is possible; and this fact, it might be claimed, makes the logical value of my proposed criterion of demarcation dubious, to say the least.

I must admit the justice of this criticism; but I need not therefore withdraw my proposal to adopt falsifiability as a criterion of demarcation. For I am going to propose (in sections 20 *f.*) that the *empirical method* shall be characterized as a method that excludes precisely those ways of evading falsification which, as my imaginary critic rightly insists, are logically admissible. According to my proposal, what characterizes the empirical method is its manner of exposing to falsification, in every conceivable way, the system to be tested. Its aim is not to save the lives of untenable systems but, on the contrary, to select the one which is by comparison the fittest, by exposing them all to the fiercest struggle for survival.

The proposed criterion of demarcation also leads us to a solution of Hume's problem of induction—of the problem of the validity of natural laws. The root of this problem is the apparent contradiction between what may be called 'the fundamental thesis of empiricism' —the thesis that experience alone can decide upon the truth or falsity of scientific statements—and Hume's realization of the inadmissibility of inductive arguments. This contradiction arises only if it is assumed that all empirical scientific statements must be 'conclusively decidable', i.e. that their verification and their falsification must both in principle be possible. If we renounce this requirement and admit as empirical also statements which are decidable in one sense only—unilaterally decidable and, more especially, falsifiable— and which may be tested by systematic attempts to falsify them, the contradiction disappears: the method of falsification presupposes no inductive inference, but only the tautological transformations of deductive logic whose validity is not in dispute.[4]

---

[4] For this see also my paper mentioned in note 1 to section 4, *now here reprinted as appendix *i; and my *Postscript*, esp. section *2.

7. *The Problem of the 'Empirical Basis'*.

If falsifiability is to be at all applicable as a criterion of demarcation, then singular statements must be available which can serve as premises in falsifying inferences. Our criterion therefore appears only to shift the problem—to lead us back from the question of the empirical character of theories to the question of the empirical character of singular statements.

Yet even so, something has been gained. For in the practice of scientific research, demarcation is sometimes of immediate urgency in connection with theoretical systems, whereas in connection with singular statements, doubts as to their empirical character rarely arise. It is true that errors of observation occur and give rise to false singular statements, but the scientist scarcely ever has occasion to describe a singular statement as non-empirical or metaphysical.

Problems of the empirical basis—that is, problems concerning the empirical character of singular statements, and how they are tested—thus play a part within the logic of science that differs somewhat from that played by most of the other problems which will concern us. For most of these stand in close relation to the *practice* of research, whilst the problem of the empirical basis belongs almost exclusively to the *theory* of knowledge. I shall have to deal with them, however, since they have given rise to many obscurities. This is especially true of the relation between *perceptual experiences* and *basic statements*. (What I call a 'basic statement' or a 'basic proposition' is a statement which can serve as a premise in an empirical falsification; in brief, a statement of a singular fact.)

Perceptual experiences have often been regarded as providing a kind of justification for basic statements. It was held that these statements are 'based upon' these experiences; that their truth becomes 'manifest by inspection' through these experiences; or that it is made 'evident' by these experiences, etc. All these expressions exhibit the perfectly sound tendency to emphasize the close connection between the basic statements and our perceptual experiences. Yet it was also rightly felt that *statements can be logically justified only by statements*. Thus the connection between the perceptions and the statements remained obscure, and was described by correspondingly obscure expressions which elucidated nothing, but slurred over the difficulties or, at best, adumbrated them through metaphors.

43

Here too a solution can be found, I believe, if we clearly separate the psychological from the logical and methodological aspects of the problem. We must distinguish between, on the one hand, *our subjective experiences or our feelings of conviction*, which can never justify any statement (though they can be made the subject of psychological investigation) and, on the other hand, the *objective logical relations* subsisting among the various systems of scientific statements, and within each of them.

The problems of the empirical basis will be discussed in some detail in sections 25 to 30. For the present I had better turn to the problem of scientific objectivity, since the terms 'objective' and 'subjective' which I have just used are in need of elucidation.

8. *Scientific Objectivity and Subjective Conviction.*

The words 'objective' and 'subjective' are philosophical terms heavily burdened with a heritage of contradictory usages and of inconclusive and interminable discussions.

My use of the terms 'objective' and 'subjective' is not unlike Kant's. He uses the word 'objective' to indicate that scientific knowledge should be *justifiable*, independently of anybody's whim: a justification is 'objective' if in principle it can be tested and understood by anybody. 'If something is valid', he writes, 'for anybody in possession of his reason, then its grounds are objective and sufficient.'[1]

Now I hold that scientific theories are never fully justifiable or verifiable, but that they are nevertheless testable. I shall therefore say that the *objectivity* of scientific statements lies in the fact that they can be *inter-subjectively tested*.[*1]

The word 'subjective' is applied by Kant to our feelings of conviction (of varying degrees).[2] To examine how these come about

---

[1] *Kritik der reinen Vernunft*, Methodenlehre, 2. Haupstück, 3. Abschnitt (2nd edition, p. 848; English Translation by N. Kemp Smith, 1933: *Critique of Pure Reason*, The Transcendental Doctrine of Method, chapter ii, section 3, p. 645).

[*1] I have since generalized this formulation; for inter-subjective *testing* is merely a very important aspect of the more general idea of inter-subjective *criticism*, or in other words, of the idea of mutual rational control by critical discussion. This more general idea, discussed at some length in my *Open Society and Its Enemies*, chapters 23 and 24, and in my *Poverty of Historicism*, section 32, is also discussed in my *Postscript*, especially in chapters *i, *ii, and *vi.

[2] *Ibid.*

is the business of psychology. They may arise, for example, 'in accordance with the laws of association'.[3] Objective reasons too may serve as 'subjective *causes* of judging',[4] in so far as we may reflect upon these reasons, and become convinced of their cogency.

Kant was perhaps the first to realize that the objectivity of scientific statements is closely connected with the construction of theories—with the use of hypotheses and universal statements. Only when certain events recur in accordance with rules or regularities, as is the case with repeatable experiments, can our observations be tested—in principle—by anyone. We do not take even our own observations quite seriously, or accept them as scientific observations, until we have repeated and tested them. Only by such repetitions can we convince ourselves that we are not dealing with a mere isolated 'coincidence', but with events which, on account of their regularity and reproducibility, are in principle inter-subjectively testable.[5]

Every experimental physicist knows those surprising and inexplicable apparent 'effects' which can perhaps even be reproduced in his laboratory for some time, but which finally disappear without trace. Of course, no physicist would say in such a case that he had made a scientific discovery (though he might try to rearrange his experiments so as to make the effect reproducible). Indeed the scientifically significant *physical effect* may be defined as that which can be regularly reproduced by anyone who carries out the appropriate experiment in the way prescribed. No serious physicist would offer for publication, as a scientific discovery, any such 'occult effect', as I propose to call it—one for whose reproduction he could give no instructions. The 'discovery' would be only too soon rejected as chimerical, simply because attempts to test it would lead to negative

---

[3] *Cf. Kritik der reinen Vernunft*, Transcendentale Elementarlehre §19 (2nd edition, p. 142; English translation by N. Kemp Smith, 1933: *Critique of Pure Reason*, Transcendental Doctrine of Elements, §19, p. 159).

[4] *Cf. Kritik der reinen Vernuft*, Methodenlehre, 2. Hauptstück, 3. Abschnitt (2nd edition, p. 849; English translation, chapter ii, section 3, p. 646).

[5] Kant realized that from the required objectivity of scientific statements it follows that they must be at any time inter-subjectively testable, and that they must therefore have the form of universal laws or theories. He formulated this discovery somewhat obscurely by his 'principle of temporal succession according to the law of causality' (which principle he believed that he could prove *a priori* by employing the reasoning here indicated). I do not postulate any such principle (*cf.* section 12); but I agree that scientific statements, since they must be inter-subjectively testable, must always have the character of universal hypotheses. * See also note *1 to section 22.

results.[6] (It follows that any controversy over the question whether events which are in principle unrepeatable and unique ever do occur cannot be decided by science: it would be a metaphysical controversy.)

We may now return to a point made in the previous section: to my thesis that a subjective experience, or a feeling of conviction, can never justify a scientific statement, and that within science it can play no part but that of the subject of an empirical (a psychological) inquiry. No matter how intense a feeling of conviction it may be, it can never justify a statement. Thus I can be utterly convinced of the truth of a statement; certain of the evidence of my perceptions; overwhelmed by the intensity of my experience: every doubt may seem to me absurd. But does this afford the slightest reason for science to accept my statement? Can any statement be justified by the fact that K. R. P. is utterly convinced of its truth? The answer is, 'No'; and any other answer would be incompatible with the idea of scientific objectivity. Even the fact, for me so firmly established, that I am experiencing this feeling of conviction, cannot appear within the field of objective science except in the form of a *psychological hypothesis* which, of course, calls for inter-subjective testing: from the conjecture that I have this feeling of conviction the psychologist may deduce, with the help of psychological and other theories, certain predictions about my behaviour; and these may be confirmed or refuted in the course of experimental tests. But from the epistemological point of view, it is quite irrelevant whether my feeling of conviction was strong or weak; whether it came from a strong or even irresistible impression of indubitable certainty (or 'self-evidence'), or merely from a doubtful surmise. None of this has any bearing on the question of how scientific statements can be justified.

Considerations like these do not of course provide an answer to the problem of the empirical basis. But at least they help us to see its main difficulty. In demanding objectivity for basic statements as well as for other scientific statements, we deprive ourselves of any logical means by which we might have hoped to reduce the truth of

[6] In the literature of physics there are to be found some instances of reports, by serious investigators, of the occurrence of effects which could not be reproduced, since further tests led to negative results. A well-known example from recent times is the unexplained positive result of Michelson's experiment observed by Miller (1921-1926) at Mount Wilson, after he himself (as well as Morley) had previously reproduced Michelson's negative result. But since later tests again gave negative results it is now customary to regard these latter as decisive, and to explain Miller's divergent result as 'due to unknown sources of error'. *See also section 22, especially footnote *1.

scientific statements to our experiences. Moreover we debar ourselves from granting any favoured status to statements which represent experiences, such as those statements which describe our perceptions (and which are sometimes called 'protocol sentences'). They can occur in science only as psychological statements; and this means, as hypotheses of a kind whose standards of inter-subjective testing (considering the present state of psychology) are certainly not very high.

Whatever may be our eventual answer to the question of the empirical basis, one thing must be clear: if we adhere to our demand that scientific statements must be objective, then those statements which belong to the empirical basis of science must also be objective, i.e. inter-subjectively testable. Yet inter-subjective testability always implies that from the statements which are to be tested, other testable statements can be deduced. Thus if the basic statements in their turn are to be inter-subjectively testable, *there can be no ultimate statements in science*: there can be no statements in science which cannot be tested, and therefore none which cannot in principle be refuted, by falsifying some of the conclusions which can be deduced from them.

We thus arrive at the following view. Systems of theories are tested by deducing from them statements of a lesser level of universality. These statements in their turn, since they are to be inter-subjectively testable, must be testable in like manner—and so *ad infinitum*.

It might be thought that this view leads to an infinite regress, and that it is therefore untenable. In section 1, when criticizing induction, I raised the objection that it may lead to an infinite regress; and it might well appear to the reader now that the very same objection can be urged against that procedure of deductive testing which I myself advocate. However, this is not so. The deductive method of testing cannot establish or justify the statements which are being tested; nor is it intended to do so. Thus there is no danger of an infinite regress. But it must be admitted that the situation to which I have drawn attention—testability *ad infinitum* and the absence of ultimate statements which are not in need of tests—does create a problem. For, clearly, tests cannot in fact be carried on *ad infinitum*: sooner or later we have to stop. Without discussing this problem here in detail, I only wish to point out that the fact that the tests cannot go on for ever does not clash with my demand that every scientific statement must be testable. For I do not demand that every scientific statement must *have in fact been tested* before it is accepted. I only demand that every such statement must be *capable* of being tested; or in other words, I refuse to accept the view that there are statements in science which we have, resignedly, to accept as true merely because it does not seem possible, for logical reasons, to test them.

CHAPTER II

ON THE PROBLEM OF A
THEORY OF SCIENTIFIC METHOD

IN accordance with my proposal made above, epistemology, or the logic of scientific discovery, should be identified with the theory of scientific method. The theory of method, in so far as it goes beyond the purely logical analysis of the relations between scientific statements, is concerned with *the choice of methods*—with decisions about the way in which scientific statements are to be dealt with. These decisions will of course depend in their turn upon the *aim* which we choose from among a number of possible aims. The decision here proposed for laying down suitable rules for what I call the 'empirical method' is closely connected with my criterion of demarcation: I propose to adopt such rules as will ensure the testability of scientific statements; which is to say, their falsifiability.

9. *Why Methodological Decisions are Indispensable.*

What are rules of scientific method, and why do we need them? Can there be a theory of such rules, a methodology?

The way in which one answers these questions will largely depend upon one's attitude to science. Those who, like the positivists, see empirical science as a system of statements which satisfy certain *logical criteria*, such as meaningfulness or verifiability, will give one answer. A very different answer will be given by those who tend to see (as I do) the distinguishing characteristic of empirical statements in their susceptibility to revision—in the fact that they can be criticized, and superseded by better ones; and who regard it as their task to analyse the characteristic ability of science to advance, and the characteristic manner in which a choice is made, in crucial cases, between conflicting systems of theories.

I am quite ready to admit that there is a need for a purely logical analysis of theories, for an analysis which takes no account of how they change and develop. But this kind of analysis does not elucidate those aspects of the empirical sciences which I, for one, so highly prize. A system such as classical mechanics may be 'scientific' to any degree you like; but those who uphold it dogmatically—believing, perhaps, that it is their business to defend such a successful system against criticism as long as it is not *conclusively disproved*—are adopting the very reverse of that critical attitude which in my view is the proper one for the scientist. In point of fact, no conclusive disproof of a theory can ever be produced; for it is always possible to say that the experimental results are not reliable, or that the discrepancies which are asserted to exist between the experimental results and the theory are only apparent and that they will disappear with the advance of our understanding. (In the struggle against Einstein, both these arguments were often used in support of Newtonian mechanics, and similar arguments abound in the field of the social sciences.) If you insist on strict proof (or strict disproof*1) in the empirical sciences, you will never benefit from experience, and never learn from it how wrong you are.

If therefore we characterize empirical science merely by the formal or logical structure of its statements, we shall not be able to exclude from it that prevalent form of metaphysics which results from elevating an obsolete scientific theory into an incontrovertible truth.

Such are my reasons for proposing that empirical science should be characterized by its methods: by our manner of dealing with scientific systems: by what we do with them and what we do to them. Thus I shall try to establish the rules, or if you will the norms, by which the scientist is guided when he is engaged in research or in discovery, in the sense here understood.

10. *The Naturalistic Approach to the Theory of Method.*
The hint I gave in the previous section as to the deep-seated

*1 I have now here added in brackets the words 'or strict disproof' to the text (a) because they are clearly implied by what is said immediately before ('no conclusive disproof of a theory can ever be produced'), and (b) because I have been constantly misinterpreted as upholding a criterion (and moreover one of *meaning* rather than of *demarcation*) based upon a doctrine of 'complete' or 'conclusive' falsifiability.

difference between my position and that of the positivists is in need of some amplification.

The positivist dislikes the idea that there should be meaningful problems outside the field of 'positive' empirical science—problems to be dealt with by a genuine philosophical theory. He dislikes the idea that there should be a genuine theory of knowledge, an epistemology or a methodology.*1 He wishes to see in the alleged philosophical problems mere 'pseudo-problems' or 'puzzles'. Now this wish of his—which, by the way, he does not express as a wish or a proposal but rather as a statement of fact*2—can always be gratified. For nothing is easier than to unmask a problem as 'meaningless' or 'pseudo'. All you have to do is to fix upon a conveniently narrow meaning for 'meaning', and you will soon be bound to say of any inconvenient question that you are unable to detect any meaning in it. Moreover, if you admit as meaningful none except problems in natural science,[1] any debate about the concept of 'meaning' will also turn out to be meaningless.[2] The dogma of meaning, once enthroned, is elevated forever above the battle. It can no longer be attacked. It has become (in Wittgenstein's own words) 'unassailable and definitive'.[3]

The controversial question whether philosophy exists, or has any right to exist, is almost as old as philosophy itself. Time and again an entirely new philosophical movement arises which finally unmasks the old philosophical problems as pseudo-problems, and which confronts the wicked nonsense of philosophy with the good sense of meaningful, positive, empirical, science. And time and again do the despised defenders of 'traditional philosophy' try to explain to the leaders of the latest positivistic assault that the main problem of philosophy is the critical analysis of the appeal to the authority of

*1 In the two years before the first publication of this book, it was the standing criticism raised by members of the Vienna Circle against my ideas that a theory of method which is neither an empirical science nor pure logic is impossible, since what lies outside these two fields must be sheer nonsense. (The same view was still maintained by Wittgenstein in 1948; cf. my paper 'The Nature of Philosophical Problems', The British Journal for the Philosophy of Science 3, 1952, note on p. 128.) Later, the standing criticism became anchored in the legend that I had proposed to replace the verifiability criterion by a falsifiability criterion of meaning. See my Postscript, especially sections *19 to *22.

*2 Some positivists have since changed this attitude; see note 6, below.

[1] Wittgenstein, Tractatus Logico-Philosophicus, Proposition 6.53.

[2] Wittgenstein at the end of the Tractatus (in which he explains the concept of meaning) writes, 'My propositions are elucidatory in this way: he who understands me finally recognizes them as senseless. . . .'

[3] Wittgenstein, op. cit., at the end of his Preface.

'experience'[4]—precisely that 'experience' which every latest discoverer of positivism is, as ever, artlessly taking for granted. To such objections, however, the positivist only replies with a shrug: they mean nothing to him, since they do not belong to empirical science, which alone is meaningful. 'Experience' for him is a programme, not a problem (unless it is studied by empirical psychology).

I do not think positivists are likely to respond any differently to my own attempts to analyse 'experience' which I interpret as the method of empirical science. For only two kinds of statement exist for them: logical tautologies and empirical statements. If methodology is not logic, then, they will conclude, it must be a branch of some empirical science—the science, say, of the behaviour of scientists at work.

This view, according to which methodology is an empirical science in its turn—a study of the actual behaviour of scientists, or of the actual procedure of 'science'—may be described as *naturalistic*. A naturalistic methodology (sometimes called an 'inductive theory of science'[5]) has its value, no doubt. A student of the logic of science may well take an interest in it, and learn from it. But what I call 'methodology' should not be taken for an empirical science. I do not believe that it is possible to decide, by using the methods of an empirical science, such controversial questions as whether science actually uses a principle of induction or not. And my doubts increase when I remember that what is to be called a 'science' and who is to be called a 'scientist' must always remain a matter of convention or decision.

I believe that questions of this kind should be treated in a different way. For example, we may consider and compare two different systems of methodological rules; one with, and one without, a principle of induction. And we may then examine whether such a principle, once introduced, can be applied without giving rise to inconsistencies; whether it helps us; and whether we really need it. It is this type of inquiry which leads me to dispense with the principle of induction: not because such a principle is as a matter of fact never

---

[4] H. Gomperz (*Weltanschauungslehre I*, 1905, p. 35) writes: 'If we consider how infinitely problematic the concept of *experience* is . . . we may well be forced to believe that . . . enthusiastic affirmation is far less appropriate in regard to it . . . than the most careful and guarded criticism. . . .'

[5] Dingler, *Physik und Hypothesis*, Versuch einer induktiven Wissenschaftslehre (1921); similarly V. Kraft, *Die Grundformen der Wissenschaftlichen Methoden* (1925).

used in science, but because I think that it is not needed; that it does not help us; and that it even gives rise to inconsistencies.

Thus I reject the naturalistic view. It is uncritical. Its upholders fail to notice that whenever they believe themselves to have discovered a fact, they have only proposed a convention.[6] Hence the convention is liable to turn into a dogma. This criticism of the naturalistic view applies not only to its criterion of meaning, but also to its idea of science, and consequently to its idea of empirical method.

11. *Methodological Rules as Conventions.*

Methodological rules are here regarded as *conventions*. They might be described as the rules of the game of empirical science. They differ from the rules of pure logic rather as do the rules of chess, which few would regard as part of *pure* logic: seeing that the rules of pure logic govern transformations of linguistic formulae, the result of an inquiry into the rules of chess could perhaps be entitled 'The Logic of Chess', but hardly 'Logic' pure and simple. (Similarly, the result of an inquiry into the rules of the game of science—that is, of scientific discovery—may be entitled 'The Logic of Scientific Discovery'.)

Two simple examples of methodological rules may be given. They will suffice to show that it would be hardly suitable to place an inquiry into method on the same level as a purely logical inquiry.

(1) The game of science is, in principle, without end. He who decides one day that scientific statements do not call for any further test, and that they can be regarded as finally verified, retires from the game.

(2) Once a hypothesis has been proposed and tested, and has proved its mettle,[*1] it may not be allowed to drop out without 'good

---

[6] (Addition made in 1934 while this book was in proof.) The view, only briefly set forth here, that it is a matter for decision what is to be called 'a genuine statement' and what 'a meaningless pseudo-statement' is one that I have held for years. (Also the view that the exclusion of metaphysics is likewise a matter for decision.) However, my present criticism of positivism (and of the naturalistic view) no longer applies, as far as I can see, to Carnap's *Logische Syntax der Sprache* (1934), in which he too adopts the standpoint that all such questions rest upon decisions (the 'principle of tolerance'). According to Carnap's preface, Wittgenstein has for years propounded a similar view in unpublished works. (*See however note *1 above.) Carnap's *Logische Syntax* was published while the present book was in proof. I regret that I was unable to discuss it in my text.

[*1] Regarding the translation 'to prove one's mettle' for '*sich bewähren*', see the footnote to the beginning of chapter x (*Corroboration*).

reason'. A 'good reason' may be, for instance: replacement of the hypothesis by another which is better testable; or the falsification of one of the consequences of the hypothesis. (The concept 'better testable' will later be analysed more fully.)

These two examples show what methodological rules look like. Clearly they are very different from the rules usually called 'logical'. Although logic may perhaps set up criteria for deciding whether a statement is testable, it certainly is not concerned with the question whether anyone exerts himself to test it.

In section 6 I tried to define empirical science with the help of the criterion of falsifiability; but as I was obliged to admit the justice of certain objections, I promised a methodological supplement to my definition. Just as chess might be defined by the rules proper to it, so empirical science may be defined by means of its methodological rules. In establishing these rules we may proceed systematically. First a supreme rule is laid down which serves as a kind of norm for deciding upon the remaining rules, and which is thus a rule of a higher type. It is the rule which says that the other rules of scientific procedure must be designed in such a way that they do not protect any statement in science against falsification.

Methodological rules are thus closely connected both with other methodological rules and with our criterion of demarcation. But the connection is not a strictly deductive or logical one.[1] It results, rather, from the fact that the rules are constructed with the aim of ensuring the applicability of our criterion of demarcation; thus their formulation and acceptance proceeds according to a practical rule of a higher type. An example of this has been given above (cf. rule 1): theories which we decide not to submit to any further test would no longer be falsifiable. It is this systematic connection between the rules which makes it appropriate to speak of a *theory* of method. Admittedly the pronouncements of this theory are, as our examples show, for the most part conventions of a fairly obvious kind. Profound truths are not to be expected of methodology. Nevertheless it may help us in many cases to clarify the logical situation, and even to solve some far-reaching problems which have hitherto proved intractable. One of these, for example, is the problem of deciding when a probability statement should be accepted or rejected. (*Cf.* section 68.)

It has often been doubted whether the various problems of the

[1] *Cf.* K. Menger. *Moral, Wille und Weltgestaltung* (1934), p. 58 *ff.*

theory of knowledge stand in any systematic relation to one another, and also whether they can be treated systematically. I hope to show in this book that these doubts are unjustified. The point is of some importance. My only reason for proposing my criterion of demarcation is that it is fruitful: that a great many points can be clarified and explained with its help. 'Definitions are dogmas; only the conclusions drawn from them can afford us any new insight', says Menger.[2] This is certainly true of the definition of the concept 'science'. It is only from the consequences of my definition of empirical science, and from the methodological decisions which depend upon this definition, that the scientist will be able to see how far it conforms to his intuitive idea of the goal of his endeavours.[*2]

The philosopher too will accept my definition as useful only if he can accept its consequences. We must satisfy him that these consequences enable us to detect inconsistencies and inadequacies in older theories of knowledge, and to trace these back to the fundamental assumptions and conventions from which they spring. But we must also satisfy him that our own proposals are not threatened by the same kind of difficulties. This method of detecting and resolving contradictions is applied also within science itself, but it is of particular importance in the theory of knowledge. It is by this method, if by any, that methodological conventions might be justified, and might prove their value.[3]

Whether philosophers will regard these methodological investigations as belonging to philosophy is, I fear, very doubtful, but this does not really matter much. Yet it may be worth mentioning in this connection that not a few doctrines which are metaphysical, and thus certainly philosophical, could be interpreted as typical hypostatizations of methodological rules. An example of this, in the shape of what is called 'the principle of causality', will be discussed in the next section. Another example which we have already encountered

---

[2] K. Menger, *Dimensionstheorie* (1928), p. 76.

[*2] See also section *15, 'The Aim of Science', of my *Postscript*.

[3] In the present work I have relegated the critical—or, if you will, the 'dialectical'—method of resolving contradictions to second place, since I have been concerned with the attempt to develop the practical methodological aspects of my views. In an as yet unpublished work I have tried to take the critical path; and I have tried to demonstrate that the problems of both the classical and the modern theory of knowledge (from Hume via Kant to Russell and Whitehead) can be traced back to the problem of demarcation, that is, to the problem of finding the criterion of the empirical character of science.

is the problem of objectivity. For the requirement of scientific objectivity can also be interpreted as a methodological rule: the rule that only such statements may be introduced in science as are inter-subjectively testable (see sections 8, 20, 27, and elsewhere). It might indeed be said that the majority of the problems of theoretical philosophy, and the most interesting ones, can be re-interpreted in this way as problems of method.

PART II

SOME STRUCTURAL COMPONENTS OF A THEORY OF EXPERIENCE

CHAPTER III

THEORIES

THE empirical sciences are systems of theories. The logic of scientific knowledge can therefore be described as a theory of theories.

Scientific theories are universal statements. Like all linguistic representations they are systems of signs or symbols. Thus I do not think it is helpful to express the difference between universal theories and singular statements by saying that the latter are 'concrete' whereas theories are *merely* symbolic formulae or symbolic schemata; for exactly the same may be said of even the most 'concrete' statements.[*1]

Theories are nets cast to catch what we call 'the world': to rationalize, to explain, and to master it. We endeavour to make the mesh ever finer and finer.

12. *Causality, Explanation, and the Deduction of Predictions.*

To give a *causal explanation* of an event means to deduce a statement which describes it, using as premises of the deduction one or more *universal laws*, together with certain singular statements, the *initial conditions*. For example, we can say that we have given a causal

---

[*1] This is a critical allusion to a view which I later described as 'instrumentalism' and which was represented in Vienna by Mach, Wittgenstein, and Schlick (*cf.* notes *4 and 7 to section 4, and note 5 to section 27). It is the view that a theory is *nothing but* a tool or an instrument for prediction. I have analysed and criticized it in my papers 'A Note on Berkeley as a Precursor of Mach', *Brit. Journ. Philos. Science* 6, 1953, pp. 26 ff.; 'Three Views Concerning Human Knowledge', in *Contemporary British Philosophy* iii, 1956, edited by H. D. Lewis, pp. 355 ff.; and more fully in my *Postscript*, sections *11 to *15 and *19 to *26. My point of view is, briefly, that our ordinary language is full of theories; that observation is always *observation in the light of theories*; and that it is only the inductivist prejudice which leads people to think that there could be a phenomenal language, free of theories, and distinguishable from a 'theoretical language'; and lastly, that the theorist is interested in explanation as such, that is to say, in testable explanatory theories: applications and predictions interest him only for theoretical reasons—because they may be used as *tests* of theories.

explanation of the breaking of a certain piece of thread if we have found that the thread has a tensile strength of 1 *lb.* and that a weight of 2 *lbs.* was put on it. If we analyse this causal explanation we shall find several constituent parts. On the one hand there is the hypothesis: 'Whenever a thread is loaded with a weight exceeding that which characterizes the tensile strength of the thread, then it will break'; a statement which has the character of a universal law of nature. On the other hand we have singular statements (in this case two) which apply only to the specific event in question: 'The weight characteristic for this thread is 1 *lb.*', and 'The weight put on this thread was 2 *lbs.*'.[*1]

We have thus two different kinds of statement, both of which are necessary ingredients of a complete causal explanation. They are (1) *universal statements*, *i.e.* hypotheses of the character of natural laws, and (2) *singular statements*, which apply to the specific event in question and which I shall call 'initial conditions'. It is from universal statements in conjunction with initial conditions that we *deduce* the singular statement, 'This thread will break'. We call this statement a specific or singular *prediction*.[*2]

The initial conditions describe what is usually called the '*cause*' of the event in question. (The fact that a load of 2 *lbs.* was put on a thread with a tensile strength of 1 *lb.* was the 'cause' of its breaking.) And the prediction describes what is usually called the '*effect*'. Both these terms I shall avoid. In physics the use of the expression '*causal explanation*' is restricted as a rule to the special case in which the universal laws have the form of laws of 'action by contact'; or more precisely, of *action at a vanishing distance*, expressed by differential equations. This restriction will not be assumed here. Furthermore, I shall not make any general assertion as to the universal applicability of this deductive method of theoretical explanation. Thus I shall not assert any '*principle of causality*' (or 'principle of universal causation').

[*1] A clearer analysis of this example—and one which distinguishes *two* laws as well as two initial conditions—would be the following: 'For every thread of a given structure S (determined by its material, thickness, etc.) there is a characteristic weight w, such that the thread will break if any weight exceeding w is suspended from it.'—'For every thread of the structure S_1, the characteristic weight w_1 equals 1 *lb.*' These are the two universal laws. The two initial conditions are, 'This is a thread of structure S_1' and, 'The weight to be put on this thread is equal to 2 *lbs.*'

[*2] The term 'prediction', as used here, comprises statements about the past ('retrodictions'), or even 'given' statements which we wish to explain ('*explicanda*'); *cf.* my *Poverty of Historicism* (1945), p. 133 of the edition of 1957, and the *Postscript*, section *15.

The 'principle of causality' is the assertion that any event whatsoever *can* be causally explained—that it *can* be deductively predicted. According to the way in which one interprets the word 'can' in this assertion, it will be either tautological (analytic), or else an assertion about reality (synthetic). For if 'can' means that it is always logically possible to construct a causal explanation, then the assertion is tautological, since for any prediction whatsoever we can always find universal statements and initial conditions from which the prediction is derivable. (Whether these universal statements have been tested and corroborated in other cases is of course quite a different question.) If, however, 'can' is meant to signify that the world is governed by strict laws, that it is so constructed that every specific event is an instance of a universal regularity or law, then the assertion is admittedly synthetic. But in this case it is *not falsifiable*, as will be seen later, in section 78. I shall, therefore, neither adopt nor reject the 'principle of causality'; I shall be content simply to exclude it, as 'metaphysical', from the sphere of science.

I shall, however, propose a methodological rule which corresponds so closely to the 'principle of causality' that the latter might be regarded as its metaphysical version. It is the simple rule that we are not to abandon the search for universal laws and for a coherent theoretical system, nor ever give up our attempts to explain causally any kind of event we can describe.[1] This rule guides the scientific investigator in his work. The view that the latest developments in physics demand the renunciation of this rule, or that physics has

---

[1] The idea of regarding the principle of causality as the expression of a rule or of a decision is due to H. Gomperz, *Das Problem der Willensfreiheit* (1907). *Cf.* Schlick, *Die Kausalität in der gegenwartigen Physik*, *Naturwissenschaften* 19, 1931, p. 154.

*I feel that I should say here more explicitly that the decision to search for causal explanation is that by which the theoretician adopts his aim—or the aim of theoretical science. His aim is to find *explanatory theories* (if possible, *true* explanatory theories); that is to say, theories which describe certain structural properties of the world, and which permit us to deduce, with the help of initial conditions, the effects to be explained. It was the purpose of the present section to explain, if only very briefly, what we mean by causal explanation. A somewhat fuller statement will be found in appendix *x, and in my *Postscript*, section *15. My explanation of explanation has been adopted by certain positivists or 'instrumentalists' who saw in it an attempt to explain it away—as the assertion that explanatory theories are *nothing but* premises for deducing predictions. I therefore wish to make it quite clear that I consider the theorist's interest in *explanation*— that is, in discovering explanatory theories—as irreducible to the practical technological interest in the deduction of predictions. The theorist's interest in *predictions*, on the other hand, is explicable as due to his interest in the problem whether his theories are true; or in other words, as due to his interest in testing his theories—in trying to find out whether they cannot be shown to be false. See also appendix *x, note 4 and text.

now established that within one field at least it is pointless to seek any longer for laws, is not accepted here.[2] This matter will be discussed in section 78.[*3]

13. Strict and Numerical Universality.

We can distinguish two kinds of universal synthetic statement: the 'strictly universal' and the 'numerically universal'. It is the *strictly universal statements* which I have had in mind so far when speaking of universal statements—of theories or natural laws. The other kind, the numerically universal statements, are in fact equivalent to certain singular statements, or to conjunctions of singular statements, and they will be classed as singular statements here.

Compare, for example, the following two statements: (a) Of all harmonic oscillators it is true that their energy never falls below a certain amount (*viz.* $h\nu/2$); and (b) Of all human beings now living on the earth it is true that their height never exceeds a certain amount (say 8 ft.). Formal logic (including symbolic logic), which is concerned only with the theory of deduction, treats these two statements alike as universal statements ('formal' or 'general' implications).[1] I think however that it is necessary to emphasize the difference between them. Statement (a) claims to be true for any place and any time. Statement (b) refers only to a finite class of specific elements within a finite individual (or particular) spatio-temporal region. Statements of this latter kind can, in principle, be replaced by a conjunction of singular statements; for given sufficient time, one can *enumerate* all the elements of the (finite) class concerned. This is why we speak in such cases of 'numerical universality'. By contrast, statement (a), about the oscillators, cannot be replaced by a con-

[2] The view here opposed is held by Schlick, among others; *op cit.* p. 155 ... that impossibility ...' (he is referring to the impossibility of exact prediction maintained by Heisenberg) '... means that it is impossible to *search for* that formula.' (*Cf.* also note 1 to section 78.)

[*3] But see now also chapters *iv to *vi of my *Postscript*.

[1] Classical logic (and similarly symbolic logic or 'logistic') distinguishes universal, particular, and singular statements. A universal statement is one referring to all the elements of some class; a particular statement is one referring to some among its elements; a singular statement is one referring to one given element—an individual. This classification is not based on reasons connected with the logic of knowledge. It was developed with an eye to the technique of inference. We can therefore identify our 'universal statements' neither with the universal statements of classical logic nor with the 'general' or 'formal' implications of logistic (*cf.* note 6 to section 14). *See now also appendix *x, and my *Postscript*, especially section *15.

junction of a finite number of singular statements about a definite spatio-temporal region; or rather, it could be so replaced only on the assumption that the world is bounded in time and that there exists only a finite number of oscillators in it. But we do not make any such assumption; in particular, we do not make any such assumption in defining the concepts of physics. Rather we regard a statement of type (a) as an *all-statement*, *i.e.* a universal assertion about an unlimited number of individuals. So interpreted it clearly cannot be replaced by a conjunction of a finite number of singular statements.

My use of the concept of a strictly universal statement (or 'all-statement') stands opposed to the view that every synthetic universal statement must in principle be translatable into a conjunction of a finite number of singular statements. Those who adhere to this view[2] insist that what I call 'strictly universal statements' can never be verified, and they therefore reject them, referring either to their criterion of meaning, which demands verifiability, or to some similar consideration.

It is clear that on any such view of natural laws which obliterates the distinction between singular and universal statements, the problem of induction would seem to be solved; for obviously, inferences from singular statements to merely numerically universal ones may be perfectly admissible. But it is equally clear that the methodological problem of induction is not affected by this solution. For the verification of a natural law could only be carried out by empirically ascertaining every single event to which the law might apply, and by finding that every such event actually conforms to the law—clearly an impossible task.

In any case, the question whether the laws of science are strictly or numerically universal cannot be settled by argument. It is one of those questions which can be settled only by an agreement or a convention. And in view of the methodological situation just referred to, I consider it both useful and fruitful to regard natural laws as synthetic and strictly universal statements ('all-statements'). This is to regard them as non-verifiable statements which can be put in the form: 'Of all points in space and time (or in all regions of space and time) it is true that . . .'. By contrast, statements which relate only to certain finite regions of space and time I call 'specific' or 'singular' statements.

---

[2] *Cf.* for instance F. Kaufmann, *Bemerkungen zum Grundlagenstreit in Logik und Mathematik, Erkenntnis* 2, 1931, p. 274.

The distinction between strictly universal statements and merely numerically universal statements (*i.e.* really a kind of singular statement) will be applied to synthetic statements only. I may, however, mention the possibility of applying this distinction to analytic statements also (for example, to certain mathematical statements).[3]

14. *Universal Concepts and Individual Concepts.*

The distinction between universal and singular *statements* is closely connected with that between *universal and individual concepts or names.*

It is usual to elucidate this distinction with the help of examples of the following kind: 'dictator', 'planet', 'H_2O' are universal concepts or universal names. 'Napoleon', 'the earth', 'the Atlantic' are singular or individual concepts or names. In these examples individual concepts or names appear to be characterized either by being proper names, or by having to be defined by means of proper names, whilst universal concepts or names can be defined without the use of proper names.

I consider the distinction between universal and individual concepts or names to be of fundamental importance. Every application of science is based upon an inference from scientific hypotheses (which are universal) to singular cases, *i.e.* upon a deduction of singular predictions. But in every singular statement individual concepts or names must occur.

The individual names that occur in the singular statements of science often appear in the guise of spatio-temporal co-ordinates. This is easily understood if we consider that the *application* of a spatio-temporal system of co-ordinates always involves reference to individual names. For we have to fix its points of origin, and this we can do only by making use of proper names (or their equivalents). The use of the names 'Greenwich' and 'The year of Christ's birth' illustrates what I mean. By this method an arbitrarily large number of individual names may be reduced to a very few.[1]

---

[3] Examples: (a) Every natural number has a successor. (b) With the exception of the numbers 11, 13, 17, and 19, all numbers between 10 and 20 are divisible.

[1] But the units of measurement of the co-ordinate system which first were also established by individual names (the rotation of the earth; the standard metre in Paris) can be defined in principle by means of universal names, for example by means of the wave-length or frequency of the monochromatic light emitted by a certain kind of atoms treated in a certain way.

Such vague and general expressions as 'this thing here', 'that thing over there', etc., can sometimes be used as individual names, perhaps in conjunction with ostensive gestures of some kind; in short, we can use signs which are not proper names but which to some extent are interchangeable with proper names or with individual co-ordinates. But universal concepts too can be indicated, if only vaguely, with the help of ostensive gestures. Thus we may point to certain individual things (or events) and then express by a phrase like 'and other similar things' (or 'and so on') our intention to regard these individuals only as representatives of some class which should properly be given a universal name. There can be no doubt that *we learn the use* of universal words, that is their *application* to individuals, by ostensive gestures and by similar means. The logical basis of applications of this kind is that individual concepts may be concepts not only of elements but also of classes, and that they may thus stand to universal concepts not only in a relation corresponding to that of an element to a class, but also in a relation corresponding to that of a sub-class to a class. For example, my dog Lux is not only an element of the class of Viennese dogs, which is an individual concept, but also an element of the (universal) class of mammals, which is a universal concept. And the Viennese dogs, in their turn, are not only a sub-class of the (individual) class of Austrian dogs, but also a sub-class of the (universal) class of mammals.

The use of the word 'mammals' as an example of a universal name might possibly cause misunderstanding. For words like 'mammal', 'dog', etc., are in their ordinary use not free from ambiguity. Whether these words are to be regarded as individual class names or as universal class names depends upon our intentions: it depends upon whether we wish to speak of a race of animals living on our planet (an individual concept), or of a kind of physical bodies with properties which can be described in universal terms. Similar ambiguities arise in connection with the use of concepts such as 'pasteurized', 'Linnean System', and 'Latinism', in so far as it is possible to eliminate the proper names to which they allude (or else, to define them with the help of these proper names).[*1]

The above examples and explanations should make it clear what

[*1] 'Pasteurized' may be defined, either, as 'treated according to the advice of M. Louis Pasteur' (or something like this), or else as 'heated to 80 degrees centigrade and kept at this temperature for ten minutes'. The first definition makes 'pasteurized' an individual concept; the second makes it a universal concept.

will here be meant by 'universal concepts' and 'individual concepts'. If I were asked for definitions I should probably have to say, as above: 'An individual concept is a concept in the definition of which proper names (or equivalent signs) are indispensable. If any reference to proper names can be completely eliminated, then the concept is a universal concept.' Yet any such definition would be of very little value, since all that it does is to reduce the idea of an individual concept or name to that of a proper name (in the sense of a name of one individual physical thing).

I believe that my usage corresponds fairly closely to the customary use of the expressions 'universal' and 'individual'. But whether or not this is so I certainly consider the distinction here made to be indispensable if we are not to blur the corresponding distinction between universal and singular statements. (There is a complete analogy between the problem of universals and the problem of induction.) The attempt to identify an individual thing *merely* by its universal properties and relations, which appear to belong to it alone and to nothing else, is foredoomed to failure. Such a procedure would describe not a single individual thing but the universal class of all those individuals to which these properties and relations belong. Even the use of a universal spatio-temporal system of co-ordinates would alter nothing.[2] For whether there are any individual things corresponding to a description by means of universal names, and if so how many, must always remain an open question.

In the same way, any attempt to define universal names with the help of individual names is bound to fail. This fact has often been overlooked, and it is widely believed that it is possible to rise by a process called 'abstraction' from individual concepts to universal concepts. This view is a near relation of inductive logic, with its passage from singular statements to universal statements. Logically, these procedures are equally impracticable.[3] It is true that one can obtain classes of individuals in this way, but these classes will still be individual concepts—concepts defined with the help of proper names. (Examples of such individual class-concepts are 'Napoleon's

---

[2] Not 'space and time' in general but individual determinations (spatial, temporal, or others) based on proper names are 'principles of individuation'.

[3] Similarly, the 'method of abstraction' used in symbolic logic is unable to accomplish the ascent from individual names to universal names. If the class defined by means of abstraction is defined extensionally with the help of individual names, then it is in its turn an individual concept.

generals', and 'the inhabitants of Paris'.) Thus we see that my distinction between universal names or concepts and individual names or concepts has nothing to do with the distinction between classes and elements. Both universal names and individual names may occur as names of some classes, and also as the names of elements of some classes.

It is therefore not possible to abolish the distinction between individual concepts and universal concepts with arguments like the following of Carnap's: '. . . this distinction is not justified', he says, because '. . . every concept can be regarded as an individual or universal concept according to the point of view adopted'. Carnap tries to support this by the assertion '. . . that (almost) *all so-called individual concepts are* (names of) *classes*, just like universal concepts'.[4] This last assertion is quite correct, as I have shown, but has nothing to do with the distinction in question.

Other workers in the field of symbolic logic (at one time called 'logistics') have similarly confused the distinction between universal names and individual names with the distinction between classes and their elements.[5] It is certainly permissible to use the term 'universal name' as a synonym for 'name of a class', and 'individual name' as a synonym for 'name of an element'; but there is little to be said for

---

[4] Carnap, *Der logische Aufbau der Welt*, p. 213. (Addition made in 1934 while the work was in proof.) In Carnap's *Logical Syntax of Language* (1934; Engl. ed. 1937), the distinction between individual names and universal names does not seem to have been considered; nor does this distinction seem to be expressible in the 'co-ordinate language' which he constructs. One might perhaps think that the 'co-ordinates', being signs of lowest type (*cf.* pp. 12 *f.*), are to be interpreted as *individual* names (and that Carnap uses a co-ordinate system defined with the help of individuals). But this interpretation will not do, since Carnap writes (p. 87; see also p. 12 Engl. ed., p. 97, para. 4) that in the language which he uses '. . . all expressions of lowest type are numerical expressions' in the sense that they denote what would fall under Peano's undefined primitive sign 'number' (*cf.* pp. 31 and 33). This makes it clear that the number signs appearing as co-ordinates are not to be thought of as proper names or individual co-ordinates, but as universals. (They are 'individual' only in a Pickwickian sense, *cf.* note 3 (b) to section 13.)

[5] The distinction drawn by Russell and Whitehead between individuals (or particulars) and universals has also nothing to do with the distinction here introduced between individual names and universal names. According to Russell's terminology, in the sentence 'Napoleon is a French general', 'Napoleon' is, as in my scheme, an individual, but 'French general' is a universal; but conversely, in the sentence 'Nitrogen is a non-metal', 'non-metal' is, as in my scheme, a universal, but 'nitrogen' is an individual. Moreover, what Russell calls 'descriptions' does not correspond to my 'individual names' since *e.g.* the class of 'geometrical points falling within my body', is for me an individual concept, but cannot be represented by means of a 'description'. *Cf.* Whitehead and Russell *Principia Mathematica* (2nd edition 1925, vol. I), Introduction to the second edition, II 1, pp. xix, *f.*

this usage. Problems cannot be solved in this way; on the other hand, this usage may very well prevent one from seeing them. The situation here is quite similar to what we met before when discussing the distinction between universal and singular statements. The instruments of symbolic logic are no more adequate for handling the problem of universals than for handling the problem of induction.[6]

15. *Strictly Universal and Existential Statements.*

It is not enough, of course, to characterize universal statements as statements in which no individual names occur. If the word 'raven' is used as a universal name, then, clearly, the statement 'all ravens are black' is a strictly universal statement. But in many other statements such as 'many ravens are black' or perhaps 'some ravens are black' or 'there are black ravens', etc., there also occur only universal names; yet we should certainly not describe such statements as universal.

Statements in which only universal names and no individual names occur will here be called 'strict' or 'pure'. Most important among them are the *strictly universal* statements which I have already discussed. In addition to these, I am especially interested in statements of the form 'there are black ravens', which may be taken to mean the same as 'there exists at least one black raven'. Such statements will be called *strictly or purely existential statements* (or *'there-is' statements*).

The negation of a strictly universal statement is always equivalent to a strictly existential statement and *vice versa*. For example, 'not all ravens are black' says the same thing as 'there exists a raven which is not black', or 'there are non-black ravens'.

The theories of natural science, and especially what we call

---

[6] The difference between universal and singular statements can also not be expressed in the system of Whitehead and Russell. It is not correct to say that the so-called 'formal' or 'general' implications must be universal statements. For every singular statement can be put in the form of a general implication. For example, the statement 'Napoleon was born in Corsica' can be expressed in the form, $(x) (x = N \rightarrow \phi x)$, in words: it is true for all values of x that, if x is identical with Napoleon, then x was born in Corsica.

A *general implication* is written, '$(x) (\phi x \rightarrow f x)$', where the universal 'operator', '(x)', can be read: 'It is true for all values of x'; 'ϕx' and '$f x$' are *propositional functions*: (e.g. 'x was born in Corsica', without its being said who x is; a propositional function can be neither true nor false). '$\rightarrow$' stands for: 'if it is true that . . . then it is true that . . .' the propositional function ϕx preceding '$\rightarrow$' may be called the *antecedent* or the *conditioning propositional function*, and $f x$ the '*consequent propositional function*' or the '*predication*'; and the *general implication*, $(x) (\phi x \rightarrow f x)$, asserts that all values of x which satisfy ϕ also satisfy f.

natural laws, have the logical form of strictly universal statements; thus they can be expressed in the form of negations of strictly existential statements or, as we may say, in the form of *non-existence statements* (or 'there-is-not' statements). For example, the law of the conservation of energy can be expressed in the form: 'There is no perpetual motion machine', or the hypothesis of the electrical elementary charge in the form: 'There is no electrical charge other than a multiple of the electrical elementary charge'.

In this formulation we see that natural laws might be compared to 'proscriptions' or 'prohibitions'. They do not assert that something exists or is the case; they deny it. They insist on the non-existence of certain things or states of affairs, proscribing or prohibiting, as it were, these things or states of affairs: they rule them out. And it is precisely because they do this that they are *falsifiable*. If we accept as true one singular statement which, as it were, infringes the prohibition by asserting the existence of a thing (or the occurrence of an event) ruled out by the law, then the law is refuted. (An instance would be, 'In such-and-such a place, there is an apparatus which is a perpetual motion machine'.)

Strictly existential statements, by contrast, cannot be falsified. No singular statement (that is to say, no 'basic statement', no statement of an observed event) can contradict the existential statement, 'There are white ravens'. Only a universal statement could do this. On the basis of the criterion of demarcation here adopted I shall therefore have to treat strictly existential statements as non-empirical or 'metaphysical'. This characterization may perhaps seem dubious at first sight and not quite in accordance with the practice of empirical science. By way of objection, it might be asserted (with justice) that there are theories even in physics which have the form of strictly existential statements. An example would be a statement, deducible from the periodic system of chemical elements, which asserts the existence of elements of certain atomic numbers. But if the hypothesis that an element of a certain atomic number exists is to be so formulated that it becomes testable, then much more is required than a purely existential statement. For example, the element with the atomic number 72 (Hafnium) was not discovered merely on the basis of an isolated purely existential statement. On the contrary, all attempts to find it were in vain until Bohr succeeded in predicting several of its properties by deducing them from his theory. But Bohr's theory and those of its conclusions which were relevant to this element and which helped to bring about its discovery are far from being isolated purely existential statements.*1 They are strictly universal statements. That my decision to regard strictly existential statements as non-empirical—because they are not falsifiable—is helpful, and also in accordance with ordinary usage, will be seen from its application to probability statements and to the problem of testing them empirically. (*Cf.* sections 66–68.)

Strict or pure statements, whether universal or existential, are not limited as to space and time. They do not refer to an individual, restricted, spatio-temporal region. This is the reason why strictly existential statements are not falsifiable. We cannot search the whole world in order to establish that something does not exist, has never existed, and will never exist. It is for precisely the same reason that strictly universal statements are not verifiable. Again, we cannot search the whole world in order to make sure that nothing exists which the law forbids. Nevertheless, both kinds of strict statements, strictly existential and strictly universal, are in principle empirically decidable; each, however, in *one way only*: they are *unilaterally decidable*. Whenever it is found that something exists here or there, a strictly existential statement may thereby be verified, or a universal one falsified.

The asymmetry here described, with its consequence, the one-sided falsifiability of the universal statements of empirical science, may now perhaps seem less dubious than it did before (in section 6). We now see that no asymmetry of any purely *logical* relationship is involved. On the contrary, the logical relationships show symmetry. Universal and existential statements are constructed symmetrically. It is only*2 the line drawn by our criterion of demarcation which produces an asymmetry.

*1 The word 'isolated' has been inserted to avoid misinterpretation of the passage though its tendency, I feel, was clear enough: an *isolated* existential statement is never falsifiable; but if taken *in a context* with other statements, an existential statement *may in some cases* add to the empirical content of the whole context: it may enrich the theory to which it belongs, and may add to its degree of falsifiability or testability. In this case, the theoretical system including the existential statement in question is to be described as scientific rather than metaphysical.

*2 The word 'only' here should not be taken too seriously. The situation is quite simple. If it is characteristic of empirical science to look upon *singular* statements as test-statements, then the asymmetry arises from the fact that, *with respect to singular statements*, universal statements are falsifiable only and existential statements verifiable only. See also section *22 of my *Postscript*.

16. *Theoretical Systems.*

Scientific theories are perpetually changing. This is not due to mere chance but might well be expected, according to our characterization of empirical science.

Perhaps this is why, as a rule, only *branches* of science—and these only temporarily—ever acquire the form of an elaborate and logically well-constructed system of theories. In spite of this, a tentative system can usually be quite well surveyed as a whole, with all its important consequences. This is very necessary; for a severe test of a system presupposes that it is at the time sufficiently definite and final in form to make it impossible for new assumptions to be smuggled in. In other words, the system must be formulated sufficiently clearly and definitely to make every new assumption easily recognizable for what it is: a modification and therefore a *revision* of the system.

This, I believe, is the reason why the form of a rigorous system is aimed at. It is the form of a so-called '*axiomatized system*'—the form which Hilbert, for example, was able to give to certain branches of theoretical physics. The attempt is made to collect all the assumptions which are needed, but no more, to form the apex of the system. They are usually called the 'axioms' (or 'postulates', or 'primitive propositions'; no claim to truth is implied in the term 'axiom' as here used). The axioms are chosen in such a way that all the other statements belonging to the theoretical system can be derived from the axioms by purely logical or mathematical transformations.

A theoretical system may be said to be axiomatized if a set of statements, the axioms, has been formulated which satisfies the following four fundamental requirements. (a) The system of axioms must be *free from contradiction* (whether self-contradiction or mutual contradiction). This is equivalent to the demand that not every arbitrarily chosen statement is deducible from it.[1] (b) The system must be *independent*, *i.e.* it must not contain any axiom deducible from the remaining axioms. (In other words, a statement is to be called an axiom only if it is not deducible within the rest of the system.) These two conditions concern the axiom system as such; as regards the relation of the axiom system to the bulk of the theory, the axioms should be (c) *sufficient* for the deduction of all statements belonging

[1] *Cf.* section 24.

to the theory which is to be axiomatized, and (d) *necessary*, for the same purpose; which means that they should contain no superfluous assumptions.[2]

In a theory thus axiomatized it is possible to investigate the mutual dependence of various parts of the system. For example, we may investigate whether a certain part of the theory is derivable from some part of the axioms. Investigations of this kind (of which more will be said in sections 63 and 64, and 75 to 77) have an important bearing on the problem of falsifiability. They make it clear why the falsification of a logically deduced statement may sometimes not affect the whole system but only some part of it, which may then be regarded as falsified. This is possible because, although the theories of physics are in general not completely axiomatized, the connections between its various parts may yet be sufficiently clear to enable us to decide which of its sub-systems are affected by some particular falsifying observation.[*1]

17. *Some Possibilities of Interpreting a System of Axioms.*

The view of classical rationalism that the 'axioms' of certain systems, *e.g.* those of Euclidean geometry, must be regarded as immediately or intuitively certain, or self-evident, will not be discussed here. I will only mention that I do not share this view. I consider two different interpretations of any system of axioms to be admissible. The axioms may be regarded either (i) as *conventions*, or they may be regarded (ii) as empirical or scientific *hypotheses*.

(i) If the axioms are regarded as *conventions* then they tie down the use or meaning of the fundamental ideas (or primitive terms, or concepts) which the axioms introduce; they determine what can and what cannot be said about these fundamental ideas. Sometimes the axioms are described as '*implicit definitions*' of the ideas which they introduce. This view can perhaps be elucidated by means of an analogy between an axiomatic system and a (consistent and soluble) system of equations.

The admissible values of the 'unknowns' (or variables) which appear in a system of equations are in some way or other determined

---

[2] Regarding these four conditions, and also the following section, see, for example, the somewhat different account in Carnap's *Abriss der Logistik* (1927), p. 70 *ff*.
[*1] The point is more fully discussed in my *Postscript*, especially section *22.

by it. Even if the system of equations does not suffice for a unique solution, it does not allow every conceivable combination of values to be substituted for the 'unknowns' (variables). Rather, the system of equations characterizes certain combinations of values or value-systems as admissible, and others as inadmissible; it distinguishes the class of admissible value systems from the class of inadmissible value systems. In a similar way, systems of concepts can be distinguished as admissible or as inadmissible by means of what might be called a 'statement-equation'. A statement-equation is obtained from a propositional function or statement-function (*cf.* note 6 to section 14); this is an incomplete statement, in which one or more 'blanks' occur. Two examples of such propositional functions or statement functions are: 'An isotope of the element x has the atomic weight 65'; or '$x + y = 12$'. Every such statement-function is transformed into a *statement* by the substitution of certain values for the blanks, x and y. The resulting statement will be either true or false, according to the values (or combination of values) substituted. Thus, in the first example, substitution of the word 'copper' or 'zinc' for 'x' yields a true statement, while other substitutions yield false ones. Now what I call a 'statement-equation' is obtained if we decide, with respect to some statement-function, to admit only such values for substitution as turn this function into a *true statement*. By means of this statement-equation a definite class of admissible value-systems is defined, namely the class of those which satisfy it. The analogy with a mathematical equation is clear. If our second example is interpreted, not as a statement-function but as a statement-equation, then it becomes an equation in the ordinary (mathematical) sense.

Since its undefined fundamental ideas or primitive terms can be regarded as blanks, an axiomatic system can, to begin with, be treated as a system of statement-functions. But if we decide that only such systems or combinations of values may be substituted as will satisfy it, then it becomes a system of statement-equations. As such it implicitly defines a class of (admissible) systems of concepts. Every system of concepts which satisfies a system of axioms can be called a *model of that system of axioms*.[*1]

The interpretation of an axiomatic system as a system of (conventions or) implicit definitions can also be expressed by saying that it amounts to the decision: only models may be admitted as

[*1] See note *2.

substitutes.*² But if a model is substituted then the result will be a system of analytic statements (since it will be true by convention). An axiomatic system interpreted in this way cannot therefore be regarded as a system of empirical or scientific hypotheses (in our sense) since it cannot be refuted by the falsification of its consequences; for these too must be analytic.

(ii) How then, it may be asked, can an axiomatic system be interpreted as a system of empirical or scientific *hypotheses*? The usual view is that the primitive terms occurring in the axiomatic system are not to be regarded as implicitly defined, but as 'extra-logical constants'. For example, such concepts as 'straight line' and 'point', which occur in every axiom system of geometry, may be interpreted as 'light ray' and 'intersection of light rays'. In this way, it is thought, the statements of the axiom system become statements about empirical objects, that is to say, synthetic statements.

At first sight, this view of the matter may appear perfectly satisfactory. It leads, however, to difficulties which are connected with the problem of the empirical basis. For it is by no means clear what would be an *empirical way of defining a concept*. It is customary to speak of 'ostensive definitions'. This means that a definite empirical meaning is assigned to a concept by *correlating it* with certain objects belonging to the real world. It is then regarded as a symbol of those objects. But it should have been clear that only individual names or concepts can be fixed by ostensively referring to 'real objects'—say, by pointing to a certain thing and uttering a name, or by attaching to it a label bearing a name, etc. Yet the concepts which are to be used in the axiomatic system should be universal names, which cannot be defined by empirical indications, pointing, etc. They can be defined if at all only explicitly, *with the help of other universal names*; otherwise they can only be left undefined. That some universal names should remain undefined is therefore quite unavoidable; and herein lies the difficulty. For these undefined concepts can always be used in the non-empirical sense (i), *i.e.* as if they were implicitly defined concepts. Yet this use must inevitably destroy the empirical character of the system. This difficulty, I believe, can only be overcome by

*² Today I should clearly distinguish between the *systems of objects* which satisfy an axiom system and the *system of names of these objects* which may be substituted in the axioms (rendering them true); and I should call only the first system a 'model'. Accordingly, I should now write: 'only names of objects which constitute a model may be admitted for substitution'.

means of a methodological decision. I shall, accordingly, adopt a rule not to use undefined concepts as if they were implicitly defined. (This point will be dealt with below in section 20.)

Here I may perhaps add that it is usually possible for the primitive concepts of an axiomatic system such as geometry to be correlated with, or interpreted by, the concepts of another system, *e.g.* physics. This possibility is particularly important when, in the course of the evolution of a science, one system of statements is being *explained* by means of a new—a more general—system of hypotheses which permits the deduction not only of statements belonging to the first system, but also of statements belonging to other systems. In such cases it may be possible to define the fundamental concepts of the new system with the help of concepts which were originally used in some of the old systems.

18. *Levels of Universality. The Modus Tollens.*

We may distinguish, within a theoretical system, statements belonging to various levels of universality. The statements on the highest level of universality are the axioms; statements on the lower levels can be deduced from them. Higher level empirical statements have always the character of hypotheses relative to the lower level statements deducible from them: they can be falsified by the falsification of these less universal statements. But in any hypothetical deductive system, these less universal statements are themselves still strictly universal statements, in the sense here understood. Thus they too must have the character of *hypotheses*—a fact which has often been overlooked in the case of lower-level universal statements. Mach, for example, calls[1] Fourier's theory of heat conduction a 'model theory of physics' for the curious reason that 'this theory is founded not on a *hypothesis* but on an *observable fact*'. However, the 'observable fact' to which Mach refers is described by him by the statement, '... the velocity of the levelling out of temperature differences, provided these differences of temperature are small, is proportional to these differences themselves'—an all-statement whose hypothetical character should be sufficiently conspicuous.

I shall say even of some singular statements that they are hypothetical, seeing that conclusions may be derived from them (with the

[1] Mach, *Principien der Wärmelehre* (1896), p. 115.

75

help of a theoretical system) such that the falsification of these conclusions may falsify the singular statements in question.

The falsifying mode of inference here referred to—the way in which the falsification of a conclusion entails the falsification of the system from which it is derived—is the *modus tollens* of classical logic. It may be described as follows:[*1]

Let p be a conclusion of a system t of statements which may consist of theories and initial conditions (for the sake of simplicity I will not distinguish between them). We may then symbolize the relation of derivability (analytical implication) of p from t by '$t \to p$' which may be read: 'p follows from t'. Assume p to be false, which we may write '$\bar{p}$', to be read 'not-p'. Given the relation of deducibility, $t \to p$, and the assumption $\bar{p}$, we can then infer $\bar{t}$ (read 'not-t'); that is, we regard t as falsified. If we denote the conjunction (simultaneous assertion) of two statements by putting a point between the symbols standing for them, we may also write the falsifying inference thus: $((t \to p).\bar{p}) \to \bar{t}$, or in words: 'If p is derivable from t, and if p is false, then t also is false'.

By means of this mode of inference we falsify *the whole system* (the theory as well as the initial conditions) which was required for the deduction of the statement p, *i.e.* of the falsifying statement. Thus it cannot be asserted of any one statement of the system that it is, or is not, specifically upset by the falsification. Only if p is *independent* of some part of the system can we say that this part is not involved in the falsification.[2] With this is connected the following

---

[*1] In connection with the present passage and two later passages (*cf.* notes *I to section 35 and *I to section 36) in which I use the symbol '$\to$', I wish to say that when writing the book, I was still in a state of confusion about the distinction between a conditional statement (if-then-statement; sometimes called, somewhat misleadingly, 'material implication') and a statement about deducibility (or a statement asserting that some conditional statement is logically true, or analytic, or that its antecedent entails its consequent)—a distinction which I was taught to understand by Alfred Tarski, a few months after the publication of the book. The problem is not very relevant to the context of the book; but the confusion should be pointed out nevertheless. (These problems are discussed more fully, for example, in my paper in *Mind* 56, 1947, pp. 193 *ff.*)

[2] Thus we cannot at first know which among the various statements of the remaining sub-system t' (of which p is not independent) we are to blame for the falsity of p; which of these statements we have to alter, and which we should retain. (I am not here discussing interchangeable statements.) It is often only the scientific instinct of the investigator (influenced, of course, by the results of testing and re-testing) that makes him guess which statements of t' he should regard as innocuous, and which he should regard as being in need of modification. Yet it is worth remembering that it is often the modification of what we are inclined to regard as obviously innocuous (because of its complete agreement with our normal habits of thought) which may produce a decisive advance. A notable example of this is Einstein's modification of the concept of simultaneity.

possibility: we may, in some cases, perhaps in consideration of the *levels of universality*, attribute the falsification to some definite hypothesis—for instance to a newly introduced hypothesis. This may happen if a well-corroborated theory, and one which continues to be further corroborated, has been deductively explained by a new hypothesis of a higher level. The attempt will have to be made to test this new hypothesis by means of some of its consequences which have not yet been tested. If any of these are falsified, then we may well attribute the falsification to the new hypothesis alone. We shall then seek, in its stead, other high-level generalizations, but we shall not feel obliged to regard the old system, of lesser generality, as having been falsified. (*Cf.* also the remarks on 'quasi-induction' in section 85.)

CHAPTER IV

FALSIFIABILITY

THE question whether there is such a thing as a falsifiable singular statement (or a 'basic statement') will be examined later. Here I shall assume a positive answer to this question; and I shall examine how far my criterion of demarcation is applicable to theoretical systems—if it is applicable at all. A critical discussion of a position usually called 'conventionalism' will raise first some problems of method, to be met by taking certain *methodological decisions*. Next I shall try to characterize the logical properties of those systems of theories which are falsifiable—falsifiable, that is, if our methodological decisions are adopted.

19. *Some Conventionalist Objections.*

Objections are bound to be raised against my proposal to adopt falsifiability as our criterion for deciding whether or not a theoretical system belongs to empirical science. They will be raised, for example, by those who are influenced by the school of thought known as 'conventionalism'.[1] Some of these objections have already been touched upon in sections 6, 11, and 17; they will now be considered a little more closely.

---

[1] The chief representatives of the school are Poincaré and Duhem (*cf. La theorie physique, son objet et sa structure*, 1906; English translation by P. P. Wiener: *The Aim and Structure of Physical Theory*, Princeton, 1954). A recent adherent is H. Dingler (among his numerous works may be mentioned: *Das Experiment*, and *Der Zusammenbruch der Wissenschaft und das Primat der Philosophie*, 1926). * The German Hugo Dingler should not be confused with the Englishman Herbert Dingle. The chief representative of Conventionalism in the English-speaking world is Eddington. It may be mentioned here that Duhem denies (Engl. transl. p. 300) the possibility of crucial experiments, because he thinks of them as verifications, while I assert the possibility of crucial *falsifying* experiments. *Cf.* also my paper 'Three Views Concerning Human Knowledge', in *Contemporary British Philosophy*, iii, 1956, and in my *Conjectures and Refutations*, 1959.

The source of the conventionalist philosophy would seem to be wonder at the austerely beautiful *simplicity of the world* as revealed in the laws of physics. Conventionalists seem to feel that this simplicity would be incomprehensible, and indeed miraculous, if we were bound to believe, with the realists, that the laws of nature reveal to us an inner, a structural, simplicity of our world beneath its outer appearance of lavish variety. Kant's idealism sought to explain this simplicity by saying that it is our own intellect which imposes its laws upon nature. Similarly, but even more boldly, the conventionalist treats this simplicity as our own creation. For him, however, it is not the effect of the laws of our intellect imposing themselves upon nature, thus making nature simple; for he does not believe that nature is simple. Only the '*laws of nature*' are simple; and these, the conventionalist holds, are our own free creations; our inventions; our arbitrary decisions and conventions. For the conventionalist, theoretical natural science is not a picture of nature but merely a logical construction. It is not the properties of the world which determine this construction; on the contrary it is this construction which determines the properties of an artificial world: a world of concepts implicitly defined by the natural laws which we have chosen. It is only *this* world of which science speaks.

According to this conventionalist point of view, laws of nature are not falsifiable by observation; for they are needed to determine what an observation and, more especially, what a scientific measurement is. It is these laws, laid down by us, which form the indispensable basis for the regulation of our clocks and the correction of our so-called 'rigid' measuring-rods. A clock is called 'accurate' and a measuring rod 'rigid' only if the movements measured with the help of these instruments satisfy the axioms of mechanics which we have decided to adopt.[2]

---

[2] This view can also be regarded as an attempt to solve the problem of induction; for the problem would vanish if natural laws were definitions, and therefore tautologies. Thus according to the views of Cornelius (*cf. Zur Kritik der wissenschaftlichen Grundbegriffe, Erkenntnis* **2**, 1931, Number 4) the statement, 'The melting point of lead is about 335° C.' is part of the definition of the concept 'lead' (suggested by inductive experience) and cannot therefore be refuted. A substance otherwise resembling lead but with a different melting point would simply not be lead. But according to my view the statement of the melting point of lead is, *qua* scientific statement, synthetic. It asserts, among other things, that an element with a given atomic structure (atomic number 82) always has this melting point, whatever name we may give to this element.

(Added to the book in proof.) Ajdukiewicz appears to agree with Cornelius (*cf. Erkenntnis* **4**, 1934, p. 100 *f.*, as well as the work there announced, *Das Weldbild und die Begriffsapparatur*); he calls his standpoint 'radical conventionalism'.

The philosophy of conventionalism deserves great credit for the way it has helped to clarify the relations between theory and experiment. It recognized the importance, so little noticed by inductivists, of the part played by our actions and operations, planned in accordance with conventions and deductive reasoning, in conducting and interpreting our scientific experiments. I regard conventionalism as a system which is self-contained and defensible. Attempts to detect inconsistencies in it are not likely to succeed. Yet in spite of all this I find it quite unacceptable. Underlying it is an idea of science, of its aims and purposes, which is entirely different from mine. Whilst I do not demand any final certainty from science (and consequently do not get it), the conventionalist seeks in science 'a system of knowledge based upon ultimate grounds', to use a phrase of Dingler's. This goal is attainable; for it is possible to interpret any given scientific system as a system of implicit definitions. And periods when science develops slowly will give little occasion for conflict—unless purely academic—to arise between scientists inclined towards conventionalism and others who may favour a view like the one I advocate. It will be quite otherwise in a time of crisis. Whenever the 'classical' system of the day is threatened by the results of new experiments which might be interpreted as falsifications according to my point of view, the system will appear unshaken to the conventionalist. He will explain away the inconsistencies which may have arisen; perhaps by blaming our inadequate mastery of the system. Or he will eliminate them by suggesting *ad hoc* the adoption of certain auxiliary hypotheses, or perhaps of certain corrections to our measuring instruments.

In such times of crisis this conflict over the aims of science will become acute. We, and those who share our attitude, will hope to make new discoveries; and we shall hope to be helped in this by a newly erected scientific system. Thus we shall take the greatest interest in the falsifying experiment. We shall hail it as a success, for it has opened up new vistas into a world of new experiences. And we shall hail it even if these new experiences should furnish us with new arguments against our own most recent theories. But the newly rising structure, the boldness of which we admire, is seen by the conventionalist as a monument to the 'total collapse of science', as Dingler puts it. In the eyes of the conventionalist one principle only can help us to select a system as the chosen one from among all other possible systems: it is the principle of selecting the simplest system—the simplest system of implicit definitions; which of course means in practice the 'classical' system of the day. (For the problem of simplicity see sections 41–45, and especially 46.)

Thus my conflict with the conventionalists is not one that can be ultimately settled merely by a detached theoretical discussion. And yet it is possible I think to extract from the conventionalist mode of thought certain interesting arguments against my criterion of demarcation; for instance the following. I admit, a conventionalist might say, that the theoretical systems of the natural sciences are not verifiable, but I assert that they are not falsifiable either. For there is always the possibility of '. . . attaining, for any chosen axiomatic system, what is called its "correspondence with reality" ';[3] and this can be done in a number of ways (some of which have been suggested above). Thus we may introduce *ad hoc* hypotheses. Or we may modify the so-called 'ostensive definitions' (or the 'explicit definitions' which may replace them as shown in section 17). Or we may adopt a sceptical attitude as to the reliability of the experimenter whose observations, which threaten our system, we may exclude from science on the ground that they are insufficiently supported, unscientific, or not objective, or even on the ground that the experimenter was a liar. (This is the sort of attitude which the physicist may sometimes quite rightly adopt towards alleged occult phenomena.) In the last resort we can always cast doubt on the acumen of the theoretician (for example if he does not believe, as does Dingler, that the theory of electricity will one day be derived from Newton's law of gravitation).

Thus, according to the conventionalist view, it is not possible to divide systems of theories into falsifiable and non-falsifiable ones; or rather, such a distinction will be ambiguous. As a consequence, our criterion of falsifiability must turn out to be useless as a criterion of demarcation.

20. *Methodological Rules.*

These objections of an imaginary conventionalist seem to me incontestable, just like the conventionalist philosophy itself. I admit that my criterion of falsifiability does not lead to an unambiguous

---

[3] Carnap, *Über die Aufgabe der Physik*, *Kantstudien* 28 (1923), p. 100.

F—L S D

classification. Indeed, it is impossible to decide, by analysing its logical form, whether a system of statements is a conventional system of irrefutable implicit definitions, or whether it is a system which is empirical in my sense; that is, a refutable system. This however only goes to show that my criterion of demarcation cannot be applied immediately to a *system of statements*—a fact I have already pointed out in sections 9 and 11. The question whether a given *system* should as such be regarded as a conventionalist or an empirical one is therefore misconceived. *Only with reference to the method applied* to a theoretical system is it at all possible to ask whether we are dealing with a conventionalist or an empirical theory. The only way to avoid conventionalism is by taking a *decision*: the decision not to apply its methods. We decide that, in the case of a threat to our system, we will not save it by any kind of *conventionalist stratagem*. Thus we shall guard against exploiting the ever open possibility just mentioned of '. . . attaining for any chosen . . . system what is called its "correspondence with reality" '.

A clear appreciation of what may be gained (and lost) by conventionalist methods was expressed, a hundred years before Poincaré, by Black who wrote: 'A nice adaptation of conditions will make almost any hypothesis agree with the phenomena. This will please the imagination but does not advance our knowledge.'[1]

In order to formulate methodological rules which prevent the adoption of conventionalist stratagems, we should have to acquaint ourselves with the various forms these stratagems may take, so as to meet each with the appropriate anti-conventionalist counter-move. Moreover we should decide that, whenever we find that a system has been rescued by a conventionalist stratagem, we shall test it afresh, and reject it, as circumstances may require.

The four main conventionalist stratagems have already been listed at the end of the previous section. The list makes no claim to completeness: it must be left to the investigator, especially in the fields of sociology and psychology (the physicist may hardly need the warning) to guard constantly against the temptation to employ new conventionalist stratagems—a temptation to which psycho-analysts, for example, often succumb.

As regards *auxiliary hypotheses* we decide to lay down the rule

[1] J. Black, *Lectures on the Elements of Chemistry*, Vol. I, Edinburgh 1803, p. 193.

that only those are acceptable whose introduction does not diminish the degree of falsifiability or testability of the system in question, but, on the contrary, increases it. (How degrees of falsifiability are to be estimated will be explained in sections 31 to 40.) If the degree of falsifiability is increased, then introducing the hypothesis has actually strengthened the theory: the system now rules out more than it did previously: it prohibits more. We can also put it like this. The introduction of an auxiliary hypothesis should always be regarded as an attempt to construct a new system; and this new system should then always be judged on the issue of whether it would, if adopted, constitute a real advance in our knowledge of the world. An example of an auxiliary hypothesis which is eminently acceptable in this sense is Pauli's exclusion principle (*cf.* section 38). An example of an unsatisfactory auxiliary hypothesis would be the contraction hypothesis of Fitzgerald and Lorentz which had no falsifiable consequences but merely served to restore the agreement between theory and experiment—mainly the findings of Michelson and Morley. An advance was here achieved only by the theory of relativity which predicted new consequences, new physical effects, and thereby opened up new possibilities for testing, and for falsifying, the theory. Our methodological rule may be qualified by the remark that we need not reject, as conventionalistic, every auxiliary hypothesis that fails to satisfy these standards. In particular, there are *singular* statements which do not really belong to the theoretical system at all. They are sometimes called 'auxiliary hypotheses', and although they are introduced to assist the theory, they are quite harmless. (An example would be the assumption that a certain observation or measurement which cannot be repeated may have been due to error. *Cf.* note 6 to section 8, and sections 27 and 68.)

In section 17 I mentioned *explicit definitions* whereby the concepts of an axiom system are given a meaning in terms of a system of lower level universality. Changes in these definitions are permissible if useful; but they must be regarded as modifications of the system, which thereafter has to be re-examined as if it were new. As regards undefined universal names, two possibilities must be distinguished: (1) There are some undefined concepts which only appear in statements of the highest level of universality, and whose use is established by the fact that we know in what logical relation other concepts stand to them. They can be eliminated in the course of deduction

83

(an example is 'energy').[2] (2) There are other undefined concepts which occur in statements of lower levels of universality also, and whose meaning is established by usage (e.g. 'movement', 'mass-point', 'position'). In connection with these, we shall forbid surreptitious alterations of usage, and otherwise proceed in conformity with our methodological decisions, as before.

As to the two remaining points in our list—which are concerned with the competence of the experimenter or theoretician—we shall adopt similar rules. Inter-subjectively testable experiments are either to be accepted, or to be rejected in the light of counter-experiments. The bare appeal to logical derivations to be discovered in the future can be disregarded.

21. *Logical Investigation of Falsifiability.*

Only in the case of systems which would be falsifiable if treated in accordance with our rules of empirical method is there any need to guard against conventionalist strategems. Let us assume that we have successfully banned these stratagems by our rules: we may now ask for a *logical* characterization of such falsifiable systems. We shall attempt to characterize the falsifiability of a theory by the logical relations holding between the theory and the class of basic statements.

The character of the singular statements which I call 'basic statements' will be discussed more fully in the next chapter, and also the question whether they, in their turn, are falsifiable. Here we shall assume that falsifiable basic statements exist. It should be borne in mind that when I speak of 'basic statements', I am not referring to a system of *accepted* statements. The system of basic statements, as I use the term, is to include, rather, *all self-consistent singular statements* of a certain logical form—all conceivable singular statements of fact, as it were. Thus the system of all basic statements will contain many statements which are mutually incompatible.

As a first attempt one might perhaps try calling a theory 'empirical' whenever singular statements can be deduced from it. This attempt fails, however, because in order to deduce singular statements from

[2] Compare, for instance, Hahn, *Logik, Mathematik, und Naturerkennen,* in *Einheits-wissenschaft* 2, 1933, p. 22 *ff.* In this connection, I only wish to say that in my view 'constituable' (*i.e.* empirically definable) terms do not exist at all. I am using in their place undefinable universal names which are established only by linguistic usage. See also end of section 25.

a theory, we always need other singular statements—the initial conditions that tell us what to substitute for the variables in the theory. As a second attempt, one might try calling a theory 'empirical' if singular statements are derivable with the help of other singular statements serving as initial conditions. But this will not do either; for even a non-empirical theory, for example a tautological one, would allow us to derive some singular statements from other singular statements. (According to the rules of logic we can for example say: From the conjunction of 'Twice two is four' and 'Here is a black raven' there follows, among other things, 'Here is a raven'.) It would not even be enough to demand that from the theory together with some initial conditions we should be able to deduce *more* than we could deduce from those initial conditions alone. This demand would indeed exclude tautological theories, but it would not exclude synthetic metaphysical statements. (For example from 'Every occurrence has a cause' and 'A catastrophe is occurring here', we can deduce 'This catastrophe has a cause'.)

In this way we are led to the demand that the theory should allow us to deduce, roughly speaking, more *empirical* singular statements than we can deduce from the initial conditions alone.*1 This

*1 Formulations equivalent to the one given here have been put forward as criteria of the *meaningfulness* of *sentences* (rather than as criteria of *demarcation* applicable to theoretical *systems*) again and again after the publication of my book, even by critics who pooh-poohed my criterion of falsifiability. But it is easily seen that, if used as a criterion of *demarcation*, our present formulation is equivalent to falsifiability. For if the basic statement b_2 does not follow from b_1, but follows from b_1 in conjunction with the theory t (this is the present formulation) then this amounts to saying that the conjunction of b_1 with the negation of b_2 contradicts the theory t. But the conjunction of b_1 with the negation of b_2 is a basic statement (*cf.* section 28). Thus our criterion demands the existence of a falsifying basic statement, *i.e.* it demands falsifiability in precisely my sense. (See also note *1 to section 82.)

As a criterion of *meaning* (or of 'weak verifiability') it breaks down, however, for various reasons. First, because the negations of some meaningful statements would become meaningless, according to this criterion. Secondly, because the conjunction of a meaningful statement and a 'meaningless pseudo-sentence' would become meaningful —which is equally absurd.

If we now try to apply these two criticisms to our criterion of *demarcation*, they both prove harmless. As to the first, see section 15 above, especially note *2 (and section *22 of my *Postscript*). As to the second, empirical theories (such as Newton's) may contain 'metaphysical' elements. But these cannot be eliminated by a hard and fast rule; though if we succeed in so presenting the theory that it becomes a conjunction of a testable and a non-testable part, we know, of course, that we can now eliminate one of its metaphysical components.

The preceding paragraph of this note may be taken as illustrating another *rule of method* (*cf.* the end of note *4 to section 80): that after having produced some criticism of a rival theory, we should always make a serious attempt to apply this or a similar criticism to our own theory.

means that we must base our definition upon a particular class of singular statements; and this is the purpose for which we need the basic statements. Seeing that it would not be very easy to say in detail how a complicated theoretical system helps in the deduction of singular or basic statements, I propose the following definition. A theory is to be called 'empirical' or 'falsifiable' if it divides the class of all possible basic statements unambiguously into the following two non-empty subclasses. First, the class of all those basic statements with which it is inconsistent (or which it rules out, or prohibits): we call this the class of the *potential falsifiers* of the theory; and secondly, the class of those basic statements which it does not contradict (or which it 'permits'). We can put this more briefly by saying: a theory is falsifiable if the class of its potential falsifiers is not empty.

It may be added that a theory makes assertions only about its potential falsifiers. (It asserts their falsity.) About the 'permitted' basic statements it says nothing. In particular, it does not say that they are true.*²

22. *Falsifiability and Falsification.*

We must clearly distinguish between falsifiability and falsification. We have introduced falsifiability solely as a criterion for the empirical character of a system of statements. As to falsification, special rules must be introduced which will determine under what conditions a system is to be regarded as falsified.

We say that a theory is falsified only if we have accepted basic statements which contradict it (*cf.* section 11, rule 2). This condition is necessary, but not sufficient; for we have seen that non-reproducible single occurrences are of no significance to science. Thus a few stray basic statements contradicting a theory will hardly induce us to reject it as falsified. We shall take it as falsified only if we discover a *reproducible effect* which refutes the theory. In other words, we only accept the falsification if a low-level empirical hypothesis which describes such an effect is proposed and corroborated. This kind of

---

*² In fact, many of the 'permitted' basic statements will, in the presence of the theory, contradict each other. (*Cf.* section 38.) For example, the universal law 'All planets move in circles' (i.e. 'Any set of positions of any one planet is co-circular') is trivially 'instantiated' by any set of no more than three positions of one planet; but two such 'instances' together will in most cases contradict the law.

hypothesis may be called a *falsifying hypothesis*.[1] The requirement that the falsifying hypothesis must be empirical, and so falsifiable, only means that it must stand in a certain logical relationship to possible basic statements; thus this requirement only concerns the logical form of the hypothesis. The rider that the hypothesis should be corroborated refers to tests which it ought to have passed—tests which confront it with accepted basic statements.[*1]

Thus the basic statements play two different rôles. On the one hand, we have used the system of all *logically possible* basic statements in order to obtain with its help the logical characterization for which we were looking—that of the form of empirical statements. On the other hand, the *accepted* basic statements are the basis for the corroboration of hypotheses. If accepted basic statements contradict a theory, then we take them as providing sufficient grounds for its falsification only if they corroborate a falsifying hypothesis at the same time.

---

[1] The falsifying hypothesis can be of a very low level of universality (obtained as it were, by generalising the individual co-ordinates of a result of observation; as an instance I might cite Mach's so-called 'fact' referred to in section 18). Even if it is to be inter-subjectively testable, it need not in fact be a strictly universal statement. Thus to falsify the statement 'All ravens are black' the inter-subjectively testable statement that there is a family of white ravens in the zoo at New York would suffice. *All this shows the urgency of replacing a falsified hypothesis by a better one. In most cases we have, before falsifying a hypothesis, another one up our sleeves; for the falsifying experiment is usually a *crucial experiment* designed to decide between the two. That is to say, it is suggested by the fact that the two hypotheses differ in some respect; and it makes use of this difference to refute (at least) one of them.

[*1] This reference to accepted basic statements may seem to contain the seeds of an infinite regress. For our problem here is this. Since a hypothesis is falsified by *accepting* a basic statement, we need *methodological rules for the acceptance of basic statements*. Now if these rules in their turn refer to accepted basic statements, we may get involved in an infinite regress. To this I reply that the rules we need are merely rules for accepting basic statements that falsify a well-tested and so far successful hypothesis; and the accepted basic statements to which the rule has recourse need not be of this character. Moreover, the rule formulated in the text is far from exhaustive; it only mentions an important aspect of the acceptance of basic statements that falsify an otherwise successful hypothesis, and it will be expanded in chapter v (especially in section 29).

Professor J. H. Woodger, in a personal communication, has raised the question: how often has an effect to be actually reproduced in order to be a *'reproducible effect'* (or a *'discovery'*)? The answer is: in some cases *not even once*. If I assert that there is a family of white ravens in the New York zoo, then I assert something which can be tested *in principle*. If somebody wishes to test it and is informed, upon arrival, that the family has died, or that it has never been heard of, it is left to him to accept or reject my falsifying basic statement. As a rule, he will have means for forming an opinion by examining witnesses, documents, etc.; that is to say, by appealing to other intersubjectively testable and reproducible facts. (*Cf.* sections 27 to 30.)

23. Occurrences and Events.

The requirement of falsifiability which was a little vague to start with has now been split into two parts. The first, the methodological postulate (*cf.* section 20), can hardly be made quite precise. The second, the logical criterion, is quite definite as soon as it is clear which statements are to be called 'basic' (*cf.* section 28). This logical criterion has so far been presented, in a somewhat formal manner, as a logical relation between statements—the theory and the basic statements. Perhaps it will make matters clearer and more intuitive if I now express my criterion in a more 'realistic' language. Although it is equivalent to the formal mode of speech, it may be a little nearer to ordinary usage.

In this 'realistic' mode of speech we can say that a singular statement (a basic statement) describes an *occurrence*. Instead of speaking of basic statements which are ruled out or prohibited by a theory, we can then say that the theory rules out certain possible occurrences, and that it will be falsified if these possible occurrences do in fact occur.

The use of this vague expression 'occurrence' is perhaps open to criticism. It has sometimes been said[1] that expressions such as 'occurrence' or 'event' should be banished altogether from epistemological discussion, and that we should not speak of 'occurrences' or 'non-occurrences', or of the 'happening' of 'events', but instead of the truth or falsity of statements. I prefer, however, to retain the expression 'occurrence'. It is easy enough to define its use so that it is unobjectionable. For we may use it in such a way that whenever we speak of an occurrence, we could speak instead of some of the singular statements which correspond to it.

When defining 'occurrence', we may remember the fact that it would be quite natural to say that two singular statements which are *logically equivalent* (*i.e.* mutually deducible) describe the same occurrence. This suggests the following definition. Let p_k be a singular

---

[1] Especially by some writers on probability; *cf.* Keynes, *A Treatise on Probability* (1921), p. 5. Keynes refers to Ancillon as the first to propose the 'formal mode of expression'; also to Boole, Czuber, and Stumpf. *Although I still regard my ('syntactical') definitions of '*occurrence*' and '*event*', given below, as adequate *for my purpose*, I do no longer believe that they are intuitively adequate; that is, I do not believe that they adequately represent our usage, or our intentions. It was Alfred Tarski who pointed out to me (in Paris, in 1935) that a 'semantic' definition would be required instead of a 'syntactical' one.

statement. (The subscript 'k' refers to the individual names or co-ordinates which occur in p_k.) Then we call the class of all statements which are equivalent to p_k the occurrence P_k. Thus we shall say that it is an occurrence, for example, *that it is now thundering here*. And we may regard this occurrence as the class of the statements 'It is now thundering here'; 'It is thundering in the 13th District of Vienna on the 10th of June 1933 at 5.15 p.m.', and of all other statements equivalent to these. The realistic formulation 'The statement p_k represents the occurrence P_k' can then be regarded as meaning the same as the somewhat trivial statement 'The statement p_k is an element of the class P_k of all statements which are equivalent to it'. Similarly, we regard the statement 'The occurrence P_k has occurred' (or 'is occurring') as meaning the same as 'p_k and all statements equivalent to it are true'.

The purpose of these rules of translation is not to assert that whoever uses, in the realistic mode of speech, the word 'occurrence' is thinking of a class of statements; their purpose is merely to give an interpretation of the realistic mode of speech which makes intelligible what is meant by saying, for example, that an occurrence P_k contradicts a theory t. This statement will now simply mean that every statement equivalent to p_k contradicts the theory t, and is thus a potential falsifier of it.

Another term, 'event', will now be introduced, to denote what may be *typical or universal* about an occurrence, or what, in an occurrence, can be described with the help of universal names. (Thus we do not understand by an event a complex, or perhaps a protracted, occurrence, whatever ordinary usage may suggest.) We define: Let P_k, P_l, ... be elements of a class of occurrences which differ *only* in respect of the individuals (the spatio-temporal positions or regions) involved; then we call this class 'the event (P)'. In accordance with this definition, we shall say, for example, of the statement 'A glass of water has just been upset here' that the class of statements which are equivalent to it is an element of the event, 'upsetting of a glass of water'.

Speaking of the singular statement p_k, which represents an occurrence P_k, one may say, in the realistic mode of speech, that this statement asserts the occurrence of the event (P) at the spatio-temporal position k. And we take this to mean the same as 'the class P_k, of the singular statements equivalent to p_k, is an element of the event (P)'.

We will now apply this terminology[2] to our problem. We can say of a theory, provided it is falsifiable, that it rules out, or prohibits, not merely one occurrence, but always *at least one event*. Thus the class of the prohibited basic statements, i.e. of the potential falsifiers of the theory, will always contain, if it is not empty, an unlimited number of basic statements; for a theory does not refer to individuals as such. We may call the singular basic statements which belong to *one* event 'homotypic', so as to point to the analogy between *equivalent* statements describing *one* occurrence, and *homotypic* statements describing one (typical) event. We can then say that every non-empty class of potential falsifiers of a theory contains at least one non-empty class of homotypic basic statements.

Let us now imagine that the class of all possible basic statements is represented by a circular area. The area of the circle can be regarded as representing something like the totality of *all possible worlds of experience*, or of all possible empirical worlds. Let us imagine, further, that each event is represented by one of the radii (or more precisely, by a very narrow area—or a very narrow sector—along one of the radii) and that any two occurrences involving the same co-ordinates (or individuals) are located at the same distance from the centre, and thus on the same concentric circle. Then we can illustrate the postulate of falsifiability by the requirement that for every empirical theory there must be at least *one* radius (or very narrow sector) in our diagram which the theory forbids.

This illustration may prove helpful in the discussion of our various problems,[*1] such as that of the metaphysical character of purely existential statements (briefly referred to in section 15). Clearly, to each of these statements there will belong one event (one radius) such that the various basic statements belonging to this event will each verify the purely existential statement. Nevertheless, the class of its potential falsifiers is empty; so from the existential statement *nothing follows* about the possible worlds of experience. (It excludes

---

[2] It is to be noted that although singular statements *represent* occurrences, universal statements do not represent events: they *exclude* them. Similarly to the concept of 'occurrence' a 'uniformity' or 'regularity' can be defined by saying that universal statements *represent* uniformities. But here we do not need any such concept, seeing that we are only interested in what universal statements *exclude*. For this reason such questions as whether uniformities (universal 'states of affairs' etc.) exist, do not concern us. *But such questions are discussed in section 79, and now also in appendix *x, and in section *15 of the *Postscript*.

[*1] The illustration will be used, more especially, in sections 31 *ff*., below.

or forbids none of the radii.) The fact that, conversely, from every basic statement a purely existential statement follows, cannot be used as an argument in support of the latter's empirical character. For every tautology also follows from every basic statement, since it follows from any statement whatsoever.

At this point I may perhaps say a word about self-contradictory statements.

Whilst tautologies, purely existential statements and other non-falsifiable statements assert, as it were, *too little* about the class of possible basic statements, self-contradictory statements assert *too much*. From a self-contradictory statement, any statement whatsoever can be validly deduced.[*2] Consequently, the class of its potential falsifiers is identical with that of all possible basic statements: it is falsified by any statement whatsoever. (One could perhaps say that this fact illustrates an advantage of our method, *i.e.* of our way of considering possible falsifiers rather than possible verifiers. For if one could verify a statement by the verification of its logical consequences, or merely make it probable in this way, then one would expect that, by the acceptance of any basic statement whatsoever, any self-contradictory statements would become confirmed, or verified, or at least probable.)

24. *Falsifiability and Consistency.*

The requirement of consistency plays a special rôle among the

---

[*2] This fact was even ten years after publication of this book not yet generally understood. The situation can be summed up as follows: a factually false statement 'materially implies' every statement (but it does not logically entail every statement). A logically false statement logically implies—or entails—every statement. It is therefore of course essential to distinguish clearly between a merely *factually false* (synthetic) statement and a *logically false* or *inconsistent* or *self-contradictory* statement; that is to say, one from which a statement of the form $p \cdot \bar{p}$ can be deducted.

That an inconsistent statement entails every statement can be shown as follows:
From Russell's 'primitive propositions' we get at once
(1) $\qquad\qquad\qquad\qquad p \rightarrow (p \lor q)$
and further, by substituting here first '$\bar{p}$' for 'p', and then '$p \rightarrow q$' for '$\bar{p} \lor q$' we get
(2) $\qquad\qquad\qquad\qquad \bar{p} \rightarrow (p \rightarrow q),$
which yields, by 'importation',
(3) $\qquad\qquad\qquad\qquad \bar{p} \cdot p \rightarrow q$
But (3) allows us to deduce, using the *modus ponens*, any statement q from any statement of the form '$\bar{p} \cdot p$', or '$p \cdot \bar{p}$'. (See also my note in *Mind* **52**, 1943, pp. 47 *ff.*) The fact that everything is deducible from an inconsistent set of premises is rightly treated as well known by P. P. Wiener (*The Philosophy of Bertrand Russell*, edited by P. A. Schilpp, 1944, p. 264); but surprisingly enough, Russell challenged this fact in his reply to Wiener (*op. cit.*, p. 695 *f.*), speaking however of '*false* propositions' where Wiener spoke of '*inconsistent* premises'.

various requirements which a theoretical system, or an axiomatic system, must satisfy. It can be regarded as the first of the requirements to be satisfied by *every* theoretical system, be it empirical or non-empirical.

In order to show the fundamental importance of this requirement it is not enough to mention the obvious fact that a self-contradictory system must be rejected because it is 'false'. We frequently work with statements which, although actually false, nevertheless yield results which are adequate for certain purposes.*[1] (An example is Nernst's approximation for the equilibrium equation of gases.) But the importance of the requirement of consistency will be appreciated if one realizes that a self-contradictory system is uninformative. It is so because any conclusion we please can be derived from it. Thus no statement is singled out, either as incompatible or as derivable, since all are derivable. A consistent system, on the other hand, divides the set of all possible statements into two: those which it contradicts and those with which it is compatible. (Among the latter are the conclusions which can be derived from it.) This is why consistency is the most general requirement for a system, whether empirical or non-empirical, if it is to be of any use at all.

Besides being consistent, an empirical system should satisfy a further condition: it must be *falsifiable*. The two conditions are to a large extent analogous.[1] Statements which do not satisfy the condition of consistency fail to differentiate between any two statements within the totality of all possible statements. Statements which do not satisfy the condition of falsifiability fail to differentiate between any two statements within the totality of all possible empirical basic statements.

---

*[1] *Cf.* my *Postscript*, section *3 (my reply to the 'second proposal'); and section *12, point (2).
  [1] *Cf.* my note in *Erkenntnis* **3**, 1933, p. 426. * This is now printed in appendix *i, below.

CHAPTER V

THE PROBLEM OF THE EMPIRICAL BASIS

WE have now reduced the question of the falsifiability of theories to that of the falsifiability of those singular statements which I have called basic statements. But what kind of singular statements are these basic statements? How can they be falsified? To the practical research worker, these questions may be of little concern. But the obscurities and misunderstandings which surround the problem make it advisable to discuss it here in some detail.

25. *Perceptual Experiences as Empirical Basis: Psychologism.*

The doctrine that the empirical sciences are reducible to sense-perceptions, and thus to our experiences, is one which many accept as obvious beyond all question. However, this doctrine stands or falls with inductive logic, and is here rejected along with it. I do not wish to deny that there is a grain of truth in the view that mathematics and logic are based on thinking, and the factual sciences on sense-perceptions. But what is true in this view has little bearing on the epistemological problem. And indeed, there is hardly a problem in epistemology which has suffered more severely from the confusion of psychology with logic than this problem of the basis of statements of experience.

The problem of the basis of experience has troubled few thinkers so deeply as Fries.[1] He taught that, if the statements of science are not to be accepted *dogmatically*, we must be able to *justify* them. If we demand justification by reasoned argument, in the logical sense, then we are committed to the view that *statements can be justified only by statements*. The demand that *all* statements are to be logically justified (described

[1] J. F. Fries, *Neue oder anthropologische Kritik der Vernunft* (1828 to 1831).

by Fries as a 'predilection for proofs') is therefore bound to lead to an *infinite regress*. Now, if we wish to avoid the danger of dogmatism as well as an infinite regress, then it seems as if we could only have recourse to *psychologism, i.e.* the doctrine that statements can be justified not only by statements but also by perceptual experience. Faced with this *trilemma* —dogmatism *vs.* infinite regress *vs.* psychologism—Fries, and with him almost all epistemologists who wished to account for our empirical knowledge, opted for psychologism. In sense-experience, he taught, we have 'immediate knowledge':[2] by this immediate knowledge, we may justify our 'mediate knowledge'—knowledge expressed in the symbolism of some language. And this mediate knowledge includes, of course, the statements of science.

Usually the problem is not explored as far as this. In the epistemologies of sensationalism and positivism it is taken for granted that empirical scientific statements 'speak of our experiences'.[3] For how could we ever reach any knowledge of facts if not through sense-perception? Merely by taking thought a man cannot add an iota to his knowledge of the world of facts. Thus perceptual experience must be the sole 'source of knowledge' of all the empirical sciences. All we know about the world of facts must therefore be expressible in the form of statements *about our experiences*. Whether this table is red or blue can be found out only by consulting our sense-experience. By the immediate feeling of conviction which it conveys, we can distinguish the true statement, the one whose terms agree with experience, from the false statement, whose terms do not agree with it. Science is merely an attempt to classify and describe this perceptual knowledge, these immediate experiences whose truth we cannot doubt; *it is the systematic presentation of our immediate convictions.*

This doctrine founders in my opinion on the problems of induction and of universals. For we can utter no scientific statement that does not go far beyond what can be known with certainty 'on the basis of immediate experience'. (This fact may be referred to as the 'transcendence inherent in any description'.) Every description uses *universal* names (or symbols, or ideas); every statement has the

---

[2] *Cf.* for example, J. Kraft, *Von Husserl zu Heidegger* (1932), p. 102 *f.* (*Second edition, 1957, p. 108 *f.*)
[3] I am following here almost word for word the expositions of Frank (*cf.* section 27, note 4) and Hahn (*cf.* section 27, note 1).

character of a theory, of a hypothesis. The statement, 'Here is a glass of water' cannot be verified by any observational experience. The reason is that the *universals* which appear in it cannot be correlated with any specific sense-experience. (An 'immediate experience' is *only once* 'immediately given'; it is unique.) By the word 'glass', for example, we denote physical bodies which exhibit a certain *law-like behaviour*, and the same holds for the word 'water'. Universals cannot be reduced to classes of experiences; they cannot be 'constituted'.[4]

26. *Concerning the So-Called 'Protocol Sentences'.*

The view which I call 'psychologism', discussed in the previous section, still underlies, it seems to me, a modern theory of the empirical basis, even though its advocates do not speak of experiences or perceptions but, instead, of 'sentences'—sentences which represent experiences. These are called *protocol sentences* by Neurath[1] and by Carnap.[2]

A similar theory had been held even earlier by Reininger. His starting-point was the question: Wherein lies the correspondence or agreement between a statement and the fact or state of affairs which it describes? He came to the conclusion that statements can be compared only with statements. According to his view, the correspondence of a statement with a fact is nothing else than a logical correspondence between statements belonging to different levels of universality: it is[3] '... the correspondence of higher level statements with statements which are of similar content, and ultimately with statements recording experiences'. (These are sometimes called 'elementary statements' by Reininger.[4])

Carnap starts with a somewhat different question. His thesis is that all philosophical investigations speak 'of the forms of speech'.[5] The logic of science has to investigate 'the forms of scientific language'.[6] It does not speak of (physical) 'objects' but of words; not of fact, but of sentences. With this, the correct, the *'formal mode of speech's*

---

[4] *Cf.* note 2 to section 20, and text. *'Constituted' is Carnap's term.Ɪ
[1] The term is due to Neurath; *cf.*, for example, *Soziologie*, in *Erkenntnis* **2**, 1932, p. 393.
[2] Carnap, *Erkenntnis* **2**, 1932, p. 432 *ff.*; *Ibid.* **3** (1932), p. 107 *ff.*
[3] Reininger, *Metaphysik der Wirklichkeit* (1931), p. 134.
[4] Reininger, *op. cit.*, p. 132.
[5] Carnap, *Erkenntnis* **2**, 1932, p. 435, *'These der Metalogik'*.
[6] Carnap, *Ibid.* **3**, 1933, p. 228.

Carnap contrasts the ordinary or, as he calls it, the 'material mode of speech'. If confusion is to be avoided, then the material mode of speech should only be used where it is possible to translate it into the correct formal mode of speech.

Now this view—with which I can agree—leads Carnap (like Reininger) to assert that we must not say, in the logic of science, that sentences are tested by comparing them with states of affairs or with experiences: we may only say that they can be tested by comparing them with other *sentences*. Yet Carnap is nevertheless really retaining the fundamental ideas of the psychologistic approach to the problem; all that he is doing is to translate them into the 'formal mode of speech'. He says that the sentences of science are tested 'with the help of protocol sentences'[7]; but since these are explained as statements or sentences 'which are not in need of confirmation but serve as a basis for all the other sentences of science', this amounts to saying—in the ordinary 'material' mode of speech—that the protocol sentences refer to the 'given': to the 'sense data'. They describe (as Carnap himself puts it) 'the contents of immediate experience, or the phenomena; and thus the simplest knowable facts'.[8] Which shows clearly enough that the theory of protocol sentences is nothing but psychologism translated into the formal mode of speech. Much the same can be said of Neurath's view[9]: he demands that in protocol sentences such words as 'perceive', 'see', etc., should occur together with the full name of the author of the protocol sentence. Protocol sentences, as the term indicates, should be *records or protocols of immediate observations, or perceptions.*

Like Reininger,[10] Neurath holds that perceptual statements recording experiences—*i.e.* 'protocol sentences'—are not irrevocable, but that they can sometimes be rejected. He opposes[11] Carnap's view (since revised by the latter[12]) that protocol sentences are ultimate, and *not in need of confirmation*. But whilst Reininger describes a method of testing his 'elementary' statements, in cases of doubt, by means

[7] Carnap, *Ibid.* **2**, 1932, p. 437.
[8] Carnap, *Ibid.*, p. 438.
[9] Neurath, *Erkenntnis* **3**, 1933, p. 205 *ff*. Neurath gives the following example, 'A complete protocol statement might run: {Otto's protocol at 3 hrs. 17 mins. [Otto's speech-thought was at 3 hrs. 16 mins.: (in the room at 3 hrs. 15 mins. was a table observed by Otto)]}.'
[10] Reininger, *op. cit.*, p. 133.
[11] Neurath, *op. cit.*, p. 209 *f*.
[12] Carnap, *Erkenntnis* **3**, 1933, p. 215 *ff.*; *cf.* note 1 to section 29.

of other statements—it is the method of deducing and testing conclusions—Neurath gives no such method. He only remarks that we can either 'delete' a protocol sentence which contradicts a system, '. . . or else accept it, and modify the system in such a way that, with the sentence added, it remains consistent'.

Neurath's view that protocol sentences are not inviolable represents, in my opinion, a notable advance. But apart from the replacement of perceptions by perception-statements—merely a translation into the formal mode of speech—the doctrine that protocol sentences may be revised is his only advance upon the theory (due to Fries) of the immediacy of perceptual knowledge. It is a step in the right direction; but it leads nowhere if it is not followed up by another step: we need a set of rules to limit the arbitrariness of 'deleting' (or else 'accepting') a protocol sentence. Neurath fails to give any such rules and thus unwittingly throws empiricism overboard. For without such rules, empirical statements are no longer distinguished from any other sort of statements. Every system becomes defensible if one is allowed (as everybody is, in Neurath's view) simply to 'delete' a protocol sentence if it is inconvenient. In this way one could not only rescue any system, in the manner of conventionalism; but given a good supply of protocol sentences, one could even confirm it, by the testimony of witnesses who have testified, or protocolled, what they have seen and heard. Neurath avoids one form of dogmatism, yet he paves the way for any arbitrary system to set itself up as an 'empirical science'.

Thus it is not quite easy to see what part the protocol sentences are supposed to play in Neurath's scheme. In Carnap's earlier view, the system of protocol sentences was the touchstone by which every assertion of an empirical science had to be judged. This is why they had to be 'irrefutable'. For they alone could overthrow sentences—sentences other than protocol sentences, of course. But if they are deprived of this function, and if they themselves can be overthrown by theories, what are they for? Since Neurath does not try to solve the problem of demarcation, it seems that his idea of protocol sentences is merely a relic— a surviving memorial of the traditional view that empirical science starts from perception.

27. *The Objectivity of the Empirical Basis.*

I propose to look at science in a way which is slightly different

97

from that favoured by the various psychologistic schools: I wish to *distinguish sharply between objective science on the one hand, and 'our knowledge' on the other.*

I readily admit that only observation can give us 'knowledge concerning facts', and that we can (as Hahn says) 'become aware of facts only by observation'.[1] But this awareness, this knowledge of ours, does not justify or establish the truth of any statement. I do not believe, therefore, that the question which epistemology must ask is, '. . . on what does our *knowledge* rest? . . . or more exactly, how can I, having had the *experience* S, justify my description of it, and defend it against doubt?'[2] This will not do, even if we change the term 'experience' into 'protocol sentence'. In my view, what epistemology has to ask is, rather: how do we test scientific statements by their deductive consequences?[*1] And *what kind* of consequences can we select for this purpose if they in their turn are to be inter-subjectively testable?

By now, this kind of objective and non-psychological approach is pretty generally accepted where logical or tautological statements are concerned. Yet not so long ago it was held that logic was a science dealing with mental processes and their laws—the laws of our thought. On this view there was no other justification to be found for logic than the alleged fact that we just could not think in any other way. A logical inference seemed to be justified because it was experienced as a necessity of thought, as a feeling of being compelled to think along certain lines. In the field of logic, this kind of psychologism is now perhaps a thing of the past. Nobody would dream of justifying the validity of a logical inference, or of defending it against doubts, by writing beside it in the margin the following protocol sentence. 'Protocol: In checking this chain of inferences today, I experienced an acute feeling of conviction.'

The position is very different when we come to *empirical statements of science*. Here everybody believes that these are grounded on experiences such as perceptions; or in the formal mode of speech, on protocol sentences. Most people would see that any attempt to base

---

[1] H. Hahn, *Logik, Mathematik und Naturerkennen*, in *Einheitswissenschaft*, **2** 1933, pp. 19 and 24.

[2] *Cf.* Carnap, for instance, *Scheinprobleme in der Philosophie* (1928), p. 15 (no italics in the original).

[*1] At present, I should formulate this question thus. How can we best *criticise* our theories (our hypotheses, our guesses), rather than defend them against doubt? Of course, *testing* was always, in my view, part of *criticising*. (*Cf.* my *Postscript*, sections *7, text between notes 5 and 6, and end of *52.)

logical statements on protocol sentences is a case of psychologism. But curiously enough, when it comes to empirical statements, the same kind of thing goes today by the name of 'physicalism'. Yet whether statements of logic are in question or statements of empirical science, I think the answer is the same: our *knowledge*, which may be described vaguely as a system of *dispositions*, and which may be of concern to psychology, may be in both cases linked with feelings of belief or of conviction: in the one case, perhaps, with the feeling of being compelled to think in a certain way; in the other with that of 'perceptual assurance'. But all this interests only the psychologist. It does not even touch upon problems like those of the logical connections between scientific statements, which alone interest the epistemologists.

(There is a widespread belief that the statement 'I see that this table here is white', possesses some profound advantage over the statement 'This table here is white', from the point of view of epistemology. But from the point of view of evaluating its possible objective tests, the first statement, in speaking about me, does not appear more secure than the second statement, which speaks about the table here.)

There is only one way to make sure of the validity of a chain of logical reasoning. This is to put it in the form in which it is most easily testable: we break it up into many small steps, each easy to check by anybody who has learnt the mathematical or logical technique of transforming sentences. If after this anybody still raises doubts then we can only beg him to point out an error in the steps of the proof, or to think the matter over again. In the case of the empirical sciences, the situation is much the same. Any empirical scientific statement can be presented (by describing experimental arrangements, etc.) in such a way that anyone who has learned the relevant technique can test it. If, as a result, he rejects the statement, then it will not satisfy us if he tells us all about his feelings of doubt or about his feelings of conviction as to his perceptions. What he must do is to formulate an assertion which contradicts our own, and give us his instructions for testing it. If he fails to do this we can only ask him to take another and perhaps a more careful look at our experiment, and think again.

An assertion which owing to its logical form is not testable can at best operate, within science, as a stimulus: it can suggest a problem. In the field of logic and mathematics, this may be exemplified by Fermat's problem, and in the field of natural history, say, by reports about sea-serpents. In such cases science does not say that the reports are unfounded; that Fermat was in error or that all the records of observed sea-serpents are lies. Instead, it suspends judgment.[3]

Science can be viewed from various standpoints, not only from that of epistemology; for example, we can look at it as a biological or as a sociological phenomenon. As such it might be described as a tool, or an instrument, comparable perhaps to some of our industrial machinery. Science may be regarded as a means of production—as the last word in 'roundabout production'.[4] Even from this point of view science is no more closely connected with 'our experience' than other instruments or means of production. And even if we look at it as gratifying our intellectual needs, its connection with our experiences does not differ in principle from that of any other objective structure. Admittedly it is not incorrect to say that science is '. . . an instrument' whose purpose is '. . . to predict from immediate or given experiences later experiences, and even as far as possible to control them'.[5] But I do not think that this talk about experiences contributes to clarity. It has hardly more point than, say, the not incorrect characterization of an oil derrick . by the assertion that its purpose is to give us certain experiences : not oil, but rather the sight and smell of oil; not money, but rather the feeling of having money.

28. *Basic Statements.*

It has already been briefly indicated what rôle the basic statements play within the epistemological theory I advocate. We need them in order to decide whether a theory is to be called falsifiable, *i.e.* empirical. (*Cf.* section 21.) And we also need them for the corroboration of falsifying hypotheses, and thus for the falsification of theories. (*Cf.* section 22.)

Basic statements must therefore satisfy the following conditions. (a) From a universal statement without initial conditions, no basic

---

[3] *Cf.* the remark on 'occult effects' in section 8.
[4] The expression is Böhm-Bawerk's ('*Produktionsumweg*').
[5] Frank, *Das Kausalgesetz und seine Grenzen* (1932), p. 1. * Concerning instrumentalism, see note *1 before section 12, and my *Postscript*, especially sections *12 to *15.

statementc an be deduced.*1 On the other hand, (b) a universal statement and a basic statement can contradict each other. Condition (b)
can only be satisfied if it is possible to derive the negation of a basic
statement from the theory which it contradicts. From this and condition (a) it follows that a basic statement must have a logical form
such that its negation cannot be a basic statement in its turn.

We have already encountered statements whose logical form is
different from that of their negations. These were universal statements
and existential statements: they are negations of one another, and they
differ in their logical form. *Singular* statements can be constructed in
an analogous way. The statement: 'There is a raven in the space-time
region k' may be said to be different in its logical form—and not
only in its linguistic form—from the statement 'There is no raven in
the space-time region k'. A statement of the form 'There is a so-and-so
in the region k' or 'such-and-such an event is occurring in the region

---

*1 When writing this, I believed that it was plain enough that from Newton's theory
alone, without initial conditions, nothing of the nature of an observation statement can
be deducible (and therefore certainly no basic statements). Unfortunately, it turned out
that this fact, and its consequences for the problem of observation statements or 'basic
statements', was not appreciated by some of the critics of my book. I may therefore add
here a few remarks.

First, nothing observable follows from any pure all-statement—'All swans are white',
say. This is easily seen if we contemplate the fact that 'All swans are white' and 'All swans
are black' do not, of course, contradict each other, but together merely imply that there
are no swans—clearly not an observation statement, and not even one that can be
'verified'. (A unilaterally falsifiable statement like 'All swans are white', by the way, has
the same logical form as 'There are no swans', for it is equivalent to 'There are no non-
white swans'.)

Now if this is admitted, it will be seen at once that the singular statements which *can*
be deduced from purely universal statements cannot be basic statements. I have in mind
statements of the form: 'If there is a swan at the place k, then there is a white swan at
the place k.' (Or, 'At k, there is either no swan or a white swan'.) We see now at once
why these 'instantial statements' (as they may be called) are not basic statements. The
reason is that these instantial statements *cannot play the role of test statements* (or of potential
falsifiers) which is precisely the role which basic statements are supposed to play. If we
were to accept instantial statements as test statements, we should obtain for any
theory (and thus both for 'All swans are white' *and* for 'All swans are black') an overwhelming number of verifications—indeed, an infinite number, once we accept as a
fact that the overwhelming part of the world is empty of swans.

Since 'instantial statements' are derivable from universal ones, their negations must
be potential falsifiers, and *may* therefore be basic statements (if the conditions stated
below in the text are satisfied). Instantial statements, *vice versa*, will then be of the form
of negated basic statements (see also note *4 to section 80). It is interesting to note that
basic statements (which are too strong to be derivable from universal laws alone)
will have a greater informative content than their instantial negations; which means
that *the content of basic statements exceeds their logical probability* (since it must exceed 1/2).

These were some of the considerations underlying my theory of the logical form
of basic statements. (See also section *43 of my *Postscript*.)

k' (*cf.* section 23) may be called a '*singular* existential statement' or a '*singular* there-is statement'. And the statement which results from negating it, *i.e.* 'There is no so-and-so in the region *k'* or 'No event of such-and-such a kind is occurring in the region *k'*, may be called a '*singular* non-existence statement', or a '*singular* there-is-not statement'.

We may now lay down the following rule concerning basic statements: *basic statements have the form of singular existential statements*. This rule means that basic statements will satisfy condition (a), since a singular existential statement can never be deduced from a strictly universal statement, *i.e.* from a strict non-existence statement. They will also satisfy condition (b), as can be seen from the fact that from every singular existential statement a purely existential statement can be derived simply by omitting any reference to any individual space-time region; and as we have seen, a purely existential statement may indeed contradict a theory.

It should be noticed that the conjunction of two basic statements, *d* and *r*, which do not contradict each other, is in turn a basic statement. Sometimes we may even obtain a basic statement by joining one basic statement to another statement which is not basic. For example, we may form the conjunction of the basic statement, *r* 'There is a pointer at the place *k'* with the singular non-existence statement $\bar{p}$, 'There is no pointer in motion at the place *k'*. For clearly, the conjunction $r \cdot \bar{p}$ ('*r and non-p*') of the two statements is equivalent to the singular existential statement 'There is a pointer at rest at the place *k'*. This has the consequence that, if we are given a theory *t* and the initial conditions *r*, from which we deduce the prediction *p*, then the statement $r \cdot \bar{p}$ will be a falsifier of the theory, and so a basic statement. (On the other hand, the conditional statement '$r \rightarrow p$' *i.e.* 'If *r* then *p'*, is no more basic than the negation $\bar{p}$, since it is equivalent to the negation of a basic statement, *viz.* to the negation of $r \cdot \bar{p}$.)

These are the formal requirements for basic statements; they are satisfied by all singular existential statements. In addition to these, a basic statement must also satisfy a material requirement—a requirement concerning the event which, as the basic statement tells us, is occurring at the place *k*. This event must be an '*observable*' event; that is to say, basic statements must be testable, inter-subjectively, by 'observation'. Since they are singular statements, this requirement can of course only refer to observers who are suitably placed in space and time (a point which I shall not elaborate).

No doubt it will now seem as though in demanding observability, I have, after all, allowed psychologism to slip back quietly into my theory. But this is not so. Admittedly, it is possible to interpret the concept of an *observable event* in a psychologistic sense. But I am using it in such a sense that it might just as well be replaced by 'an event involving position and movement of macroscopic physical bodies'. Or we might lay it down, more precisely, that every basic statement must either be itself a statement about relative positions of physical bodies, or that it must be equivalent to some basic statement of this 'mechanistic' or 'materialistic' kind. (That this stipulation is practicable is connected with the fact that a theory which is inter-subjectively testable will also be inter-sensually[1] testable. This is to say that tests involving the perception of one of our senses can, in principle, be replaced by tests involving other senses.) Thus the charge that, in appealing to observability, I have stealthily readmitted psychologism would have no more force than the charge that I have admitted mechanism or materialism. This shows that my theory is really quite neutral and that neither of these labels should be pinned to it. I say all this only so as to save the term 'observable', as I use it, from the stigma of psychologism. (Observations and perceptions may by psychological, but observability is not.) I have no intention of *defining* the term 'observable' or 'observable event', though I am quite ready to elucidate it by means of either psychologistic or mechanistic examples. I think that it should be introduced as an undefined term which becomes sufficiently precise in use: as a primitive concept whose use the epistemologist has to learn, much as he has to learn the use of the term 'symbol', or as the physicist has to learn the use of the term 'mass-point'.

Basic statements are therefore—in the material mode of speech—statements asserting that an observable event is occurring in a certain individual region of space and time. The various terms used in this definition, with the exception of the primitive term 'observable', have been explained more precisely in section 23; 'observable' is undefined, but can also be explained fairly precisely, as we have seen here.

---

[1] Carnap, *Erkenntnis* **2**, 1932, p. 445.

29. *The Relativity of Basic Statements. Resolution of Fries' Trilemma.*

Every test of a theory, whether resulting in its corroboration or falsification, must stop at some basic statement or other which we *decide to accept.* If we do not come to any decision, and do not accept some basic statement or other, then the test will have led nowhere. But considered from a logical point of view, the situation is never such that it compels us to stop at this particular basic statement rather than at that, or else give up the test altogether. For any basic statement can again in its turn be subjected to tests, using as a touchstone any of the basic statements which can be deduced from it with the help of some theory, either the one under test, or another. This procedure has no natural end.[1] Thus if the test is to lead us anywhere, nothing remains but to stop at some point or other and say that we are satisfied, for the time being.

It is fairly easy to see that we arrive in this way at a procedure according to which we stop only at a kind of statement that is especially easy to test. For it means that we are stopping at statements about whose acceptance or rejection the various investigators are likely to reach agreement. And if they do not agree, they will simply continue with the tests, or else start them all over again. If this too leads to no result, then we might say that the statements in question were not inter-subjectively testable, or that we were not, after all, dealing with observable events. If some day it should no longer be possible for scientific observers to reach agreement about basic statements this would amount to a failure of language as a means of universal communication. It would amount to a new 'Babel of Tongues': scientific discovery would be reduced to absurdity. In this new Babel, the soaring edifice of science would soon lie in ruins.

Just as a logical proof has reached a satisfactory shape when the difficult work is over, and everything can be easily checked, so, after science has done its work of deduction or explanation, we stop at

---

[1] *Cf.* Carnap, *Erkenntnis* **3**, 1933, p. 224. I can accept this report by Carnap of my theory, save for a few not too important details. These are, first, the suggestion that basic statements (called by Carnap 'protocol statements') are the starting points from which science is built up; secondly, the remark (p. 225) that a protocol statement can be confirmed 'with such and such degree of certainty'; thirdly that 'statements about perceptions' constitute 'equally valid links in the chain' and that it is these statements about perception to which we 'appeal in critical cases'. *Cf.* the quotation in the text to the next note. I wish to take this opportunity of thanking Professor Carnap for his friendly words about my unpublished work, at the place mentioned.

basic statements which are easily testable. Statements about personal experiences—*i.e.* protocol sentences—are clearly *not* of this kind; thus they will not be very suitable to serve as statements at which we stop. We do of course make use of records or protocols, such as certificates of tests issued by a department of scientific and industrial research. These, if the need arises, can be re-examined. Thus it may become necessary, for example, to test the reaction-times of the experts who carry out the tests (*i.e.* to determine their personal equations). But in general, and especially '. . . in critical cases' we do stop at easily testable statements, and *not*, as Carnap recommends, at perception or protocol sentences; *i.e.* we *do not* '. . . stop just at these . . . because the inter-subjective testing of statements about perceptions . . . is relatively complicated and difficult'.[2]

What is our position now in regard to Fries' trilemma, the choice between dogmatism, infinite regress, and psychologism? (*Cf.* section 25.) The basic statements at which we stop, which we decide to accept as satisfactory, and as sufficiently tested, have admittedly the character of *dogmas*, but only in so far as we may desist from justifying them by further arguments (or by further tests). But this kind of dogmatism is innocuous since, should the need arise, these statements can easily be tested further. I admit that this too makes the chain of deduction in principle infinite. But this kind of '*infinite regress*' is also innocuous since in our theory there is no question of trying to prove any statements by means of it. And finally, as to *psychologism*: I admit, again, that the decision to accept a basic statement, and to be satisfied with it, is causally connected with our experiences—especially with our *perceptual experiences*. But we do not attempt to *justify* basic statements by these experiences. Experiences can *motivate a decision*, and hence an acceptance or a rejection of a statement, but a basic statement cannot be *justified* by them—no more than by thumping the table.[3]

---

[2] *Cf.* the previous note. * This paper of Carnap's contained the first published report of my theory of tests; and the view here quoted from it was there erroneously attributed to me.

[3] It seems to me that the view here upheld is closer to that of the 'critical' (Kantian) school of philosophy (perhaps in the form represented by Fries) than to positivism. Fries in his theory of our 'predilection for proofs' emphasizes that the (logical) relations holding between statements are quite different from the relation between statements and sense experiences; positivism on the other hand always tries to abolish the distinction: either all science is made part of my knowing, 'my' sense experience (monism of sense data); or sense experiences are made part of the objective scientific network of arguments in the form of protocol statements (monism of statements).

30. *Theory and Experiment.*

Basic statements are accepted as the result of a decision or agreement; and to that extent they are conventions. The decisions are reached in accordance with a procedure governed by rules. Of special importance among these is a rule which tells us that we should not accept *stray basic statements*—i.e. logically disconnected ones—but that we should accept basic statements in the course of testing *theories*; of raising searching questions about these theories, to be answered by the acceptance of basic statements.

Thus the real situation is quite different from the one visualized by the naïve empiricist, or the believer in inductive logic. He thinks that we begin by collecting and arranging our experiences, and so ascend the ladder of science. Or, to use the more formal mode of speech, that if we wish to build up a science, we have first to collect protocol sentences. But if I am ordered: 'Record what you are now experiencing' I shall hardly know how to obey this ambiguous order. Am I to report that I am writing; that I hear a bell ringing; a newsboy shouting; a loudspeaker droning; or am I to report, perhaps, that these noises irritate me? And even if the order could be obeyed: however rich a collection of statements might be assembled in this way, it could never add up to a *science*. A science needs points of view, and theoretical problems.

Agreement upon the acceptance or rejection of basic statements is reached, as a rule, on the occasion of *applying* a theory; the agreement, in fact, is part of an application which puts the theory to the test. Coming to an agreement upon basic statements is, like other kinds of applications, to perform a purposeful action, guided by various theoretical considerations.

We are now, I think, in a position to solve such problems as, for instance, Whitehead's problem of how it is that the tactile breakfast should always be served along with the visual breakfast, and the tactile *Times* with the visible and the audibly rustling *Times*. The inductive logician who believes that all science starts from stray elementary perceptions must be puzzled by such regular coincidences; they must seem to him entirely 'accidental'. He is prevented from explaining regularity by theories, because he is committed to the view that theories are nothing but statements of regular coincidences.

But according to the position reached here, the connections between our various experiences are explicable, and deducible, in terms of *theories* which we are engaged in testing. (Our theories do not lead us to expect that along with the visible moon we shall be served a tactile moon; nor do we expect to be bothered by an auditory nightmare.) One question, certainly, does remain—a question which obviously cannot be answered by any falsifiable theory and which is therefore 'metaphysical': how is it that we are so often lucky in the theories we construct—how is it that there are 'natural laws'?*[1]

All these considerations are important for the epistemological *theory of experiment*. The theoretician puts certain definite questions to the experimenter, and the latter, by his experiments, tries to elicit a decisive answer to these questions, and to no others. All other questions he tries hard to exclude. (Here the relative independence of sub-systems of a theory may be important.) Thus he makes his test with respect to this one question '. . . as sensitive as possible, but as insensitive as possible with respect to all other associated questions. . . . Part of this work consists in screening off all possible sources of error.'[1] But it is a mistake to suppose that the experimenter proceeds in this way 'in order to lighten the task of the theoretician',[2] or perhaps in order to furnish the theoretician with a basis for inductive generalizations. On the contrary, the theoretician must long before have done his work, or at least what is the most important part of his work: he must have formulated his question as sharply as possible. Thus it is he who shows the experimenter the way. But even the experimenter is not in the main engaged in making exact observations; his work, too, is largely of a theoretical kind. Theory dominates the experimental work from its initial planning up to the finishing touches in the laboratory.*[2]

This is well illustrated by cases in which the theoretician succeeded in predicting an observable effect which was later experimentally produced; perhaps the most beautiful instance is de Broglie's pre-

---

*[1] This question will be discussed in section 79 and in appendix *x; see also my *Postscript*, especially sections *15 and *16.

[1] H. Weyl, *Philosophie der Mathematik und Naturwissenschaft* (1927), p. 113; English Edition: *Philosophy of Mathematics and Natural Science*, Princeton, 1949, p. 116.

[2] Weyl, *ibid.*

*[2] I now feel that I should have emphasized in this place a view which can be found elsewhere in the book (see for example the fourth and the last paragraphs of section 19). I mean the view that observations, and even more so observation statements and statements of experimental results, are always *interpretations* of the facts observed; that they are *interpretations in the light of theories*. This is one of the main reasons why it is always deceptively easy to find *verifications* of a theory, and why we have to adopt a highly critical attitude towards our theories if we do not wish to argue in circles: the attitude of trying to *falsify* them.

diction of the wave-character of matter, first confirmed experimentally by Davisson and Germer.*[3] It is illustrated perhaps even better by cases in which experiments had a conspicuous influence upon the progress of theory. What compels the theorist to search for a better theory, in these cases, is almost always the experimental *falsification* of a theory, so far accepted and corroborated: it is, again, the outcome of tests guided by theory. Famous examples are the Michelson-Morley experiment which led to the theory of relativity, and the falsification, by Lummer and Pringsheim, of the radiation formula of Rayleigh and Jeans, and of that of Wien, which led to the quantum theory. Accidental discoveries occur too, of course, but they are comparatively rare. Mach[3] rightly speaks in such cases of a 'correction of scientific opinions by accidental circumstances' (thus acknowledging the significance of theories in spite of himself).

It may now be possible for us to answer the question: How and why do we accept one theory in preference to others?

The preference is certainly not due to anything like an experiential justification of the statements composing the theory; it is not due to a logical reduction of the theory to experience. We choose the theory which best holds its own in competition with other theories; the one which, by natural selection, proves itself the fittest to survive. This will be the one which not only has hitherto stood up to the severest tests, but the one which is also testable in the most rigorous way. A theory is a tool which we test by applying it, and which we judge as to its fitness by the results of its applications.*[4]

From a logical point of view, the testing of a theory depends upon basic statements whose acceptance or rejection, in its turn, depends upon our *decisions*. Thus it is *decisions* which settle the fate of theories. To this extent my answer to the question, 'how do we select a theory?' resembles that given by the conventionalist; and like him I say that this choice is in part determined by considerations of utility. But in spite of this, there is a vast difference between my views and his. For I hold that what characterizes the empirical method is just this:

---

*[3] The story is briefly and excellently told by Max Born in *Albert Einstein, Philosopher-Scientist*, edited by P. A. Schilpp, 1949, p. 174. There are better illustrations, such as Adams' and Leverier's discovery of Neptune, or that of Hertzean waves.

[3] Mach, *Die Prinzipien der Wärmelehre* (1896), p. 438.

*[4] For a criticism of the 'instrumentalist' view see however the references in note *[1] before section 12 (p 59), and in the starred addition to note 1, section 12.

that the convention or decision does not immediately determine our acceptance of *universal* statements but that, on the contrary, it enters into our acceptance of the *singular* statements—that is, the basic statements.

For the conventionalist, the acceptance of universal statements is governed by his principle of *simplicity*: he selects that system which is the simplest. I, by contrast, propose that the first thing to be taken into account should be the severity of tests. (There is a close connection between what I call 'simplicity' and the severity of tests; yet my idea of simplicity differs widely from that of the conventionalist; see section 46.) And I hold that what ultimately decides the fate of a theory is the result of a test, *i.e.* an agreement about basic statements. With the conventionalist I hold that the choice of any particular theory is an act, a practical matter. But for me the choice is decisively influenced by the application of the theory and the acceptance of the basic statements in connection with this application; whereas for the conventionalist, aesthetic motives are decisive.

Thus I differ from the conventionalist in holding that the statements decided by agreement are *not universal but singular*. And I differ from the positivist in holding that basic statements are not justifiable by our immediate experiences, but are, from the logical point of view, accepted by an act, by a free decision. (From the psychological point of view this may perhaps be a purposeful and well-adapted reaction.)

This important distinction, between a *justification* and a *decision* —a decision reached in accordance with a procedure governed by rules—might be clarified, perhaps, with the help of an analogy: the old procedure of trial by jury.

The *verdict* of the jury (*vere dictum* = spoken truly), like that of the experimenter, is an answer to a question of fact (*quid facti?*) which must be put to the jury in the sharpest, the most definite form. But what question is asked, and how it is put, will depend very largely on the legal situation, *i.e.* on the prevailing system of criminal law (corresponding to a system of theories). By its decision, the jury accepts, by agreement, a statement about a factual occurrence—a basic statement, as it were. The significance of this decision lies in the fact that from it, together with the universal statements of the system (of criminal law) certain consequences can be deduced. In other words, the decision forms the basis for the *application* of the system; the verdict plays the part of a 'true statement of fact'. But it is clear that the statement need not be true merely because the jury has accepted it. This fact is acknowledged in the rule allowing a verdict to be quashed or revised.

The verdict is reached in accordance with a procedure which is governed by rules. These rules are based on certain fundamental principles which are chiefly, if not solely, designed to result in the discovery of objective truth. They sometimes leave room not only for subjective convictions but even for subjective bias. Yet even if we disregard these special aspects of the older procedure and imagine a procedure based solely on the aim of promoting the discovery of objective truth, it would still be the case that the verdict of the jury never justifies, or gives grounds for, the truth of what it asserts.

Neither can the subjective convictions of the jurors be held to justify the decision reached; although there is, of course, a close causal connection between them and the decision reached—a connection which might be stated by psychological laws; thus these convictions may be called the 'motives' of the decision. The fact that the convictions are not justifications is connected with the fact that different rules may regulate the jury's procedure (for example, simple or qualified majority). This shows that the relationship between the convictions of the jurors and their verdict may vary greatly.

In contrast to the verdict of the jury, the *judgment* of the judge is 'reasoned'; it needs, and contains, a justification. The judge tries to justify it by, or deduce it logically from, other statements: the statements of the legal system, combined with the verdict that plays the rôle of initial conditions. This is why the judgment may be challenged on logical grounds. The jury's decision, on the other hand, can only be challenged by questioning whether it has been reached in accordance with the accepted rules of procedure; *i.e.* formally, but not as to its content. (A justification of the content of a decision is significantly called a 'motivated report', rather than a 'logically justified report'.)

The analogy between this procedure and that by which we decide basic statements is clear. It throws light, for example, upon their relativity, and the way in which they depend upon questions raised by the theory. In the case of the trial by jury, it would be clearly impossible to *apply* the 'theory' unless there is first a verdict arrived at by decision; yet the verdict has to be found in a procedure that conforms to, and thus applies, part of the general legal code. The case is analogous to that of basic statements. Their acceptance is part of the application of a theoretical system; and it is only this application which makes any further applications of the theoretical system possible.

The empirical basis of objective science has thus nothing 'absolute' about it.[4] Science does not rest upon rock-bottom. The bold structure of its theories rises, as it were, above a swamp. It is like a building erected on piles. The piles are driven down from above into the swamp, but not down to any natural or 'given' base; and when we cease our attempts to drive our piles into a deeper layer, it is not because we have reached firm ground. We simply stop when we are satisfied that they are firm enough to carry the structure, at least for the time being.

[4] Weyl (*op. cit.*, p. 83, English edition p. 116) writes: ' . . . this pair of opposites, *subjective-absolute* and *objective-relative* seems to me to contain one of the most profound epistemological truths which can be gathered from the study of nature. Whoever wants the absolute must get subjectivity—ego-centricity—into the bargain, and whoever longs for objectivity cannot avoid the problem of relativism.' And before this we find, 'What is immediately experienced is *subjective and absolute* . . . ; the objective world, on the other hand, which natural science seeks to precipitate in pure crystalline form . . . is relative.' Born expresses himself in similar terms (*Die Relativitätstheorie Einsteins und ihre physikalschen Grundlagen*, 3rd edition, 1922, Introduction). Fundamentally, this view is Kant's theory of objectivity consistently developed (*cf.* section 8 and note 5 to that section). Reininger also refers to this situation. He writes in *Das Psycho-Physische Problem* (1916), p. 29, 'Metaphysics *as science* is impossible . . . because although the absolute is indeed experienced, and for that reason can be intuitively felt, it yet refuses to be expressed in words. For "*Spricht* die Seele, so spricht, ach ! schon die *Seele* nicht mehr". (If the soul *speaks* then alas it is no longer the *soul* that speaks.)'

CHAPTER VI

DEGREES OF TESTABILITY

THEORIES may be more, or less, severely testable; that is to say, more, or less, easily falsifiable. The degree of their testability is of significance for the selection of theories.

In this chapter, I shall compare the various degrees of testability or falsifiability of theories through comparing the classes of their potential falsifiers. This investigation is quite independent of the question whether or not it is possible to distinguish in an absolute sense between falsifiable and non-falsifiable theories. Indeed one might say of the present chapter that it 'relativises' the requirement of falsifiability by showing falsifiability to be a matter of degree.

31. *A Programme and an Illustration.*

A theory is falsifiable, as we saw in section 23, if there exists at least one non-empty class of homotypic basic statements which are forbidden by it; that is, if the class of its potential falsifiers is not empty. If, as in section 23, we represent the class of all possible basic statements by a circular area, and the possible events by the radii of the circle, then we can say: At least *one* radius—or perhaps better, one narrow sector whose width may represent the fact that the event is to be 'observable'—must be incompatible with the theory and ruled out by it. One might then represent the potential falsifiers of various theories by sectors of various widths. And according to the greater and lesser width of the sectors ruled out by them, theories might then be said to have more, or less, potential falsifiers. (The question whether this 'more' or 'less' could be made at all precise will be left open for the moment.) It might then be said, further, that if the class of potential falsifiers of one theory is 'larger' than that of another, there will be more opportunities for the first theory to be refuted by experience; thus compared with the second theory, the first theory may be said to be 'falsifiable in a higher degree'. This also means that the first theory *says more* about the world of experience than the second theory, for it rules out a larger class of basic statements. Although the class of permitted statements will thereby become smaller, this does not affect our argument; for we have seen that the theory does not assert anything about this class. Thus it can be said that the amount of empirical information conveyed by a theory, or its *empirical content*, increases with its degree of falsifiability.

Let us now imagine that we are given a theory, and that the sector representing the basic statements which it forbids becomes wider and wider. Ultimately the basic statements *not* forbidden by the theory will be represented by a narrow remaining sector. (If the theory is to be consistent, then some such sector must remain.) A theory like this would obviously be very easy to falsify, since it allows the empirical world only a narrow range of possibilities; for it rules out almost all conceivable, *i.e.* logically possible, events. It asserts so much about the world of experience, its empirical content is so great, that there is, as it were, little chance for it to escape falsification.

Now theoretical science aims, precisely, at obtaining theories which are easily falsifiable in this sense. It aims at restricting the range of permitted events to a minimum; and, if this can be done at all, to such a degree that any further restriction would lead to an actual empirical falsification of the theory. If we could be successful in obtaining a theory such as this, then this theory would describe 'our particular world' as precisely as a theory can; for it would single out the world of 'our experience' from the class of all logically possible worlds of experience with the greatest precision attainable by theoretical science. All the events or classes of occurrences which we actually encounter and observe, and only these, would be characterized as 'permitted'.[*1]

32. *How are Classes of Potential Falsifiers to be Compared ?*

The classes of potential falsifiers are infinite classes. The intuitive

[*1] For further remarks concerning the aims of science, see appendix *x, and section *15 of the *Postscript*.

'more' and 'less' which can be applied without special safeguards to finite classes cannot similarly be applied to infinite classes.

We cannot easily get round this difficulty; not even if, instead of the forbidden basic statements or *occurrences*, we consider, for the purpose of comparison, classes of forbidden *events*, in order to ascertain which of them contains 'more' forbidden events. For the number of events forbidden by an empirical theory is also infinite, as may be seen from the fact that the conjunction of a forbidden event with any other event (whether forbidden or not) is again a forbidden event.

I shall consider three ways of giving a precise meaning, even in the case of infinite classes, to the intuitive 'more' or 'less', in order to find out whether any of them may be used for the purpose of comparing classes of forbidden events.

(1) The concept of the *cardinality (or power) of a class*. This concept cannot help us to solve our problem, since it can easily be shown that the classes of potential falsifiers have the same cardinal number for all theories.[1]

(2) *The concept of dimension.* The vague intuitive idea that a cube in some way contains more points than, say, a straight line can be clearly formulated in logically unexceptionable terms by the set-theoretical concept of dimension. This distinguishes classes or sets of points according to the wealth of the 'neighbourhood relations' between their elements: sets of higher dimension have more abundant neighbourhood relations. The concept of dimension which allows us to compare classes of 'higher' and 'lower' dimension, will be used here to tackle the problem of comparing degrees of testability. This is possible because basic statements, combined by conjunction with other basic statements, again yield basic statements which, however, are 'more highly composite' than their components; and this degree of composition of basic statements may be linked with the concept of dimension. However, it is not the composition of the forbidden events but that of the permitted ones which will have to be used. The reason is that the events forbidden by a theory can be of any degree of composition; on the other hand, some of the permitted statements are

---

[1] Tarski has proved that under certain assumptions every class of statements is denumerable (*cf. Monatshefte f. Mathem. u. Physik* **40**, 1933, p. 100, note 10). * The concept of measure is inapplicable for similar reasons (*i.e.* because the set of all statements of a language is denumerable).

permitted merely because of their form or, more precisely, because their degree of composition is too low to enable them to contradict the theory in question; and this fact can be used for comparing dimensions.[*1]

(3) The *subclass relation*. Let all elements of a class α be also elements of a class β, so that α is a subclass of β (in symbols: $\alpha \subset \beta$). Then either all elements of β are in their turn also elements of α—in which case the two classes are said to have the same extension, or to be identical—or there are elements of β which do not belong to α. In the latter case the elements of β which do not belong to α form 'the difference class' or the *complement* of α with respect to β, and α is a *proper subclass* of β. The subclass relation corresponds very well to the intuitive 'more' and 'less', but it suffers from the disadvantage that this relation can only be used to compare the two classes if one includes the other. If therefore two classes of potential falsifiers intersect, without one being included in the other, or if they have no common elements, then the degree of falsifiability of the corresponding theories cannot be compared with the help of the subclass relation: they are non-comparable with respect to this relation.

33. *Degrees of Falsifiability Compared by Means of the Subclass Relation.*
The following definitions are introduced provisionally, to be improved later in the course of our discussion of the dimensions of theories.[*1]

(1) A statement x is said to be 'falsifiable in a higher degree' or 'better testable' than a statement y, or in symbols: $Fsb(x) > Fsb(y)$, if and only if the class of potential falsifiers of x includes the class of the potential falsifiers of y as a *proper subclass*.

(2) If the classes of potential falsifiers of the two statements x and y are identical, then they have the same degree of falsifiability, i.e. $Fsb(x) = Fsb(y)$.

[*1] The German term '*komplex*' has been translated here and in similar passages by '*composite*' rather than by '*complex*'. The reason is that it does not denote, like the English 'complex', the opposite of 'simple'. The opposite of 'simple' ('*einfach*') is denoted, rather, by the German '*kompliziert*'. (*Cf.* the first paragraph of section 41 where '*kompliziert*' is translated by 'complex'.) In view of the fact that *degree of simplicity* is one of the major topics of this book, it would have been misleading to speak here (and in section 38) of *degree of complexity*. I therefore decided to use the term '*degree of composition*' which seems to fit the context very well.

[*1] See section 38, and the appendices i, *vii, and *viii.

(3) If neither of the classes of potential falsifiers of the two statements includes the other as a proper subclass, then the two statements have non-comparable degrees of falsifiability $(Fsb(x) \parallel Fsb(y))$.

If (1) applies, there will always be a non-empty complement class. In the case of universal statements, this complement class must be infinite. It is not possible, therefore, for the two (strictly universal) theories to differ in that one of them forbids a finite number of single occurrences permitted by the other.

The classes of potential falsifiers of all tautological and metaphysical statements are empty. In accordance with (2) they are, therefore, identical. (For empty classes are subclasses of all classes, and hence also of empty classes, so that all empty classes are identical; which may be expressed by saying that there exists only *one* empty class.) If we denote an empirical statement by 'e', and a tautology or a metaphysical statement (*e.g.* a purely existential statement) by 't' or 'm' respectively, then we may ascribe to tautological and metaphysical statements a zero degree of falsifiability and we can write: $Fsb(t) = Fsb(m) = 0$, and $Fsb(e) > 0$.

A self-contradictory statement (which we may denote by 'c') may be said to have the class of all logically possible basic statements as its class of potential falsifiers. This means that any statement whatsoever is comparable with a self-contradictory statement as to its degree of falsifiability. We have $Fsb(c) > Fsb(e) > 0$.[*2] If we arbitrarily put $Fsb(c) = 1$, i.e. arbitrarily assign the number 1 to the degree of falsifiability of a self-contradictory statement, then we may even define an empirical statement e by the condition $1 > Fsb(e) > 0$. In accordance with this formula, $Fsb(e)$ always falls within the interval between 0 and 1, excluding these limits, i.e. within the 'open interval' bounded by these numbers. By excluding contradiction and tautology (as well as metaphysical statements) the formula expresses at the same time both *the requirement of consistency and that of falsifiability.*

34. *The Structure of the Subclass Relation. Logical Probability.*

We have defined the comparison of the degree of falsifiability of two statements with the help of the subclass relation; it therefore

[*2] See however now appendix *vii.

shares all the structural properties of the latter. The question of comparability can be elucidated with the help of a diagram (fig. 1), in which certain subclass relations are depicted on the left, and the corresponding testability relations on the right. The Arabic numerals

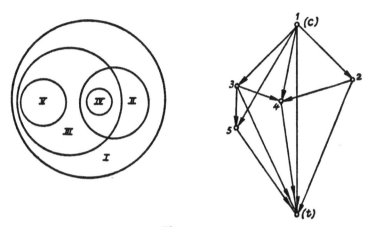

Figure 1

on the right correspond to the Roman numerals on the left in such a way that a given Roman numeral denotes the class of the potential falsifiers of that statement which is denoted by the corresponding Arabic numeral. The arrows in the diagram showing the degrees of testability run from the better testable or better falsifiable statements to those which are not so well testable. (They therefore correspond fairly precisely to derivability-arrows; see section 35.)

It will be seen from the diagram that various sequences of subclasses can be distinguished and traced, for example the sequence I–II–IV or I–III–V, and that these could be made more 'dense' by introducing new intermediate classes. All these sequences begin in this particular case with 1 and end with the empty class, since the latter is included in every class. (The empty class cannot be depicted in our diagram on the left, just because it is a subclass of every class and would therefore have to appear, so to speak, everywhere.) If we choose to identify class 1 with the class of all possible basic statements, then 1 becomes the contradiction (c); and 0 (corresponding to the empty class) may then denote the tautology (t). It is possible to pass from 1 to the empty class, or from (*c*) to (*t*) by various paths; some of these, as can be seen from the right hand diagram, may cross one another. We may therefore say that the structure of the relation is that of a lattice (a 'lattice of sequences' ordered by the arrow, or the subclass relation). There are nodal points (*e.g.* statements 4 and 5) in which the lattice is partially connected. The relation is totally connected only in the universal class and in the empty class, corresponding to the contradiction *c* and tautology *t*.

Is it possible to arrange the degrees of falsifiability of various statements on one scale, *i.e.* to correlate, with the various statements, numbers which order them according to their falsifiability? Clearly, we cannot possibly order all statements in this way;*¹ for if we did, we should be arbitrarily making the non-comparable statements comparable. There is, however, nothing to prevent us from picking out one of the sequences from the lattice, and indicating the order of its statements by numbers. In so doing we should have to proceed in such a way that a statement which lies nearer to the contradiction *c* is always given a higher number than one which lies nearer to the tautology *t*. Since we have already assigned the numbers 0 and 1 to tautology and contradiction respectively, we should have to assign *proper fractions* to the empirical statements of the selected sequence.

I do not really intend, however, to single out one of the sequences. Also, the assignment of numbers to the statements of the sequence would be entirely arbitrary. Nevertheless, the fact that it is possible to assign such fractions is of great interest, especially because of the light it throws upon the connection between degree of falsifiability and the idea of *probability*. Whenever we can compare the degrees of falsifiability of two statements, we can say that the one which is the less falsifiable is also the more probable, by virtue of its logical

---

*¹ I still believe that the attempt to make all statements comparable by introducing a *metric* must contain an arbitrary, extra-logical element. This is quite obvious in the case of statements such as 'All adult men are more than two feet high' (or 'All adult men are less than nine feet high'); that is to say, statements with predicates stating a measurable property. For it can be shown that the metric of content or falsifiability would have to be a function of the metric of the predicate; and the latter must always contain an arbitrary, or at any rate an extra-logical element. Of course, we may construct artificial languages for which we lay down a metric. But the resulting measure will not be purely logical, however 'obvious' the measure may appear as long as only discrete, qualitative yes-or-no predicates (as opposed to quantitative, measurable ones) are admitted. See also appendix *ix, the Second and Third Notes.

form. This probability I call*² 'logical probability';¹ it must not be confused with that numerical probability which is employed in the theory of games of chance, and in statistics. *The logical probability of a statement is complementary to its degree of falsifiability*: it increases with decreasing degree of falsifiability. The logical probability 1 corresponds to the degree 0 of falsifiability, and *vice versa*. The better testable statement, *i.e.* the one with the higher degree of falsifiability, is the one which is logically less probable; and the statement which is less well testable is the one which is logically more probable.

As will be shown in section 72, *numerical* probability can be linked with logical probability, and thus with degree of falsifiability. It is possible to interpret numerical probability as applying to a subsequence (picked out from the logical probability relation) for which a *system of measurement* can be defined, on the basis of frequency estimates.

These observations on the comparison of degrees of falsifiability do not hold only for universal statements, or for systems of theories; they can be extended so as to apply to singular statements. Thus they hold, for example, for theories in conjunction with initial conditions. In this case the class of potential falsifiers must not be mistaken for a class of events—for a class of homotypic basic statements—since it is a class of occurrences. (This remark has some bearing on the connection between logical and numerical probability which will be analysed in section 72.)

35. *Empirical Content, Entailment, and Degrees of Falsifiability.*

It was said in section 31 that what I call the *empirical content* of a statement increases with its degree of falsifiability: the more a statement forbids, the more it says about the world of experience. (*Cf.*

---

*² I now (since 1938; *cf.* appendix *ii) use the term 'absolute logical probability' rather than 'logical probability' in order to distinguish it from 'relative logical probability (or 'conditional logical probability'). See also appendices *iv, and *vii to *ix.

¹ To this idea of logical probability (inverted testability) corresponds Bolzano's idea of validity, especially when he applies it to the *comparison of statements*. For example, he describes the major propositions in a derivability relation as the statements of lesser validity, the consequents as those of greater validity (*Wissenschaftslehre*, 1837, Vol. II, §157, No. 1). The relation of his concept of validity to that of probability is explained by Bolzano in *op. cit.* §147. *Cf.* also Keynes, *A Treatise on Probability* (1921), p. 224. The examples there given show that my comparison of logical probabilities is identical with Keynes's 'comparison of the probability which we ascribe *a priori* to a generalization'. See also notes 1 to section 36 and 1 to section 83.

section 6.) What I call 'empirical content' is closely related to, but not identical with, the concept 'content' as defined, for instance, by Carnap.[1] For the latter I will use the term 'logical content', to distinguish it from that of *empirical* content.

I define the *empirical content* of a statement p as the class of its potential falsifiers (*cf.* section 31). The *logical content* is defined, with the help of the concept of derivability, as the class of all non-tautological statements which are derivable from the statement in question. (It may be called its 'consequence class'.) So the logical content of p is *at least equal* to (*i.e.* greater than or equal to) that of a statement q, if q is derivable from p (or, in symbols, if '$p \rightarrow q$'[*1]). If the derivability is mutual (in symbols, '$p \leftrightarrow q$'[*1]) then p and q are said to be of equal content.[2] If q is derivable from p, but not p from q, then the consequence class of q must be a proper sub-set of the consequence class of p; and p then possesses the larger consequence class, and thereby the greater logical content (or logical force[*2]).

It is a consequence of my definition of *empirical content* that the comparison of the logical and of the empirical contents of two statements p and q leads to the same result if the statements compared contain no metaphysical elements. We shall therefore require the following: (a) two statements of equal logical content must also have equal empirical content; (b) a statement p whose logical content is greater than that of a statement q must also have greater empirical content, or at least equal empirical content; and finally, (c) if the empirical content of a statement p is greater than that of a statement q, then the logical content must be greater or else non-comparable. The qualification in (b) 'or at least equal empirical content' had to be added because p might be, for example, a conjunction of q with some purely existential statement, or with some other kind of metaphysical statement to which we must ascribe a certain logical content; for in

[1] Carnap, *Erkenntnis* **2**, 1932, p. 458.

[*1] '$p \rightarrow q$' means, according to this explanation, that the conditional statement with the antecedent p and the consequent q is *tautological*, or logically true. (At the time of writing the text, I was not clear on this point; nor did I understand the significance of the fact that an assertion about deducibility was a meta-linguistic one. See also note *1 to section 18, above.) Thus '$p \rightarrow q$' may be read here: 'p entails q'.

[2] Carnap, *op. cit.*, says: 'The metalogical term "equal in content" is defined as "mutually derivable".' Carnap's *Logische Syntax der Sprache* (1934) and his *Die Aufgabe der Wissenschaftslogik* (1934) were published too late to be considered here.

[*2] If the logical content of p exceeds that of q, then we say also that p is *logically stronger* than q, or that its *logical force* exceeds that of q.

this case the empirical content of p will not be greater than that of q. Corresponding considerations make it necessary to add to (c) the qualification 'or else non-comparable'.[*3]

In comparing degrees of testability or of empirical content we shall therefore as a rule—*i.e.* in the case of purely empirical statements —arrive at the same results as in comparing logical content, or derivability-relations. Thus it will be possible to base the comparison of degrees of falsifiability to a large extent upon derivability relations. Both relations show the form of lattices which are totally connected in the self-contradiction and in the tautology (*cf.* section 34). This may be expressed by saying that a self-contradiction entails every statement and that a tautology is entailed by every statement. Moreover, *empirical* statements, as we have seen, can be characterized as those whose degree of falsifiability falls into the open interval which is bounded by the degrees of falsifiability of self-contradictions on the one side, and of tautologies on the other. Similarly, *synthetic* statements in general (including those which are non-empirical) are placed, by the entailment relation, in the open interval between self-contradiction and tautology.

To the positivist thesis that all non-empirical (metaphysical) statements are 'meaningless' there would thus correspond the thesis that my distinction between *empirical* and *synthetic* statements, or between *empirical* and *logical* content, is superfluous; for all synthetic statements would have to be empirical—all that are genuine, that is, and no mere pseudo-statements. But this way of using words, though feasible, seems to me more likely to confuse the issue than to clarify it.

Thus I regard the comparison of the empirical content of two statements as equivalent to the comparison of their degrees of falsifiability. This makes our methodological rule that those theories should be given preference which can be most severely tested (*cf.* the anti-conventionalist rules in section 20) equivalent to a rule favouring theories with the highest possible empirical content.

36. *Levels of Universality and Degrees of Precision.*

There are other methodological demands which may be reduced to the demand for the highest possible empirical content. Two of these are outstanding: the demand for the highest attainable level

[*3] See, again, appendix *vii.

(or degree) of *universality*, and the demand for the highest attainable degree of *precision*.

With this in mind we may examine the following conceivable natural laws:

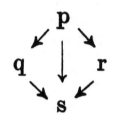

p: All heavenly bodies which move in closed orbits move in circles: or more briefly: All *orbits of heavenly bodies are circles.*

q: All *orbits of planets* are *circles.*

r: All *orbits of heavenly bodies* are *ellipses.*

s: All *orbits of planets* are *ellipses.*

The deducibility relations holding between these four statements are shown by the arrows in our diagram. From *p* all the others follow; from *q* follows *s*, which also follows from *r*; so that *s* follows from all the others.

Moving from *p* to *q* the *degree of universality* decreases; and *q* says less than *p* because the orbits of planets form a proper subclass of the orbits of heavenly bodies. Consequently *p* is more easily falsified than *q*: if *q* is falsified, so is *p*, but not *vice versa*. Moving from *p* to *r*, the *degree of precision* (of the predicate) decreases: circles are a proper subclass of ellipses; and if *r* is falsified, so is *p*, but not *vice versa*. Corresponding remarks apply to the other moves: moving from *p* to *s*, the degree of both universality and precision decreases; from *q* to *s* precision decreases; and from *r* to *s*, universality. To a higher degree of universality *or* precision corresponds a greater (logical or) empirical content, and thus a higher degree of testability.

Both universal and singular statements can be written in the form of a 'universal conditional statement' (or a 'general implication' as it is often called). If we put our four laws in this form, then we can perhaps see more easily and accurately how the degrees of universality and the degrees of precision of two statements may be compared.

A universal conditional statement (*cf.* note 6 to section 14) may be written in the form: '$(x) (\varphi x \rightarrow fx)$' or in words: 'All values of *x* which satisfy the statement function φx also satisfy the statement function fx.' The statement *s* from our diagram yields the following example: '(x) (*x* is an orbit of a planet $\rightarrow$ *x* is an ellipse)' which means:

'Whatever x may be, if x is an orbit of a planet then x is an ellipse.' Let p and q be two statements written in this 'normal' form; then we can say that p is of greater universality than q if the antecedent statement function of p (which may be denoted by '$\varphi_p x$') is tautologically implied by (or logically deducible from), but not equivalent to, the corresponding statement function of q (which may be denoted by '$\varphi_q x$'); or in other words, if '(x) $(\varphi_q x \to \varphi_p x)$' is *tautological* (or logically true). Similarly we shall say that p has greater precision than q if '(x) $(f_p x \to f_q x)$' is tautological, that is if the predicate (or the consequent statement function) of p is narrower than that of q, which means that the predicate of p entails that of q. *[1]

This definition may be extended to statement functions with more than one variable. Elementary logical transformations lead from it to the derivability relations which we have asserted, and which may be expressed by the following rule:[1] If of two statements both their universality *and* their precision are comparable, then the less universal or less precise is derivable from the more universal or more precise; unless, of course, the one is more universal and the other more precise (as in the case of q and r in my diagram).[2]

We could now say that our methodological decision—sometimes metaphysically interpreted as the principle of causality—is to leave nothing unexplained, *i.e.* always to try to deduce statements from others of higher universality. This decision is derived from the demand for the highest attainable degree of universality and precision, and it can be reduced to the demand, or rule, that preference should be given to those theories which can be most severely tested.*[2]

---

*[1] It will be seen that in the present section (in contrast to sections 18 and 35), the arrow is used to express a conditional rather than the entailment relation; *cf.* also note *1 to section 18.

[1] We can write: $[(\phi_q x \to \phi_p x) \cdot (f_p x \to f_q x)] \to [(\phi_p x \to f_p x) \to (\phi_q x \to f_q x)]$ or for short: $[(\phi_q \to \phi_p) \cdot (f_p \to f_q)] \to (p \to q)$. *The elementary character of this formula, maintained in the text, becomes clear if we write: '$[(a \to b) \cdot (c \to d)] \to [(b \to c) \to (a \to d)]$'. We then put, in accordance with the text, 'p' for '$b \to c$' and 'q' for '$a \to d$', etc.

[2] What I call higher universality in a statement corresponds roughly to what classical logic might call the greater 'extension of the subject'; and what I call greater precision corresponds to the smaller extension, or the 'restriction of the predicate'. The rule concerning the derivability relation, which we have just discussed, can be regarded as clarifying and combining the classical *dictum de omni et nullo* and the *nota-notae* principle, the 'fundamental principle of mediate predication'. Cf. Bolzano, *Wissenschaftslehre* II (1837), §263, Nos. 1 and 4; Külpe, *Vorlesungen über Logik* (edited by Selz, 1923), §34, 5, and 7.

*[2] See now also section *15 and chapter *iv of my *Postscript*, especially section *76, text to note 5.

37. *Logical Ranges. Notes on the Theory of Measurement.*

If a statement p is more easy to falsify than a statement q, because it is of a higher level of universality or precision, then the class of the basic statements permitted by p is a proper subclass of the class of the basic statements permitted by q. The subclass-relationship holding between classes of permitted statements is the opposite of that holding between classes of forbidden statements (potential falsifiers): the two relationships may be said to be inverse (or perhaps complementary). The class of basic statements permitted by a statement may be called its '*range*'.[1] The 'range' which a statement allows to reality is, as it were, the amount of 'free play' (or the degree of freedom) which it allows to reality. Range and empirical content (*cf.* section 35) are converse (or complementary) concepts. Accordingly, the ranges of two statements are related to each other in the same way as are their logical probabilities (*cf.* sections 34 and 72).

I have introduced the concept of range because it helps us to handle certain questions connected with *degree of precision in measurement*. Assume that the consequences of two theories differ so little in all fields of application that the very small differences between the calculated observable events cannot be detected, owing to the fact that the degree of precision attainable in our measurements is not sufficiently high. It will then be impossible to decide by experiment between the two theories, without first improving our technique of measurement.[*1] This shows that the prevailing technique of measurement determines a certain range—a region within which discrepancies between the observations are permitted by the theory.

Thus the rule that theories should have the highest attainable degree of testability (and thus allow only the narrowest range) entails the demand that the degree of precision in measurement should be raised as much as possible.

It is often said that all measurement consists in the determination of coincidences of points. But any such determination can only be correct within limits. There are no coincidences of points in a strict

---

[1] The concept of range (*Spielraum*) was introduced by von Kries (1886); similar ideas are found in Bolzano. Waismann (*Erkenntnis* I, 1930, p. 228 *ff.*) attempts to combine the theory of range with the frequency theory; *cf.* section 72. * Keynes gives (*Treatise*, p. 88) 'field' as a translation of '*Spielraum*', here translated as 'range'; he also uses (p.224) 'scope' for what in my view amounts to precisely the same thing.

[*1] This is a point which, I believe, was wrongly interpreted by Duhem. See his *Aim and Structure of Physical Theory*, p. 137 *ff.*

sense.[*2] Two physical 'points'—a mark, say, on the measuring-rod, and another on a body to be measured—can at best be brought into close proximity; they cannot coincide, that is, coalesce into *one* point. However trite this remark might be in another context, it is important for the question of precision in measurement. For it reminds us that measurement should be described in the following terms. We find that the point of the body to be measured lies *between* two gradations or marks on the measuring-rod or, say, that the pointer of our measuring apparatus lies *between* two gradations on the scale. We can then either regard these gradations or marks as our two optimal limits of error, or proceed to estimate the position of (say) the pointer within the interval of the gradations, and so obtain a more accurate result. One may describe this latter case by saying that we take the pointer to lie between two imaginary gradation marks. Thus an interval, a range, always remains. It is the custom of physicists to estimate this interval for every measurement. (Thus following Millikan they give, for example, the elementary charge of the electron, measured in electrostatic units, as $e = 4.774 . 10^{-10}$, adding that the range of imprecision is $\pm 0.005 . 10^{-10}$.) But this raises a problem. What can be the purpose of replacing, as it were, one mark on a scale by *two*—to wit, the two bounds of the interval—when for each of these two bounds there must again arise the same question: what are the limits of accuracy for the bounds of the interval?

Giving the bounds of the interval is clearly useless unless these two bounds in turn can be fixed with a degree of precision greatly exceeding what we can hope to attain for the original measurement; fixed, that is, within their own intervals of imprecision which should thus be smaller, by several orders of magnitude, than the interval they determine for the value of the original measurement. In other words, the bounds of the interval are not sharp bounds but are really very small intervals, the bounds of which are in their turn still much smaller intervals, and so on. In this way we arrive at the idea of what may be called the 'unsharp bounds' or '*condensation bounds*' of the interval.

These considerations do not presuppose the mathematical theory of errors, nor the theory of probability. It is rather the other way round; by analysing the idea of a measuring interval they furnish a background without which the statistical theory of errors makes very

[*2] Note that I am speaking here of measuring, not of counting. (The difference between these two is closely related to that between real numbers and rational numbers.)

little sense. If we measure a magnitude many times, we obtain values which are distributed with different densities over an interval—the interval of precision depending upon the prevailing measuring technique. Only if we know what we are seeking—namely the condensation bounds of this interval—can we apply to these values the theory of errors, and determine the bounds of the interval.[*3]

Now all this sheds some light, I think, on the *superiority of methods that employ measurements* over *purely qualitative methods*. It is true that even in the case of qualitative estimates, such as an estimate of the pitch of a musical sound, it may sometimes be possible to give an interval of accuracy for the estimates; but in the absence of measurements, any such interval can be only very vague, since in such cases the concept of condensation bounds cannot be applied. This concept is applicable only where we can speak of orders of magnitude, and therefore only where methods of measurement are defined. I shall make further use of the concept of condensation bounds of intervals of precision in section 68, in connection with the theory of probability.

38. *Degrees of Testability Compared by Reference to Dimensions.*

Till now we have discussed the comparison of theories with respect to their degrees of testability only in so far as they can be compared with the help of the subclass-relation. In some cases this method is quite successful in guiding our choice between theories. Thus we may now say that Pauli's exclusion principle, mentioned by way of example in section 20, indeed turns out to be highly satisfactory as an auxiliary hypothesis. For it greatly increases the degree of precision and, with it, the degree of testability, of the older quantum theory (like the corresponding statement of the new quantum theory which asserts that anti-symmetrical states are realized by electrons, and symmetrical ones by uncharged, and by certain multiply charged, particles).

For many purposes, however, comparison by means of the subclass relation does not suffice. Thus Frank, for example, has pointed

---

[*3] These considerations are closely connected with, and supported by, some of the results discussed under points 8 ff of my 'Third Note', reprinted in appendix *ix. See also section *15 of the *Postscript* for the significance of measurement for the 'depth' of theories.

out that statements of a high level of universality—such as the principle of the conservation of energy in Planck's formulation—are apt to become tautological, and to lose their empirical content, unless the initial conditions can be determined '... by *few* measurements, ... *i.e.* by means of a small number of magnitudes characteristic of the state of the system'.[1] The question of the number of parameters which have to be ascertained, and to be substituted in the formulae, cannot be elucidated with the help of the sub-class relation, in spite of the fact that it is evidently closely connected with the problem of testability and falsifiability, and their degrees. The fewer the magnitudes which are needed for determining the initial conditions, the less composite[*1] will be the basic statements which suffice for the falsification of the theory; for a falsifying basic statement consists of the conjunction of the initial conditions with the negation of the derived prediction (*cf.* section 28). Thus it may be possible to compare theories as to their degree of testability by ascertaining the minimum degree of composition which a basic statement must have if it is to be able to contradict the theory; provided always that we can find a way to compare basic statements in order to ascertain whether they are more (or less) composite, *i.e.* compounds of a greater (or a smaller) number of basic statements of a simpler kind. All basic statements, whatever their content, whose degree of composition does not reach the requisite minimum, would be permitted by the theory simply because of their low degree of composition.

But any such programme is faced with difficulties. For generally it is not easy to tell, merely by inspecting it, whether a statement is composite, *i.e.* equivalent to a conjunction of simpler statements. In all statements there occur universal names, and by analysing these one can often break down the statement into conjunctive components. (For example, the statement 'There is a glass of water at the place *k*' might perhaps be analysed, and broken down into the two statements 'There is a glass containing a fluid at the place *k*' and 'There is water at the place *k*'.) There is no hope of finding any natural end to hte dissection of statements by this method, especially since we can always introduce new universals defined for the purpose of making a further dissection possible.

With a view to rendering comparable the degrees of composition

---

[1] *Cf.* Frank, *Das Kausalgesetz und seine Grenzen* (1931), e.g. p. 24.
[*1] For the term 'composite', see note *1 to section 32.

of all basic statements, it might be suggested that we should choose a certain class of statements as the *elementary* or *atomic* ones,[2] from which all other statements could then be obtained by conjunction and other logical operations. If successful, we should have defined in this way an 'absolute zero' of composition, and the composition of any statement could then be expressed, as it were, in absolute degrees of composition.[*2] But for the reason given above, such a procedure would have to be regarded as highly unsuitable; for it would impose serious restrictions upon the free use of scientific language.[*3]

Yet it is still possible to compare the degrees of composition of basic statements, and thereby also those of other statements. This can be done by selecting arbitrarily a class of *relatively* atomic statements, which we take as a basis for comparison. Such a class of relatively atomic statements can be defined by means of a *generating schema or matrix* (for example, 'There is a measuring apparatus for . . . at the place . . . , the pointer of which lies between the gradation mark . . . and . . .'). We can then define as relatively atomic, and thus as equi-composite, the class of all statements obtained from this kind of matrix (or statement function) by the substitution of definite values. The class of these statements, together with all the conjunctions which can be formed from them may be called a '*field*'. A conjunction of n different relatively atomic statements of a field may be called an 'n-tuple of the field';

[2] 'Elementary propositions' in Wittgenstein, *Tractatus Logico-Philosophicus*, Proposition 5: 'Propositions are truth-functions of elementary propositions'. 'Atomic propositions' (as opposed to the composite 'molecular propositions') in Whitehead and Russell's *Principia Mathematica* Vol. I. Introduction to 2nd edition (1925), pp. xv f. C. K. Ogden translated Wittgenstein's term 'Elementarsatz' as 'elementary proposition' (*cf. Tractatus* 4.21), while Bertrand Russell in his Preface to the *Tractatus* (1922), p. 13, translated it as 'atomic proposition'. The latter term has become more popular.

[*2] Absolute degrees of composition would determine, of course, absolute degrees of content, and thus of absolute logical improbability. The programme here indicated of introducing improbability, and thus probability, by singling out a certain class of absolutely atomic statements (earlier sketched, for example, by Wittgenstein) has more recently been elaborated by Carnap in his *Logical Foundations of Probability*, 1950, in order to construct a theory of induction. See also the remarks on model languages in my *Preface to the English Edition* where I allude to the fact that the third model language (Carnap's language system) does not admit measurable properties. (Nor does it in its present form allow the introduction of a temporal or spatial order.)

[*3] The words 'scientific language' were here used quite naïvely, and should not be interpreted in the technical sense of what is today called a 'language system'. On the contrary, my main point was that we should remember the fact that scientists cannot use a 'language system' since they have constantly to change their language, with every new step they take. 'Matter', or 'atom', after Rutherford, and 'matter', or 'energy', after Einstein, meant something different from what they meant before: the meaning of these concepts is a function of the—constantly changing—*theory*.

and we can say that the degree of its composition is equal to the number n.

If there exists, for a theory t, a field of singular (but not necessarily basic) statements such that, for some number d, the theory t cannot be falsified by any d-tuple of the field, although it can be falsified by certain $d+1$-tuples, then we call d the *characteristic number* of the theory with respect to that field. All statements of the field whose degree of composition is less than d, or equal to d, are then compatible with the theory, and permitted by it, irrespective of their content.

Now it is possible to base the comparison of the degree of testability of theories upon this characteristic number d. But in order to avoid inconsistencies which might arise through the use of different fields, it is necessary to use a somewhat narrower concept than that of a field, namely that of a *field of application*. If a theory t is given, we say that a field is a *field of application of the theory t* if there exists a characteristic number d of the theory t with respect to this field, and if, in addition, it satisfies certain further conditions (which are explained in appendix 1).

The characteristic number d of a theory t, with respect to a field of application, I call the *dimension* of t with respect to this field of application. The expression 'dimension' suggests itself because we can think of all possible n-tuples of the field as spatially arranged (in a configuration space of infinite dimensions). If, for example, $d = 3$, then those statements which are admissible because their composition is too low form a three-dimensional sub-space of this configuration. Transition from $d = 3$ to $d = 2$ corresponds to the transition from a solid to a surface. The smaller the dimension d, the more severely restricted is the class of those permitted statements which, regardless of their content, cannot contradict the theory owing to their low degree of composition; and the higher will be the degree of falsifiability of the theory.

The concept of the field of application has not been restricted to basic statements, but singular statements of all kinds have been allowed to be statements belonging to a field of application. But by comparing their dimensions with the help of the field, we can estimate the degree of composition of the basic statements. (We assume that to highly composite singular statements there correspond highly composite basic statements.) It thus can be assumed that to a theory of higher dimension, there corresponds a class of basic statements of higher dimension, such that all statements of this class are permitted by the theory, irrespective of what they assert.

This answers the question of how the two methods of comparing degrees of testability are related—the one by means of the dimension of a theory, and the other by means of the subclass relation. There will be cases in which neither, or only one, of the two methods is applicable. In such cases there is of course no room for conflict between the two methods. But if in a particular case both methods are applicable, then it may conceivably happen that two theories of equal dimensions may yet have different degrees of falsifiability if assessed by the method based upon the subclass relation. In such cases, the verdict of the latter method should be accepted, since it would prove to be the more sensitive method. In all other cases in which both methods are applicable, they must lead to the same result; for it can be shown, with the help of a simple theorem of the theory of dimension, that the dimension of a class must be greater than, or equal to, that of its subclasses.[3]

39. *The Dimension of a Set of Curves.*

Sometimes we can identify what I have called the 'field of application' of a theory quite simply with the *field of its graphic representation, i.e.* the area of a graph-paper on which we represent the theory by graphs: each point of this field of graphic representation can be taken to correspond to one relatively atomic statement. The dimension of the theory with respect to this field (defined in appendix 1) is then identical with the dimension of the set of curves corresponding to the theory. I shall discuss these relations with the help of the two statements q and s of section 36. (Our comparison of dimensions applies to statements with different predicates.) The hypothesis q—that all planetary orbits are circles—is three-dimensional: for its falsification at least four singular statements of the field are necessary, corresponding to four points of its graphic representation. The hypothesis s, that all planetary orbits are ellipses, is five-dimensional, since for its falsification at least six singular statements are necessary, corresponding to six points of the graph. We saw

---

[3] *Cf.* Menger, *Dimensionstheorie* (1928), p. 81. *The conditions under which this theorem holds can be assumed to be always satisfied by the 'spaces' with which we are concerned here.

in section 36 that q is more easily falsifiable than s: since all circles are ellipses, it was possible to base the comparison on the subclass relation. But the use of dimensions enables us to compare theories which previously we were unable to compare. For example, we can now compare a circle-hypothesis with a parabola-hypothesis (which is four dimensional). Each of the words 'circle', 'ellipse', 'parabola' denotes a class or *set of curves*; and each of these sets has the dimension d if d points are necessary and sufficient to single out, or characterize, one particular curve of the set. In algebraic representation, the dimension of the set of curves depends upon the number of *parameters* whose values we may freely choose. We can therefore say that the number of freely determinable parameters of a set of curves by which a theory is represented is characteristic for the degree of falsifiability (or testability) of that theory.

In connection with the statements q and s in my example I should like to make some methodological comments on Kepler's discovery of his laws.[*1]

I do not wish to suggest that the belief in perfection—the heuristic principle that guided Kepler to his discovery—was inspired, consciously or unconsciously, by methodological considerations regarding degrees of falsifiability. But I do believe that Kepler owed his success in part to the fact that the circle-hypothesis with which he started was relatively easy to falsify. Had Kepler started with a hypothesis which owing to its logical form was not so easily testable as the circle hypothesis, he might well have got no result at all, considering the difficulties of calculations whose very basis was 'in the air'— adrift in the skies, as it were, and moving in a way unknown. The unequivocal *negative* result which Kepler reached by the falsification of his circle hypothesis was in fact his first real success. His method had been vindicated sufficiently for him to proceed further; especially since even this first attempt had already yielded certain approximations.

No doubt, Kepler's laws might have been found in another way. But I think it was no mere accident that this was the way which led to success. It corresponds to *the method of elimination* which is applicable only if the theory is sufficiently easy to falsify—sufficiently *precise* to be capable of clashing with observational experience.

[*1] The views here developed were accepted, with acknowledgments, by W. C. Kneale, *Probability and Induction* (1949), p. 230, and J. G. Kemeny, 'The Use of Simplicity in Induction', *Philos. Review* 57, 1953; see his footnote on p. 404.

40. *Two Ways of Reducing the Number of Dimensions of a Set of Curves.*
Quite different sets of curves may have the same dimension. The
set of all circles, for example, is three-dimensional; but the set of all
circles passing through a given point is a two-dimensional set (like the set
of straight lines). If we demand that the circles should all pass through
two given points, then we get a one-dimensional set, and so on. Each
additional demand that all curves of a set should pass through one more
given point reduces the dimensions of the set by one.

zero dimensional classes[1]	one dimensional classes	two dimensional classes	three dimensional classes	four dimensional classes
—	—	straight line	circle	parabola
—	straight line through one given point	circle through one given point	parabola through one given point	conic through one given point
straight line through two given points	circle through two given points	parabola through two given points	conic through two given points	—
circle through three given points	parabola through three given points	conic through three given points	—	—

The number of dimensions can also be reduced by methods other
than that of increasing the number of given points. For example the
set of elipses with given ratio of the axes is four-dimensional (as is
that of parabolas), and so is the set of ellipses with given numerical
eccentricity. The transition from the ellipse to the circle, of course,
is equivalent to specifying an eccentricity (the eccentricity o) or a
particular ratio of the axes (unity).

As we are interested in assessing degrees of falsifiability of theories
we will now ask whether the various methods of reducing the
number of dimensions are equivalent for our purposes, or whether we

[1] We could also, of course, begin with the empty (over-determined) minus-one-dimensional class.

should examine more closely their relative merits. Now the stipulation that a curve should pass through a certain *singular point* (or small region) will often be linked up with, or correspond to, the acceptance of a certain *singular statement*, *i.e.* of an initial condition. On the other hand, the transition from, say, an ellipse-hypothesis to a circle-hypothesis, will obviously correspond to a reduction of the dimension of *the theory itself*. But how are these two methods of reducing the dimensions to be kept apart? We may give the name *'material reduction'* to that method of reducing dimensions which does *not* operate with stipulations as to the 'form' or 'shape' of the curve; that is, to reductions through specifying one or more points, for example, or by some equivalent specification. The other method, in which the form or shape of the curve becomes more narrowly specified as, for example, when we pass from ellipse to circle, or from circle to straight line, *etc.*, I will call the method of *'formal reduction'* of the number of dimensions.

It is not very easy, however, to get this distinction sharp. This may be seen as follows. Reducing the dimensions of a theory means, in algebraic terms, replacing a parameter by a constant. Now it is not quite clear how we can distinguish between different methods of replacing a parameter by a constant. The *formal reduction*, by passing from the general equation of an ellipse to the equation of a circle, can be described as equating one parameter to zero, and a second parameter to one. But if another parameter (the absolute term) is equated to zero, then this would mean a *material reduction*, namely the specification of a point of the ellipse. I think, however, that it is possible to make the distinction clear, if we see its connection with the problem of universal names. For material reduction introduces an individual name, formal reduction a universal name, into the definition of the relevant set of curves.

Let us imagine that we are given a certain individual plane, perhaps by 'ostensive definition'. The set of all ellipses in this plane can be defined by means of the general equation of the ellipse; the set of circles, by the general equation of the circle. These definitions are *independent of where*, in the plane, we *draw the (Cartesian) co-ordinates* to which they relate; consequently they are independent of the choice of the origin, and the orientation, of the co-ordinates. A specific system of co-ordinates can be determined only by individual names; say, by ostensively specifying its origin and orientation. Since the definition of the set of ellipses (or circles) is the same for all Cartesian co-ordinates, it is independent of the specification of these individual names: it is *invariant* with respect to all co-ordinate transformations of the Euclidean group (displacements and similarity transformations).

If, on the other hand, one wishes to define a set of ellipses (or circles) which have a specific, an individual point of the plane in common, then we must operate with an equation which is not invariant with respect to the transformations of the Euclidean group, but relates to a singular, *i.e.* an individually or ostensively specified, co-ordinate system. Thus it is connected with individual names.[2]

The transformations can be arranged in a hierarchy. A definition which is invariant with respect to a more general group of transformations is also invariant with respect to more special ones. For each definition of a set of curves, there is one—the most general—transformation group which is characteristic of it. Now we can say: The definition D_1 of a set of a curve is called 'equally general' to (or more general than) a definition D_2 of a set of curves if it is invariant with respect to the same transformation group as is D_2 (or a more general one). A reduction of the dimension of a set of curves may now be called *formal* if the reduction does not diminish the generality of the definition; otherwise it may be called *material*.

If we compare the degree of falsifiability of two theories by considering their dimensions, we shall clearly have to take into account their *generality*, *i.e.* their invariance with respect to co-ordinate transformations, along with their dimensions.

The procedure will, of course, have to be different according to whether the theory, like Kepler's theory, in fact makes geometrical statements about the world or whether it is 'geometrical' only in that it may be represented by a graph—such as, for example, the graph which represents the dependence of pressure upon temperature. It would be inappropriate to require of this latter kind of theory, or of the corresponding set of curves, that its definition should be invariant with respect to, say, rotations of the co-ordinate system; for in these cases, the different co-ordinates may represent entirely different things (the one pressure and the other temperature).

---

[2] On the relations between transformation groups and 'individualization' *cf.* Weyl, *Philosophie der Mathematik u. Naturwissenschaft* (1927), p. 59, English edition p. 73 *f.*, where reference is made to Klein's *Erlanger Programm*.

This concludes my exposition of the methods whereby degrees of falsifiability are to be compared. I believe that these methods can help us to elucidate epistemological questions, such as the *problem of simplicity* which will be our next concern. But there are other problems which are placed in a new light by our examination of degrees of falsifiability, as we shall see; especially the problem of the so-called 'probability of hypotheses' or of *corroboration*.

CHAPTER VII

SIMPLICITY

THERE seems to be little agreement as to the importance of the so-called 'problem of simplicity'. Weyl said, not long ago, that 'the problem of simplicity is of central importance for the epistemology of the natural sciences'.[1] Yet it seems that interest in the problem has lately declined; perhaps because, especially after Weyl's penetrating analysis, there seemed to be so little chance of solving it.

Until quite recently the idea of simplicity has been used uncritically, as though it were quite obvious what simplicity is, and why it should be valuable. Not a few philosophers of science have given the concept of simplicity a place of crucial importance in their theories, without even noticing the difficulties to which it gives rise. For example, the followers of Mach, Kirchhoff, and Avenarius have tried to replace the idea of a causal explanation by that of the 'simplest description'. Without the adjective 'simplest' or a similar word this doctrine would say nothing. As it is supposed to explain why we prefer a description of the world with the help of theories to one with the help of singular statements, it seems to presuppose that theories are simpler than singular statements. Yet few have ever attempted to explain why they should be simpler, or what is meant, more precisely, by simplicity.

If, moreover, we assume that theories are to be used for the sake of simplicity then, clearly, we should use the simplest theories. This is how Poincaré, for whom the choice of theories is a matter of convention, comes to formulate his principle for the selection of theories: he chooses the *simplest* of the possible conventions. But which are the simplest?

[1] *Cf.* Weyl, *op. cit.*, p. 115 *f.*; English edition p. 155. See also section 42 below.

41. *Elimination of the Aesthetic and the Pragmatic Concepts of Simplicity.*
The word 'simplicity' is used in very many different senses. Schrödinger's theory, for instance, is of great simplicity in a methodological sense, but in another sense it might well be called 'complex'. We can say of a problem that its solution is not simple but difficult, or of a presentation or an exposition that it is not simple but intricate.

To begin with, I shall exclude from our discussion the application of the term 'simplicity' to anything like a presentation or an exposition. It is sometimes said of two expositions of one and the same mathematical proof that the one is simpler or more elegant than the other. This is a distinction which has little interest from the point of view of the theory of knowledge; it does not fall within the province of logic, but merely indicates a preference of an *aesthetic or pragmatic* character. The situation is similar when people say that one task may be 'carried out by simpler means' than another, meaning that it can be done more easily or that, in order to do it, less training or less knowledge is needed. In all such cases the word 'simple' can be easily eliminated; its use is extra-logical.

42. *The Methodological Problem of Simplicity.*
What, if anything, remains after we have eliminated the aesthetic and the pragmatic ideas of simplicity? Is there a concept of simplicity which is of importance for the logician? Is it possible to distinguish theories that are logically not equivalent according to their degrees of simplicity?

The answer to this question may well seem doubtful, seeing how little successful have been most attempts to define this concept. Schlick, for one, gives a negative answer. He says: 'Simplicity is . . . a concept indicative of preferences which are partly practical, partly aesthetic in character.'[1] And it is notable that he gives this answer when writing of the concept which interests us here, and which I shall call the *epistemological concept of simplicity*; for he continues: 'Even if we are unable to explain what is really meant by "simplicity" here, we must yet recognize the fact that any scientist who has succeeded in representing a series of observations by means of a very simple formula (*e.g.* by a linear, quadratic, or exponential function) is immediately convinced that he has discovered a law.'

---

[1] Schlick, *Naturwissenschaften* 19, 1931, p. 148.   * I have translated Schlick's term '*pragmatischer*' freely.

Schlick discusses the possibility of defining the concept of law-like regularity, and especially the distinction between 'law' and 'chance', with the help of the concept of simplicity. He finally dismisses it with the remark that 'simplicity is obviously a wholly relative and vague concept; no strict definition of causality can be obtained with its help; nor can law and chance be precisely distinguished'.[2] From this passage it becomes clear what the concept of simplicity is actually expected to achieve: it is to provide a measure of the degree of law-likeness or regularity of events. A similar view is voiced by Feigl when he speaks of the 'idea defining the degree of regularity or of law-likeness with the help of the concept of simplicity'.[3]

The epistemological idea of simplicity plays a special part in theories of inductive logic, for example in connection with the problem of the 'simplest curve'. Believers in inductive logic assume that we arrive at natural laws by generalization from particular observations. If we think of the various results in a series of observations as points plotted in a co-ordinate system, then the graphic representation of the law will be a curve passing through all these points. But through a finite number of points we can always draw an unlimited number of curves of the most diverse form. Since therefore the law is not uniquely determined by the observations, inductive logic is confronted with the problem of deciding which curve, among all these possible curves, is to be chosen.

The usual answer is, 'choose the simplest curve'. Wittgenstein, for example, says: 'The process of induction consists in assuming the *simplest* law that can be made to harmonize with our experience.'[4] In choosing the simplest law, it is usually tacitly assumed that a linear function, say, is simpler than a quadratic one, a circle simpler than an ellipse, etc. But no reasons are given either for choosing this particular hierarchy of simplicities in preference to any other, or for believing that 'simple' laws have advantages over the less simple—apart from aesthetic and practical ones.[5] Schlick and Feigl mention[6] an unpublished paper of Natkin who, according to Schlick's account,

---

[2] Schlick, *ibid.*

[3] Feigl, *Theorie und Erfahrung in der Physik* (1931), p. 25.

[4] Wittgenstein, *op. cit.*, Proposition 6.363.

[5] Wittgenstein's remark on the simplicity of logic (*op. cit.*, Proposition 5.4541) which sets 'the standard of simplicity' gives no clue. Reichenbach's 'principle of the simplest curve' (*Mathematische Zeitschrift* **34**, 1932, p. 616) rests on his Axiom of Induction (which I believe to be untenable), and also affords no help.

[6] In the places referred to.

proposes to call one curve simpler than another if its average curvature is smaller; or, according to Feigl's account, if it deviates less from a straight line. (The two accounts are not equivalent.) This definition seems to agree pretty well with our intuitions; but it somehow misses the crucial point; it would, for example, make certain parts (the asymptotic parts) of a hyperbola much simpler than a circle, etc. And really, I do not think that the question can be settled by such 'artifices' (as Schlick calls them). Moreover, it would remain a mystery why we should give preference to simplicity if defined in this particular way.

Weyl discusses and rejects a very interesting attempt to base simplicity on probability. 'Assume, for example, that twenty co-ordinated pairs of values (x, y) of the same function, $y = f(x)$ lie (within the expected accuracy) on a straight line, when plotted on square graph paper. We shall then conjecture that we are faced here with a rigorous natural law, and that y depends linearly upon x. And we shall conjecture this because of the *simplicity* of the straight line, or because it would be so *extremely improbable* that just these twenty pairs of arbitrarily chosen observations should lie very nearly on a straight line, had the law in question been a different one. If now we use the straight line for interpolation and extrapolation, we obtain predictions which go beyond what the observations tell us. However, this analysis is open to criticism. It will always be possible to define all kinds of mathematical functions which ... will be satisfied by the twenty observations; and some of these functions will deviate considerably from the straight line. And for every single one of these we may claim that it would be *extremely improbable* that the twenty observations should lie just on this curve, unless it represented the true law. It is thus essential, after all, that the function, or rather the class of functions, should be offered to us, a priori, by mathematics because of its mathematical simplicity. It should be noted that this class of functions must not depend upon as many parameters as the number of observations to be satisfied.'[7] Weyl's remark that 'the class of functions should be offered to us a priori, by mathematics, because of its mathematical simplicity', and his reference to the number of

[7] Weyl, *op. cit.*, p. 116; English edition, p. 156. * When writing my book I did not know (and Weyl, no doubt, did not know when writing his) that Harold Jeffreys and Dorothy Wrinch had suggested, six years before Weyl, that we should measure the simplicity of a function by the paucity of its freely adjustable parameters. (See their joint paper in *Phil. Mag.* **42**, 1921, pp. 369 *ff.*) I wish to take this opportunity to make full acknowledgement to these authors.

parameters agree with my view (to be developed in section 43). But Weyl does not say what 'mathematical simplicity' is; and above all, he does not say what *logical or epistemological advantages* the simpler law is supposed to possess, compared with one that is more complex.[8]

The various passages so far quoted are very important, because of their bearing upon our present aim—the analysis of the epistemological concept of simplicity. For this concept is not yet precisely determined. It is therefore possible to reject any attempt (such as mine) to make this concept precise by saying that the concept of simplicity in which epistemologists are interested is really quite a different one. To such objections I could answer that I do not attach the slightest importance to the *word* 'simplicity'. The term was not introduced by me, and I am aware of its disadvantages. All I do assert is that the concept of simplicity which I am going to clarify helps to answer those very questions which, as my quotations show, have so often been raised by philosophers of science in connection with their 'problem of simplicity'.

43. *Simplicity and Degree of Falsifiability.*

The epistemological questions which arise in connection with the concept of simplicity can all be answered if we equate this concept with *degree of falsifiability*. This assertion is likely to meet with opposition;[*1] and so I shall try, first, to make it intuitively more acceptable.

---

[8] Weyl's further comments on the connection between simplicity and corroboration are also relevant in this connection; they are largely in agreement with my own views expressed in section 82, although my line of approach and my arguments are quite different; *cf.* note 1 to section 82, * and the new note here following (note *1 to section 43).

[*1] It was gratifying to find that this theory of simplicity (including the ideas of section 40) has been accepted at least by one epistemologist—by William Kneale, who writes in his book *Probability and Induction*, 1949, p. 229 *f* : '. . . it is easy to see that the hypothesis which is simplest in this sense is also that which we can hope to eliminate most quickly if it is false. . . . In short, the policy of assuming always the simplest hypothesis which accords with the known facts is that which will enable us to get rid of false hypotheses most quickly.' Kneale adds a footnote in which he refers to p. 116 of Weyl's book, and also to mine. But I cannot detect on this page—of which I quoted the relevant portions in the text—or anywhere else in Weyl's great book (or in any other) even a trace of the view that the simplicity of a theory is connected with its falsifiability, *i.e.* with the ease of its elimination. Nor would I have written (as I did near the end of the preceding section) that Weyl 'does not say what *logical or epistemological advantages* the simpler law is supposed to possess' had Weyl (or anybody else known to me) anticipated my theory.

The facts are these. In his profound discussion of the problem (here quoted in section 42, text to note 7) Weyl mentions first the intuitive view that a simple curve—say, a straight line—has an advantage over a more complex curve *because it might be considered a highly improbable accident if all the observations would fit such a simple curve.* But instead of

I have already shown that theories of a lower dimension are more easily falsifiable than those of a higher dimension. A law having the form of a function of the first degree, for instance, is more easily falsifiable than one expressible by means of a function of the second degree. But the latter still belongs to the best falsifiable ones among the laws whose mathematical form is that of an algebraic function. This agrees well with Schlick's remark concerning simplicity: 'We should certainly be inclined to regard a function of the first degree as simpler than one of the second degree, though the latter also doubtless represents a perfectly good law. . . .'[1]

The degree of universality and of precision of a theory increases with its degree of falsifiability, as we have seen. Thus we may perhaps identify the *degree of strictness* of a theory—the degree, as it were, in which a theory imposes the rigour of law upon nature—with its degree of falsifiability; which shows that the latter does just what Schlick and Feigl expected the concept of simplicity to do. I may add that the distinction which Schlick hoped to make between law and chance can also be clarified with the help of the idea of degrees of falsifiability: probability statements about sequences having chance-like characteristics turn out to be of infinite dimension (*cf.* section 65); not simple but complex (*cf.* section 58 and latter part of 59); and falsifiable only under special safeguards (section 68).

The comparison of degrees of testability has been discussed at length in sections 31 to 40. Some of the examples and other details

---

[1] Schlick, *Naturwissenschaften* 19, 1931, p. 148 (*cf.* note 1 to the preceding section).

following up this intuitive view (which I think would have led him to see that the simpler theory is the better testable one), Weyl *rejects* it as not standing up to rational criticism: he points out that the same could be said of *any given* curve, however complex. (This argument is correct, but it does no longer hold if we consider the *potential falsifiers* —and their degree of composition—rather than the verifying instances.) Weyl then proceeds to discuss the paucity of the parameters as a criterion of simplicity, without connecting this in any way either with the intuitive view just rejected by him, or with anything which, like testability, or content, might explain our epistemological preference for the simpler theory.

Weyl's characterization of the simplicity of a curve by the paucity of its parameters was anticipated in 1921 by Harold Jeffreys and Dorothy Wrinch (*Phil. Mag.* 42, 369 ff.). But if Weyl merely failed to see what is now 'easy to see' (according to Kneale), Jeffreys actually saw—and still sees—the very opposite: he attributes to the simpler law the greater prior probability instead of the greater prior improbability. (Thus Jeffreys's and Kneale's views together may illustrate Schopenhauer's remark that the solution of a problem often first looks like a paradox and later like a truism.) I wish to add here that I have further developed my views on simplicity, and that in doing so I have tried hard and, I hope, not quite without success, to learn something from Kneale. *Cf.* appendix *x and section *15 of my *Postscript*.

given there can be easily transferred to the problem of simplicity. This holds especially for the degree of universality of a theory: a more universal statement can take the place of many less universal ones, and for that reason has often been called 'simpler'. The concept of the dimension of a theory may be said to give precision to Weyl's idea of using the number of parameters to determine the concept of simplicity.*2 And by means of our distinction between a formal and a material reduction of the dimension of a theory (*cf.* section 40), certain possible objections to Weyl's theory can be met. One of these is the objection that the set of ellipses whose axes stand in a given ratio, and whose numerical eccentricity is given, has exactly as many parameters as the set of circles, although it is obviously less 'simple'.

Above all, our theory explains *why simplicity is so highly desirable.* To understand this there is no need for us to assume a 'principle of economy of thought' or anything of the kind. Simple statements, if knowledge is our object, are to be prized more highly than less simple ones *because they tell us more; because their empirical content is greater; and because they are better testable.*

44. *Geometrical Shape and Functional Form.*

Our view of the concept of simplicity enables us to resolve a number of contradictions which up to now have made it doubtful whether this concept was of any use.

Few would regard the *geometrical shape* of, say, a logarithmic curve as particularly simple; but a *law* which can be represented by a logarithmic function is usually regarded as a simple one. Similarly

---

*2 As mentioned in notes 7 to section 42 and *1 to the present section, it was Harold Jeffreys and Dorothy Wrinch who first proposed to measure the simplicity of a function by the paucity of its freely adjustable parameters. But they also proposed to attach to the simpler hypothesis a greater prior probability. Thus their views can be presented by the schema

simplicity = paucity of parameters = high prior probability.

It so happens that I approached the matter from an entirely different angle. I was interested in assessing degrees of testability, and I found first that testability can be measured by 'logical' improbability (which corresponds exactly to Jeffreys' 'prior' improbability). Then I found that testability, and thus prior improbability, can be equated with paucity of parameters; and only at the end, I equated high testability with high simplicity. Thus my view can be presented by the schema: testability =

high prior improbability = paucity of parameters = simplicity.

It will be seen that these two schemata coincide in part; but on the decisive point— probability *vs.* improbability—they stand in direct opposition. See also appendix *viii.

a *sine function* is commonly said to be simple, even though the geo-metrical shape of the *sine curve* is perhaps not so very simple.

Difficulties like this can be cleared up if we remember the con-nection between the number of parameters and the degree of falsi-fiability, and if we distinguish between the formal and the material reduction of dimensions. (We must also remember the rôle of invariance with respect to transformations of the co-ordinate systems.) If we speak of the *geometrical form or shape* of a curve, then what we demand is invariance with respect to all transformations belonging to the group of displacements, and we may demand invariance with respect to similarity transformations; for we do not think of a geometrical figure or shape as being tied to a definite *position*. Conse-quently, if we think of the shape of a one-parametric logarithmic curve ($y = log_a x$) as lying anywhere in a plane, then it would have *five* parameters (if we allow for similarity transformations). It would thus be by no means a particularly simple curve. If, on the other hand, a *theory or law* is represented by a logarithmic curve, then co-ordinate transformations of the kind described are irrelevant. In such cases, there is no point in either rotations or parallel displacements or similarity transformations. For a logarithmic curve as a rule is a graphic representation in which the co-ordinates cannot be inter-changed. (For example, the x-axis might represent atmospheric pressure, and the y-axis height above sea-level.) For this reason, similarity transformations are equally without any significance here. Analogous considerations hold for *sine* oscillations along a particular axis, for example, the time axis; and for many other cases.

45. *The Simplicity of Euclidean Geometry.*

One of the issues which played a major rôle in most of the dis-cussions of the theory of relativity was the simplicity of Euclidean geometry. Nobody ever doubted that Euclidean geometry as such was simpler than any non-Euclidean geometry with given constant curvature—not to mention non-Euclidean geometries with curvatures varying from place to place.

At first sight the kind of simplicity here involved seems to have little to do with degrees of falsifiability. But if the statements at issue are formulated as empirical hypotheses, then we find that the two concepts, simplicity and falsifiability, coincide in this case also.

Let us consider what experiments may help us to test the hypothesis, 'In our world, we have to employ a certain metrical geometry with such and such a radius of curvature.' A test will be possible only if we identify certain geometrical entities with certain physical objects —for instance straight lines with light rays; or points with the intersection of threads. If such an identification (a correlating definition, or perhaps an ostensive definition; *cf.* section 17) is adopted, then it can be shown that the hypothesis of the validity of an Euclidean light-ray-geometry is falsifiable to a higher degree than any of the competing hypotheses which assert the validity of some non-Euclidean geometry. For if we measure the sum of the angles of a light-ray triangle, then any significant deviation from 180 degrees will falsify the Euclidean hypothesis, The hypothesis of a Bolyai-Lobatschewski geometry with given curvature, on the other hand, would be compatible with any particular measurement not exceeding 180 degrees. Moreover, to falsify this hypothesis it would be necessary to measure not only the sum of the angles, but also the (absolute) size of the triangle; and this means that in addition to angles, a further unit of measurement, such as a unit of area, would have to be defined. Thus we see that more measurements are needed for a falsification; that the hypothesis is compatible with greater variations in the results of measurements; and that it is therefore more difficult to falsify: it is falsifiable to a lesser degree. To put it in another way, Euclidean geometry is the only metric geometry with a definite curvature in which similarity transformations are possible. In consequence, Euclidean geometrical figures can be invariant with respect to more transformations; that is, they can be of lower dimension: they can be simpler.

46. *Conventionalism and the Concept of Simplicity.*

What the conventionalist calls 'simplicity' does *not* correspond to what I call 'simplicity'. It is the central idea of the conventionalist, and also his starting point, that no theory is unambiguously determined by experience; a point with which I agree. He believes that he must therefore choose the 'simplest' theory. But since the conventionalist does not treat his theories as falsifiable systems but rather as conventional stipulations, he obviously means by 'simplicity' something different from degree of falsifiability.

The conventionalist concept of simplicity turns out to be indeed partly aesthetic and partly practical. Thus the following comment by Schlick (*cf.* section 42) applies to the conventionalist concept of simplicity, but not to mine: 'It is certain that one can only define the concept of simplicity by a convention which must always be arbitrary.'[1] It is curious that conventionalists themselves have overlooked the conventional character of their own fundamental concept—that of simplicity. That they must have overlooked it is clear, for otherwise they would have noticed that their appeal to simpilcity could never save them from arbitrariness, once they had chosen the way of arbitrary convention.

From my point of view, a system must be described as *complex in the highest degree* if, in accordance with conventionalist practice, one holds fast to it as a system established forever which one is determined to rescue, whenever it is in danger, by the introduction of auxiliary hypotheses. For the degree of falsifiability of a system thus protected is equal to *zero*. Thus we are led back, by our concept of simplicity, to the methodological rules of section 20; and especially also to that rule or principle which restrains us from indulgence in *ad hoc* hypotheses and auxiliary hypotheses: to the principle of parsimony in the use of hypotheses.

---

[1] Schlick, *ibid.*, p. 148.

CHAPTER VIII

PROBABILITY

In this chapter I shall only deal with the *probability of events* and the problems it raises. They arise in connection with the theory of games of chance, and with the probabilistic laws of physics. I shall leave the problems of what may be called the *probability of hypotheses*—such questions as whether a frequently tested hypothesis is more probable than one which has been little tested—to be discussed in sections 79 to 85 under the title of 'Corroboration'.

Ideas involving the theory of probability play a decisive part in modern physics. Yet we still lack a satisfactory, consistent definition of probability; or, what amounts to much the same, we still lack a satisfactory axiomatic system for the calculus of probability. The relations between probability and experience are also still in need of clarification. In investigating this problem we shall discover what will at first seem an almost insuperable objection to my methodological views. For although probability statements play such a vitally important rôle in empirical science, they turn out to be in principle *impervious to strict falsification*. Yet this very stumbling block will become a touchstone upon which to test my theory, in order to find out what it is worth.

Thus we are confronted with two tasks. *The first is to provide new foundations for the calculus of probability.* This I shall try to do by developing the theory of probability as a frequency theory, along the lines followed by Richard von Mises, but without the use of what he calls the 'axiom of convergence' (or 'limit axiom'), and with a somewhat weakened 'axiom of randomness'. *The second task is to elucidate the relations between probability and experience.* This means solving what I call *the problem of decidability of probability statements*.

My hope is that these investigations will help to relieve the present unsatisfactory situation in which physicists make much use of probabilities without being able to say, consistently, what they mean by 'probability'.[*1]

47. *The Problem of Interpreting Probability Statements.*

I shall begin by distinguishing two kinds of probability statements: those which state a probability in terms of numbers—which I will call *numerical* probability statements—and those which do not.

Thus the statement, 'The probability of throwing eleven with two (true) dice is $1/18$', would be an example of a numerical probability statement. Non-numerical probability statements can be of various kinds. 'It is very probable that we shall obtain a homogeneous mixture by mixing water and alcohol', illustrates one kind of statement which, suitably interpreted, might perhaps be transformed into a numerical probability statement. (For example, 'The probability of obtaining ... is very near to 1'.) A very different kind of non-numerical probability statement would be, for instance, 'The discovery of a physical effect which contradicts the quantum theory is highly improbable'; a statement which, I believe, cannot be transformed into a numerical probability statement, or put on a par with one, without distorting its meaning. I shall deal first with *numerical* probability statements; non-numerical ones, which I think less important, will be considered afterwards.

In connection with every numerical probability statement, the

[*1] Within the theory of probability, I have made since 1934 three kinds of changes.

(1) The introduction of a formal (axiomatic) calculus of probabilities which can be interpreted in many ways—for example, in the sense of the logical and of the frequency interpretations discussed in this book, and also of the propensity interpretation discussed in my *Postscript*.

(2) A simplification of the frequency theory of probability through carrying out, more fully and more directly than in 1934, that programme for reconstructing the frequency theory which underlies the present chapter.

(3) The replacement of the objective interpretation of probability in terms of frequency by another objective interpretation—the *propensity interpretation*—and the replacement of the calculus of frequencies by the neo-classical (or measure-theoretical) formalism.

The first two of these changes date back to 1938 and are indicated in the book itself (*i.e.* in this volume): the first by some new appendices, *ii to *v, and the second—the one which affects the argument of the present chapter—by a number of new footnotes to this chapter, and by the new appendix *vi. The main change is described here in footnote *1 to section 57.

The third change (which I first introduced, tentatively, in 1953) is explained and developed in the *Postscript*, where it is also applied to the problems of quantum theory.

question arises: 'How are we to interpret a statement of this kind and, in particular, the numerical assertion it makes?'

48. *Subjective and Objective Interpretations.*

The classical (Laplacean) theory of probability defines the numerical value of a probability as the quotient obtained by dividing the number of favourable cases by the number of equally possible cases. We might disregard the logical objections which have been raised against this definition,[1] such as that 'equally possible' is only another expression for 'equally probable'. But even then we could hardly accept this definition as providing an unambiguously applicable interpretation. For there are latent in it several different interpretations which I will classify as subjective and objective.

A *subjective interpretation* of probability theory is suggested by the frequent use of expressions with a psychological flavour, like 'mathematical *expectation*' or, say, 'normal law of *error*', *etc.*; in its original form it is *psychologistic*. It treats the degree of probability as a measure of the feelings of certainty or uncertainty, of belief or doubt, which may be aroused in us by certain assertions or conjectures. In connection with some non-numerical statements, the word 'probable' may be quite satisfactorily translated in this way; but an interpretation along these lines does not seem to me very satisfactory for numerical probability statements.

A newer variant of the subjective interpretation,[*1] however, deserves more serious consideration here. This interprets probability statements not psychologically but *logically*, as assertions about what may be called the 'logical proximity'[2] of statements. Statements, as we all know, can stand in various logical relations to one another,

---

[1] *Cf.* for example von Mises, *Warscheinlichkeit, Statistik und Wahreit* (1928), p. 62 *ff.*; 2nd edition (1936), p. 84 *ff.*; English translation by J. Neyman, D. Sholl, and E. Rabinowitsch, *Probability, Statistics and Truth* (1939), p. 98 *ff.* * Although the classical definition is often called 'Laplacean' (also in this book), it is at least as old as De Moivre's *Doctrine of Chances*, 1718. For an early objection against the phrase 'equally possible', see C. S. Pierce, *Collected Papers* 2, 1932 (first published 1878), p. 417, para. 2, 673.

[*1] The reasons why I count the logical interpretation as a variant of the *subjective* interpretation are more fully discussed in chapter *ii of the *Postscript*, where the subjective interpretation is criticised in detail. *Cf.* also appendix *ix.

[2] Waismann, *Logische Analyse des Wahrscheinlichkeitsbegriffs*, Erkenntnis I, 1930, p. 237: 'Probability so defined is then, as it were, a measure of the logical proximity, the deductive connection between the two statements'. *Cf.* also Wittgenstein, *op. cit.*, proposition 5.15 *ff.*

like derivability, incompatibility, or mutual independence; and the logico-subjective theory, of which Keynes[3] is the principal exponent, treats the *probability relation* as a special kind of logical relationship between two statements. The two extreme cases of this probability relation are derivability and contradiction: a statement q 'gives',[4] it is said, to another statement p the probability 1 if p follows from q. In case p and q contradict each other the probability given by q to p is zero. Between these extremes lie other probability relations which, roughly speaking, may be interpreted in the following way: The numerical probability of a statement p (given q) is the greater the less its content goes beyond what is already contained in that statement q upon which the probability of p depends (and which 'gives' to p a probability).

The kinship between this and the psychologistic theory may be seen from the fact that Keynes defines probability as the 'degree of rational belief'. By this he means the amount of trust it is proper to accord to a statement p in the light of the information or knowledge which we get from that statement q which 'gives' probability to p.

A third interpretation, the *objective interpretation*, treats every numerical probability statement as a statement about the *relative frequency* with which an event of a certain kind occurs within a *sequence of occurrences*.[5]

According to this interpretation, the statement 'The probability of the next throw with this die being a five equals 1/6' is not really an assertion about the next throw; rather, it is an assertion about a whole *class of throws* of which the next throw is merely an element. The statement in question says no more than that the relative frequency of fives, within this class of throws, equals 1/6.

According to this view, numerical probability statements are only admissible if we can give a *frequency interpretation* of them. Those probability statements for which a frequency interpretation cannot

[3] J. M. Keynes, *A Treatise on Probability* (1921), p. 95 *ff*.
[4] Wittgenstein, *op. cit.*, Proposition 5.152: 'If p follows from q, the proposition q gives to the proposition p the probability 1. The certainty of logical conclusion is a limiting case of probability.'
[5] For the older frequency theory *cf.* the critique of Keynes, *op. cit.*, p. 95 *ff.*, where special reference is made to Venn's *The Logic of Chance*. For Whitehead's view *cf.* section 80 (note 2). Chief representatives of the new frequency theory are: R. von Mises (*cf.* note 1 to section 50), Dörge, Kamke, Reichenbach and Tornier. ∗ A new objective interpretation, very closely related to the frequency theory, but differing from it even in its mathematical formalism, is the *propensity interpretation*, introduced in sections ∗53 *ff.* of my *Postscript*.

be given, and especially the non-numerical probability statements, are usually shunned by the frequency theorists.

In the following pages I shall attempt to construct anew the theory of probability as a (modified) *frequency theory*. Thus I declare my faith in an *objective interpretation*; chiefly because I believe that only an objective theory can explain the application of the probability calculus within empirical science. Admittedly, the subjective theory is able to give a consistent solution to the problem of how to decide probability statements; and it is, in general, faced by fewer logical difficulties than is the objective theory. But its solution is that probability statements are non-empirical; that they are tautologies. And this solution turns out to be utterly unacceptable when we remember the use which physics makes of the theory of probability. (I reject that variant of the subjective theory which holds that objective frequency statements should be derived from subjective assumptions—perhaps using Bernoulli's theorem as a 'bridge':[6] I regard this programme for logical reasons as unrealizable.)

49. *The Fundamental Problem of the Theory of Chance.*

The most important application of the theory of probability is to what we may call 'chance-like' or 'random' events, or occurrences. These seem to be characterized by a peculiar kind of incalculability which makes one disposed to believe—after many unsuccessful attempts—that all known rational methods of prediction must fail in their case. We have, as it were, the feeling that not a scientist but only a prophet could predict them. And yet, it is just this incalculability that makes us conclude that the calculus of probability can be applied to these events.

This somewhat paradoxical conclusion from incalculability to calculability (*i.e.* to the applicability of a certain calculus) ceases, it is true, to be paradoxical if we accept the subjective theory. But this way of avoiding the paradox is extremely unsatisfactory. For it entails the view that the probability calculus is not a method of calculating predictions, in contradistinction to all the other methods of empirical science. It is, according to the subjective theory, merely a

---

[6] Keynes's greatest error; *cf.* section 62, below, especially note 3. * I have not changed my view on this point even though I now believe that Bernoulli's theorem may serve as a 'bridge' *within* an objective theory—as a bridge from propensities to statistics. See also appendix *ix and sections *55 to *57 of my *Postscript*.

method for carrying out logical transformations of what we already know; or rather what we do *not* know; for it is just when we lack knowledge that we carry out these transformations.[1] This conception dissolves the paradox indeed, but *it does not explain how a statement of ignorance, interpreted as a frequency statement, can be empirically tested and corroborated*. Yet this is precisely our problem. How can we explain the fact that from incalculability—that is, from ignorance—we may draw conclusions which we can interpret as statements about empirical frequencies, and which we then find brilliantly corroborated in practice?

Even the frequency theory has not up to now been able to give a satisfactory solution of this problem—the *fundamental problem of the theory of chance*, as I shall call it. It will be shown in section 67 that this problem is connected with the 'axiom of convergence' which is an integral part of the theory in its present form. But it is possible to find a satisfactory solution within the framework of the frequency theory, after this axiom has been eliminated. It will be found by analysing the assumptions which allow us to argue from the irregular succession of single occurrences to the regularity or stability of their frequencies.

50. *The Frequency Theory of von Mises.*

A frequency theory which provides a foundation for all the principal theorems of the calculus of probability was first proposed by Richard von Mises.[1] His fundamental ideas are as follows.

The calculus of probability is a theory of certain chance-like or random sequences of events or occurrences, *i.e.* of repetitive events such as a series of throws with a die. These sequences are defined as 'chance-like' or 'random' by means of two axiomatic conditions:

[1] Waismann, *Erkenntnis* I, 1930, p. 238, says: 'There is no other reason for introducing the concept of probability than the incompleteness of our knowledge'. A similar view is held by C. Stumpf (*Sitzungsbericht der Bayerischen Akademie der Wissenschaften*, phil.-hist. Klasse, 1892, p. 41). * I believe that this widely held view is responsible for the worst confusions. This will be shown in detail in my *Postscript*, chapters *ii and *v.

[1] R. von Mises, *Fundamentalsätze der Wahrscheinlichkeitsrechnung, Mathematische Zeitschrift* 4, 1919, p. 1; *Grundlagen der Wahrscheinlichsrechnung, Mathematische Zeitschrift* 5, 1919, p. 52; *Wahrscheinlichkeit, Statistik, und Wahrheit* (1928), 2nd edition 1936, English translation by J. Neyman, D. Sholl, and E. Rabinowitsch: *Probability, Statistics and Truth*, 1939; *Wahrscheinlichkeitsrechnung und ihre Anwendung in der Statistik und theoretischen Physik* (*Vorlesungen über angewandte Mathematik* I), 1931.

the *axiom of convergence* (or the *limit-axiom*) and the *axiom of randomness*. If a sequence of events satisfies both of these conditions it is called by von Mises a 'collective'.

A collective is, roughly speaking, a sequence of events or occurrences which is capable in principle of being continued indefinitely; for example a sequence of throws made with a supposedly indestructible die. Each of these events has a certain character or *property*; for example, the throw may show a five and so have *the property five*. If we take all those throws having the property five which have appeared up to a certain element of the sequence, and divide their number by the total number of throws up to that element (*i.e.* its ordinal number in the sequence) then we obtain the *relative frequency* of fives up to that element. If we determine the relative frequency of fives up to every element of the sequence, then we obtain in this way a new sequence—the *sequence of the relative frequencies of fives*. This sequence of frequencies is distinct from the original sequence of events to which it corresponds, and which may be called the 'event-sequence' or the 'property-sequence'.

As a simple example of a collective I choose what we may call an '*alternative*'. By this term we denote a sequence of events supposed to have *two properties only*—such as a sequence of tosses of a coin. The one property (heads) will be denoted by '1', and the other (tails) by '0'. A sequence of events (or sequence of properties) may then be represented as follows:

$$0 \quad 1 \quad 1 \quad 0 \quad 0 \quad 0 \quad 1 \quad 1 \quad 1 \quad 0 \quad 1 \quad 0 \quad 1 \quad 0 \ \ldots \ldots \text{(A)}$$

Corresponding to this 'alternative'—or, more precisely, correlated with the property '1' of this alternative—is the following sequence of relative frequencies, or 'frequency-sequence':[2]

$$0 \ \ \frac{1}{2} \ \frac{2}{3} \ \frac{2}{4} \ \frac{2}{5} \ \frac{2}{6} \ \frac{3}{7} \ \frac{4}{8} \ \frac{5}{9} \ \frac{5}{10} \ \frac{6}{11} \ \frac{6}{12} \ \frac{7}{13} \ \frac{7}{14} \ \ldots \ldots \text{(A')}$$

[2] We can correlate with every sequence of properties as many distinct sequences of relative frequencies as there are properties defined in the sequence. Thus in the case of an alternative there will be two distinct sequences. Yet these two sequences are derivable from one another, since they are complementary (corresponding terms add up to 1). For this reason I shall, for brevity, refer to 'the (one) sequence of relative frequencies correlated with the alternative (α)', by which I shall always mean the sequence of frequencies correlated with the property '1' of this alternative (α).

Now the *axiom of convergence* (or 'limit-axiom') postulates that, as the event-sequence becomes longer and longer, the frequency-sequence shall tend towards a definite *limit*. This axiom is used by von Mises because we have to make sure of *one fixed frequency value* with which we can work (even though the actual frequencies have fluctuating values). In any collective there are at least two properties; and if we are given the limits of the frequencies corresponding to *all* the properties of a collective, then we are given what is called its '*distribution*'.

The axiom of randomness or, as it is sometimes called, 'the principle of the excluded gambling system', is designed to give mathematical expression to the chance-like character of the sequence. Clearly, a gambler would be able to improve his chances by the use of a gambling system if sequences of penny tosses showed regularities such as, say, a fairly regular appearance of tails after every run of three heads. Now the axiom of randomness postulates of all collectives that there does not exist a gambling system that can be successfully applied to them. It postulates that, whatever gambling system we may choose for selecting supposedly favourable tosses, we shall find that, if gambling is continued long enough, the relative frequencies in the sequence of *tosses supposed to be favourable* will approach the same limit as those in the sequence of *all tosses*. Thus a sequence for which there exists a gambling system by means of which the gambler can improve his chances is not a collective in the sense of von Mises.

Probability, for von Mises, is thus another term for 'limit of relative frequency in a collective'. The idea of probability is therefore applicable *only to sequences of events*; a restriction likely to be quite unacceptable from a point of view such as Keynes's. To critics objecting to the narrowness of his interpretation, von Mises replied by stressing the difference between the scientific use of probability, for example in physics, and the popular uses of it. He pointed out that it would be a mistake to demand that a properly defined scientific term has to correspond in all respects to inexact, pre-scientific usage.

The *task of the calculus of probability* consists, according to von Mises, simply and solely in this: to infer certain 'derived collectives' with 'derived distributions' from certain given 'initial collectives' with certain given 'initial distributions'; in short, to calculate probabilities which are not given from probabilities which are given.

The distinctive features of his theory are summarized by von

Mises in four points:[3] the concept of the collective precedes that of
probability; the latter is defined as the limit of the relative frequencies;
an axiom of randomness is formulated; and the task of the calculus
of probability is defined.

51. *Plan for a New Theory of Probability.*

The two axioms or postulates formulated by von Mises in order to
define the concept of a collective have met with strong criticism—
criticism which is not, I think, without some justification. In particular,
objections have been raised against combining the axiom of convergence with the axiom of randomness[1] on the ground that it is
inadmissible to apply the mathematical concept of a limit, or of
convergence, to a sequence which by definition (that is, because of
the axiom of randomness) must not be subject to any mathematical
rule or law. For the mathematical limit is nothing but a *characteristic
property of the mathematical rule or law by which the sequence is determined.* It is merely a property of this rule or law if, for any chosen
fraction arbitrarily close to zero, there is an element in the sequence
such that all elements following it deviate by less than that fraction
from some definite value—which is then called their limit.

To meet such objections it has been proposed to refrain from
combining the axiom of convergence with that of randomness, and to
postulate only convergence, *i.e.* the existence of a limit. As to the
axiom of randomness, the proposal was either to abandon it altogether
(Kamke) or to replace it by a weaker requirement (Reichenbach).
These suggestions presuppose that it is the axiom of randomness
which is the cause of the trouble.

In contrast to these views, I am inclined to blame the axiom of
convergence no less than the axiom of randomness. Thus I think that
there are two tasks to be performed: the improvement of the axiom
of randomness—mainly a mathematical problem; and the complete
elimination of the axiom of convergence—a matter of particular
concern for the epistemologist.[2] (*Cf.* section 66.)

[3] *Cf.* von Mises, *Wahrscheinlichkeitsrechnung* (p. 1931), 22.
[1] Waismann, *Erkenntnis* I, 1930, p. 232.
[2] This concern is expressed by Schlick, *Naturwissenschaften* 19, 1931. * I still believe
that these two tasks are important. Although I almost succeeded in the book in achieving
what I set out to do, the two tasks were satisfactorily completed only in the new
appendix *vi.

In what follows I propose to deal first with the mathematical, and afterwards with the epistemological question.

The first of these two tasks, the reconstruction of the mathematical theory,[3] has as its main aim the derivation of Bernoulli's theorem— the first 'Law of Great Numbers'—from *a modified axiom of randomness*; modified, namely, so as to demand no more than is needed to achieve this aim. Or to be more precise, my aim is the derivation of the Binomial Formula (sometimes called 'Newton's Formula'), in what I call its 'third form'. For from this formula, Bernoulli's theorem and the other limit theorems of probability theory can be obtained in the usual way.

My plan is to work out first *a frequency theory for finite classes*, and to develop the theory, within this frame, as far as possible—that is, up to the derivation of the ('first') Binomial Formula. This frequency theory for finite classes turns out to be a quite elementary part of the theory of classes. It will be developed merely in order to obtain a basis for discussing the axiom of randomness.

Next I shall proceed to *infinite sequences, i.e.* to sequences of events which can be continued indefinitely, by the old method of introducing an axiom of convergence, since we need something like it for our discussion of the axiom of randomness. And after deriving and examining Bernoulli's theorem, I shall consider *how the axiom of convergence might be eliminated*, and what sort of axiomatic system we should be left with as the result.

In the course of the mathematical derivation I shall use *three* different frequency symbols: F'' is to symbolize relative frequency in finite classes; F' is to symbolize the limit of the relative frequencies of an infinite frequency-sequence; and finally F, is to symbolize objective probability, *i.e.* relative frequency in an 'irregular' or 'random' or 'chance-like' sequence.

52. *Relative Frequency within a Finite Class.*

Let us consider a class α of a *finite* number of occurrences, for example the class of throws made yesterday with this particular die. This class α, which is assumed to be *non-empty*, serves, as it were, as a frame of reference, and will be called a (finite) *reference-class*. The

---

[3] A full account of the mathematical construction will be published separately. * Cf. the new appendix *vi.

number of elements belonging to α, i.e. its cardinal number, is denoted by '$N(\alpha)$', to be read 'the number of α'. Now let there be another class, β, which may be finite or not. We will call β our property-class: it may be, for example, the class of *all* throws which show a five, or (as we shall say) which have the property five.

The class of those elements which belong to both α and β, for example the class of throws made yesterday with this particular die and having the property five, is called the product-class of α and β, and is denoted by '$\alpha.\beta$', to be read 'α and β'. Since α.β is a subclass of α, it can at most contain a finite number of elements (it may be empty). The number of elements in α.β is denoted by '$N(\alpha.\beta)$'.

Whilst we symbolize (finite) *numbers* of elements by N, the *relative frequencies* are symbolized by F''. For example, 'the relative frequency of the property β within the finite reference-class α' is written '$_\alpha F''(\beta)$', which may be read 'the α-frequency of β'. We can now define

$$_\alpha F''(\beta) = \frac{N(\alpha.\beta)}{N(\alpha)} \qquad \text{(Definition 1)}$$

In terms of our example this would mean: 'The relative frequency of fives among yesterday's throws with this die is, by definition, equal to the quotient obtained by dividing the number of fives, thrown yesterday with this die, by the total number of yesterday's throws with this die.'[*1]

From this rather trivial definition, the theorems of the *calculus of frequency in finite classes* can very easily be derived (more especially, the general multiplication theorem; the theorem of addition; and the theorems of division, *i.e.* Bayes's rules. *Cf.* appendix ii). Of the theorems of this calculus of frequency, and of the calculus of probability in general, it is characteristic that cardinal numbers (N-numbers) never appear in them, but only relative frequencies, *i.e.* ratios, or F-numbers. The N-numbers only occur in the proofs of a few fundamental theorems which are directly deduced from the definition; but they do not occur in the theorems themselves.[*2]

[*1] Definition 1 is of course related to the classical definition of probability as the ratio of the favourable cases to the equally possible cases; but it should be clearly distinguished from the latter definition: there is no assumption involved here that the elements of α are 'equally possible'.

[*2] By selecting a set of F-formulae from which the other F-formulae can be derived, we obtain a formal *axiom system for probability*; compare the appendices ii, *ii, *iv, and *v.

How this is to be understood will be shown here with the help of one very simple example. (Further examples will be found in appendix ii.) Let us denote the class of all elements which do not belong to β by '$\bar{\beta}$' (read: 'the complement of β' or simply: 'non-β'). Then we may write

$$_{\alpha}F''(\beta) + _{\alpha}F''(\bar{\beta}) = 1$$

While this theorem only contains F-numbers, its proof makes use of N-numbers. For the theorem follows from the definition (1) with the help of a simple theorem from the calculus of classes which asserts that $N(\alpha.\beta) + N(\alpha.\bar{\beta}) = N(\alpha)$.

53. *Selection, Independence, Insensitiveness, Irrelevance.*

Among the operations which can be performed with relative frequencies in finite classes, the operation of *selection*[1] is of special importance for what follows.

Let a finite reference-class α be given, for example the class of buttons in a box, and two property-classes, β (say, the red buttons) and γ (say, the large buttons). We may now take the product-class $\alpha.\beta$ as a *new reference-class*, and raise the question of the value of $_{\alpha.\beta}F''(\gamma)$, *i.e.* of the frequency of γ within the new reference-class.[2] The new reference-class $\alpha.\beta$ may be called 'the result of selecting β-elements from α', or the 'selection from α according to the property β'; for we may think of it as being obtained by selecting from α all those elements (buttons) which have the property β (red).

Now it is just possible that γ may occur in the new reference-class, $\alpha.\beta$, with the same relative frequency as in the original reference-class α; *i.e.* it may be true that

$$_{\alpha.\beta}F''(\gamma) = _{\alpha}F''(\gamma)$$

In this case we say (following Hausdorf[3]) that the properties β and γ are '*mutually independent*, within the reference-class α'. The relation of independence is a three-termed relation and is symmetrical in the

---

[1] Von Mises' term is '*choice*' ('*Auswahl*').

[2] The answer to this question is given by the general division theorem (*cf.* appendix ii).

[3] Hausdorff, *Berichte über die Verhandlungen der sächsischen Ges. d. Wissenschaften,* Leipzig, mathem.-physik. Klasse **53**, 1901, p. 158.

properties β and γ.⁴ If two properties β and γ are (mutually) independent within a reference-class α we can also say that the property γ is, within α, *insensitive* to the selection of β-elements; or perhaps that the reference-class α is, with respect to this property γ, insensitive to a selection according to the property β.

The mutual independence, or insensitiveness, of β and γ within α could also—from the point of view of the subjective theory—be interpreted as follows: If we are informed that a particular element of the class α has the property β, then this information is *irrelevant* if β and γ are mutually independent within α; irrelevant namely, to the question whether this element also has the property γ, or not.*¹ If, on the other hand, we know that γ occurs more often (or less often) in the subclass α.β (which has been selected from α according to β), then the information that an element has the property β is *relevant* to the question whether this element also has the property γ or not.⁵

54. *Finite Sequences. Ordinal Selection and Neighbourhood Selection.*

Let us suppose that the elements of a finite reference-class α are *numbered* (for instance that a number is written on each button in the box), and that they are arranged in a *sequence*, in accordance with these ordinal numbers. In such a sequence we can distinguish two kinds of selection which have special importance, namely selection according to the ordinal number of an element, or briefly, ordinal selection, and selection according to its neighbourhood.

⁴ It is even triply symmetrical, *i.e.* for α, β and γ, if we assume β and γ also to be *finite*. For the proof of the symmetry assertion *cf.* appendix ii, (1ˢ) and (1₈). ∗ The condition of finitude for triple symmetry asserted in this note is insufficient. I may have intended to express the condition that β and γ are bounded by the finite reference class α, or, most likely, that α should be our finite universe of discourse. (These are sufficient conditions.) The insufficiency of the condition, as formulated in my note, is shown by the following counter-example. Take a universe of 5 buttons; 4 are round (α); 2 are round and black (αβ); 2 are round and large (αγ); 1 is round, black, and large (αβγ); and 1 is square, black, and large (ᾱβγ). Then we do not have triple symmetry since $_αF''(γ) \neq {_β}F''(γ)$.

*¹ Thus any information about the possession of properties is relevant, or irrelevant, if and only if the properties in question are, respectively, dependent or independent. Relevance can thus be defined in terms of dependence, but the reverse is not the case. (*Cf.* the next footnote, and note ∗1 to section 55.)

⁵ Keynes objected to the frequency theory because he believed that it was impossible to define *relevance* in its terms; *cf. op. cit.*, p. 103 *ff.* ∗In fact, the subjective theory cannot define (objective) *independence*, which is a serious objection as I show in my *Postscript*, chapter ∗ii, especially sections ∗40 to ∗43.

Ordinal selection consists in making a selection, from the sequence α, in accordance with a property β which depends upon the ordinal number of the element (whose selection is to be decided on). For example β may be the property *even*, so that we select from α all those elements whose ordinal number is even. The elements thus selected form a *selected sub-sequence*. Should a property γ be independent of an ordinal selection according to β, then we can also say that the *ordinal selection* is independent with respect to γ; or we can say that the sequence α is, with respect to γ, insensitive to a selection of β-elements.

Neighbourhood selection is made possible by the fact that, in ordering the elements in a numbered sequence, certain neighbourhood relations are created. This allows us, for example, to select all those members whose immediate predecessor has the property γ; or, say, those whose first and second predecessors, or whose second successor, have the property γ; and so on.

Thus if we have a sequence of events—say tosses of a coin—we have to distinguish two kinds of properties: its primary properties such as 'heads' or 'tails', which belong to each element independently of its position in the sequence; and its secondary properties such as 'even' or 'successor of tails', etc., which an element acquires by virtue of its position in the sequence.

A sequence with two primary properties has been called 'alternative'. As von Mises has shown, it is possible to develop (if we are careful) the essentials of the theory of probability as a theory of alternatives, without sacrificing generality. Denoting the two primary properties of an alternative by the figures '1' and '0', every alternative can be represented as a sequence of ones and zeros.

Now the structure of an alternative can be *regular*, or it can be more or less *irregular*. In what follows we will study this regularity or irregularity of certain finite alternatives more closely.[*1]

53. *n-Freedom in Finite Sequences.*

Let us take a finite alternative α, for example one consisting of a thousand ones and zeros regularly arranged as follows:

$$1\ 1\ 0\ 0\ 1\ 1\ 0\ 0\ 1\ 1\ 0\ 0\ 1\ 1\ 0\ 0\ 1\ 1\ 0\ 0\ \ldots \qquad (\alpha)$$

[*1] I suggest that sections 55 to 64, or perhaps only 56 to 64, be skipped at first reading. It may even be advisable to turn from here, or from the end of section 55, direct to chapter x.

In this alternative we have equal distribution, *i.e.* the relative frequencies of the ones and the zeros are equal. If we denote the relative frequency of the property 1 by '$F''(1)$' and that of 0 by '$F''(0)$', we can write:

$$_\alpha F''(1) = {_\alpha}F''(0) = \tfrac{1}{2} \qquad\qquad (1)$$

We now select from α all terms with the neighbourhood-property of *immediately succeeding a one* (within the sequence α). If we denote this property by 'β', we may call the selected sub-sequence '$\alpha.\beta$'. It will have the structure:

$$1 \quad 0 \quad 1 \quad 0 \quad 1 \quad 0 \quad 1 \quad 0 \quad 1 \quad 0 \;\ldots \qquad\qquad (\alpha.\beta)$$

This sequence is again an alternative with equal distribution. Moreover, neither the relative frequency of the ones nor that of the zeros has changed; *i.e.* we have

$$_{\alpha.\beta}F''(1) = {_\alpha}F''(1) ; \quad {_{\alpha.\beta}}F''(0) = {_\alpha}F''(0) . \qquad\qquad (2)$$

In the terminology introduced in section 53, we can say that the primary properties of the alternative α are *insensitive* to selection according to the property β; or, more briefly, that α is insensitive to selection according to β.

Since every element of α has either the property β (that of being the successor of a one) or that of being the successor of a zero, we can denote the latter property by '$\bar{\beta}$'. If we now select the members having the property $\bar{\beta}$ we obtain the alternative:

$$0 \quad 1 \quad 0 \quad 1 \quad 0 \quad 1 \quad 0 \quad 1 \quad 0 \;.. \qquad\qquad (\alpha.\bar{\beta})$$

This sequence shows a very slight deviation from equal distribution in so far as it begins and ends with zero (since α itself ends with '0, 0' on account of its equal distribution). If α contains 2000 elements, then $\alpha.\bar{\beta}$ will contain 500 zeros, and only 499 ones. Such deviations from equal distribution (or from other distributions) which arise only on account of the first or last elements can be made as small as we please by making the sequence sufficiently long. For this reason they will be neglected in what follows; especially since our investigations are to be extended to infinite sequences, where these deviations vanish. Accordingly, we shall say that the alternative $\alpha.\bar{\beta}$ has equal distribution, and that the alternative α is *insensitive* to the selection of elements having the property $\bar{\beta}$. As a consequence, α, or rather the relative frequency of the primary properties of α, is insensitive to both, a selection according to β and according to β̄; and we may therefore say that α is insensitive to *every* selection according to the property of the *immediate predecessor*.

Clearly, this insensitivity is due to certain aspects of the structure of the alternative α; aspects which may distinguish it from other alternatives. For example, the alternatives α.β and α.β̄ are *not* insensitive to selection according to the property of a predecessor.

We can now investigate the alternative α in order to see whether it is insensitive to other selections, especially to selection according to the property of a *pair* of predecessors. We can, for example, select from α all those elements which are successors of a pair 1,1. And we see at once that α is *not* insensitive to the selection of the successor of any of the four possible pairs 1,1; 1,0; 0,1; 0,0. In none of these cases have the resulting sub-sequences equal distribution; on the contrary, they all consist of uninterrupted *blocks* (or '*iterations*'), *i.e.* of nothing but ones, or of nothing but zeros.

The fact that α is insensitive to selection according to single predecessors, but not insensitive to selection according to pairs of predecessors, might be expressed, from the point of view of the subjective theory, as follows. Information about the property of one predecessor of any element in α is irrelevant to the question of the property of this element. On the other hand, information about the properties of its pair of predecessors is of the highest relevance; for given the law according to which α is constructed, it enables us to *predict* the property of the element in question: the information about the properties of its pair of predecessors furnishes us, so to speak, with the initial conditions needed for deducing the prediction. (The law according to which α is constructed requires a pair of properties as initial conditions; thus it is 'two-dimensional' with respect to these properties. The specification of *one* property is 'irrelevant' only in being composite in an insufficient degree to serve as an initial condition. *Cf.* section 38.*[1]*)

Remembering how closely the idea of causality—of *cause and effect*—is related to the deduction of predictions, I shall now make use of the following terms. The assertion previously made about the

---

*[1] This is another indication of the fact that the terms 'relevant' and 'irrelevant', figuring so largely in the subjective theory, are grossly misleading. For if *p* is irrelevant, and likewise *q*, it is a little surprising to learn that *p.q* may be of the highest relevance. See also appendix *ix, especially points 5 and 6 of the first note.

alternative α, 'α is insensitive to selection according to a *single* predecessor', I shall now express by saying, 'α is free from any after-effect of *single* predecessors' or briefly, 'α is 1-free'. And instead of saying as before, that α is (or is not) 'insensitive to selection according to *pairs* of predecessors', I shall now say: 'α is (not) free from the after-effects of *pairs* of predecessors', or briefly, 'α is (not) 2-free'.*²

Using the 1-free alternative α as our prototype we can now easily construct other sequences, again with equal distribution, which are not only free from the after effects of one predecessor, *i.e.* 1-free (like α), but which are, in addition, free from the after effects of a pair of predecessors, *i.e.*, 2-free; and after this, we can go on to sequences which are 3-free, etc. In this way we are led to a general idea which is fundamental for what follows. It is the idea of freedom from the after-effects of all the predecessors up to some number n; or, as we shall say, of n-freedom. More precisely, we shall call a sequence 'n-free' if, and only if, the relative frequencies of its primary properties are 'n-insensitive', *i.e.* insensitive to selection according to single predecessors *and* according to pairs of predecessors *and* according to triplets of predecessors ... *and* according to n-tuples of predecessors.[1]

An alternative α which is 1-free can be constructed by repeating the *generating period*

$$1 \quad 1 \quad 0 \quad 0 \ldots \tag{A}$$

any number of times. Similarly we obtain a 2-free alternative with equal distribution if we take

$$1 \quad 0 \quad 1 \quad 1 \quad 1 \quad 0 \quad 0 \quad 0 \ldots \tag{B}$$

as its generating period. A 3-free alternative is obtained from the generating period

$$1 \quad 0 \quad 1 \quad 1 \quad 0 \quad 0 \quad 0 \quad 0 \quad 1 \quad 1 \quad 1 \quad 1 \quad 0 \quad 1 \quad 0 \quad 0 \ldots \tag{C}$$

*² The general idea of distinguishing neighbourhoods according to their size, and of operating with well-defined neighbourhood-selections was introduced by me. But the term 'free from after-effect' ('*nachwirkungsfrei*') is due to Reichenbach. Reichenbach, however, used it at the time only in the absolute sense of 'insensitive to selection according to *any* preceding group of elements'. The idea of introducing a *recursively definable* concept of 1-freedom, 2-freedom, ... and n-freedom, and of thus utilizing the recursive method for analysing neighbourhood selections and especially for *constructing random sequences* is mine. (I have used the same recursive method also for defining the mutual independence of n events.) This method is quite different from Reichenbach's, although it uses one of his terms in a modified sense. See also footnote 4 to section 58, and especially footnote 2 to section 60, below.

[1] As Dr. K. Schiff has pointed out to me, it is possible to simplify this definition. It is enough to demand insensitivity to selection of any predecessor n-tuple (for a given n). Insensitivity to selection of $n-1$-tuples (etc.) can then be proved easily.

and a 4-free alternative is obtained from the generating period

0 1 1 0 0 0 1 1 1 0 1 0 1 0 0 1 0 0 0 0 0 1 0 1 1 1 1 1 1 0 0 1 1 ... (D)

It will be seen that the intuitive impression of being faced with an irregular sequence becomes stronger with the growth of the number n of its n-freedom.

The generating period of an n-free alternative with equal distribution must contain at least 2^{n+1} elements. The periods given as examples can, of course, begin at different places; (C) for example can begin with its fourth element, so that we obtain, in place of (C)

1 0 0 0 0 1 1 1 1 0 1 0 0 1 0 1 ... (C')

There are other transformations which leave the n-freedom of a sequence unchanged. A method of constructing generating periods of n-free sequences for every number n will be described elsewhere.[*3]

If to the generating period of an n-free alternative we add the first n elements of the next period, then we obtain a sequence of the length $2^{n+1} + n$. This has, among others, the following property: every arrangement of $n + 1$ zeros and ones, *i e.* every possible $n + 1$-tuple, occurs in it at least once.[*4]

56. *Sequences of Segments. The First Form of the Binomial Formula.*

Given a finite sequence α, we call a sub-sequence of α consisting of n consecutive elements a 'segment of α of length n'; or, more briefly, an 'n-segment of α'. If, in addition to the sequence α, we are given some definite number n, then we can arrange the n-segments of α in a sequence—the *sequence of n-segments* of α. Given a sequence α, we may construct a new sequence, of n-segments of α, in such a way that we begin with the segment of the first n elements of α.

[*3] *Cf.* note *1 to appendix iv. The result is a sequence of the length $2^n + n - 1$ such that by omitting its last $n - 1$ elements, we obtain a generating period for an *m*-free alternative, with $m = n - 1$.

[*4] The following definition, applicable to any given long but finite alternative A, with equidistribution, seems appropriate. Let N be the length of A, and let n be the greatest integer such that $2^n + 1 \leqslant N$. Then A is said to be *perfectly random* if and only if the relative number of occurrences of any given pair, triplet, ..., *m*-tuplet (up to $m = n$) deviates from that of any other pair, triplet, ..., *m*-tuplet, by not more than, say, $m/N^{\frac{1}{2}}$ respectively. This characterization makes it possible to say of a given alternative A that it is approximately random; and it even allows us to define a degree of approximation. A more elaborate Definition may be based upon the method (of maximizing my E-function) described under point 8 *ff.* of my Third Note reprinted in appendix *ix.

Next comes the segment of the elements 2 to $n+1$ of α. In general, we take as the xth element of the new sequence the segment consisting of the elements x to $x+n-1$ of α. The new sequence so obtained may be called the 'sequence of the overlapping n-segments of α'. This name indicates that any two consecutive elements (*i.e.* segments) of the new sequence overlap in such a way that they have $n-1$ elements of the original sequence α in common.

Now we can obtain, by selection, other n-sequences from a sequence of overlapping segments; especially *sequences of adjoining n-segments*.

A sequence of adjoining n-segments contains only such n-segments as immediately follow each other in α without overlapping. It may begin, for example, with the n-segments of the elements numbered 1 to n, of the original sequence α, followed by that of the elements $n+1$ to $2n$, $2n+1$ to $3n$, and so on. In general, a sequence of adjoining segments will begin with the kth element of α and its segments will contain the elements of α numbered k to $n+k-1, n+k$ to $2n+k-1$, $2n+k$ to $3n+k-1$, and so on.

In what follows, sequences of overlapping n-segments of α will be denoted by '$\alpha_{(n)}$', and sequences of adjoining n-segments by 'α_n'.

Let us now consider the sequences of overlapping segments $\alpha_{(n)}$ a little more closely. Every element of such a sequence is an n-segment of α. As a primary property of an element of $\alpha_{(n)}$, we might consider, for instance, the ordered n-tuple of zeros and ones of which the segment consists. Or we could, more simply, regard *the number of its ones* as the primary property of the element (disregarding the *order* of the ones and zeros). If we denote the number of ones by 'm' then, clearly, we have $m \leqslant n$.

Now from every sequence $\alpha_{(n)}$ we again get an *alternative* if we select a particular m ($m \leqslant n$), ascribing the property 'm' to each element of the sequence $\alpha_{(n)}$ which has exactly m ones (and therefore $n-m$ zeros) and the property '$\overline{m}$' (non-m) to all other elements of $\alpha_{(n)}$. Every element of $\alpha_{(n)}$ must then have one or the other of these two properties.

Let us now imagine again that we are given a finite alternative α with the primary properties '1' and '0'. Assume that the frequency of the ones, $_{\alpha}F''(1)$, is equal to p, and that the frequency of the zeros, $_{\alpha}F''(0)$, is equal to q. (We do not assume that the distribution is equal, *i.e.* that $p=q$.)

Now let this alternative α be at least n—1-free (n being an arbitrarily chosen natural number). We can then ask the following question: What is the frequency with which the property m occurs in the sequence $\alpha_{(n)}$? Or in other words, what will be the value of $_{\alpha_{(n)}}F''(m)$?

Without assuming anything beyond the fact that α is at least $n-1$-free, we can settle this question[1] by elementary arithmetic. The answer is contained in the following formula, the proof of which will be found in appendix iii:

$$_{\alpha_{(n)}}F''(m) = {}^nC_m\, p^m q^{n-m} \qquad (1)$$

The right-hand side of the 'binomial' formula (1) was given—in another connection—by Newton. (It is therefore sometimes called Newton's formula.) I shall call it the 'first form of the binomial formula'.[*1]

With the derivation of this formula, I now leave the frequency theory as far as it deals with *finite* reference-classes. The formula will provide us with a foundation for our discussion of the axiom of randomness.

57. *Infinite Sequences. Hypothetical Estimates of Frequency.*

It is quite easy to extend the results obtained for n-free finite sequences to infinite n-free sequences which are defined by a *generating period* (*cf.* section 55). An infinite sequence of elements playing the rôle of the reference-class to which our relative frequencies are related may be called a 'reference-sequence'. It more or less corresponds to a 'Collective' in von Mises's sense.[*1]

---

[1] The corresponding problem in connection with infinite sequences of adjoining segments I call 'Bernoulli's problem' (following von Mises, *Wahrscheinlichkeitsrechnung*, 1931, p. 128); and in connection with infinite sequences of overlapping segments call it 'the quasi-Bernoulli problem' (*cf.* note 1 to section 60). Thus the problem here discussed would be the *quasi-Bernoulli problem for finite sequences*.

[*1] In the original text, I used the term 'Newton's formula'; but since this seems to be rarely used in English, I decided to translate it by 'binomial formula'.

[*1] I come here to the point where I failed to carry out fully my intuitive programme—that of analysing randomness as far as it is possible within the region of *finite* sequences, and of proceeding to *infinite* reference sequences (in which we need *limits* of relative frequencies) only afterwards, with the aim of obtaining a theory in which the existence of frequency limits follows from the random character of the sequence. I could have carried out this programme very easily by constructing, as my next step (finite) *shortest n-free sequences* for a growing n, as I did in my old appendix iv. It can then be easily shown that if, in these shortest sequences, n is allowed to grow without bounds, the sequences become infinite, and the frequencies turn without further assumption into frequency limits. (See note *2 to appendix iv, and my new appendix *vi.) All this would have simplified the

The concept of n-freedom presupposes that of relative frequency; for what its definition requires to be insensitive—insensitive to selection according to certain predecessors—is the *relative frequency* with which a property occurs. In our theorems dealing with infinite sequences I shall employ, but only provisionally (up to section 64), the idea of a *limit of relative frequencies* (denoted by F'), to take the place of *relative frequency in finite classes* (F''). The use of this concept gives rise to no problem so long as we confine ourselves to reference-sequences which are constructed *according to some mathematical rule*. We can always determine for such sequences whether the corresponding sequence of relative frequencies is convergent or not. The idea of a limit of relative frequencies leads to trouble only in the case of sequences for which no mathematical rule is given, but only an empirical rule (linking, for example the sequence with tosses of a coin); for in these cases the concept of limit is not defined (*cf.* section 51).

An example of a mathematical rule for constructing a sequence is the following: 'The nth element of the sequence α shall be o if, and only if, n is divisible by four'. This defines the infinite alternative

$$\text{I} \quad \text{I} \quad \text{I} \quad \text{O} \quad \text{I} \quad \text{I} \quad \text{I} \quad \text{O} \quad \ldots \qquad (\alpha)$$

with the limits of the relative frequencies: $_\alpha F'(\text{I}) = 3/4$; and $_\alpha F'(\text{o}) = 1/4$. Sequences which are defined in this way by means of a mathematical rule I shall call, for brevity, '*mathematical sequences*'.

By contrast, a rule for constructing an *empirical sequence* would be, for instance: 'The nth element of the sequence α shall be o if, and only if, the nth toss of the coin c shows tails.' But empirical rules need not always define sequences of a random character. For example, I should describe the following rule as empirical: 'The nth element of the sequence shall be I if, and only if, the nth second (counting from some zero instant) finds the pendulum p to the left of this mark.'

---

next sections which, however, retain their significance. But it would have solved completely and without further assumption the problems of sections 63 and 64; for since the existence of limits becomes demonstrable, points of accumulation need no longer be mentioned.

These improvements, however, remain all within the framework of the pure frequency theory: except in so far as they define an ideal standard of objective disorder, they become unnecessary if we adopt a propensity interpretation of the neo-classical (measure-theoretical) formalism, as explained in section *53 *ff* of my *Postscript*. But even then it remains necessary to speak of frequency hypotheses—of hypothetical estimates and their statistical tests; and thus the present section remains relevant, as does much in the succeeding sections, down to section 64.

The example shows that it may sometimes be possible to replace an empirical rule by a mathematical one—for example on the basis of certain hypotheses and measurements relating to some pendulum. In this way, we may find a mathematical sequence approximating to our empirical sequence with a degree of precision which may or may not satisfy us, according to our purposes. Of particular interest in our present context is the possibility (which our example could be used to establish) of obtaining a mathematical sequence whose various *frequencies* approximate to those of a certain empirical sequence.

In dividing sequences into mathematical and empirical ones I am making use of a distinction that may be called 'intensional' rather than 'extensional'. For if we are given a sequence 'extensionally', *i.e.* by listing its elements singly, one after the other—so that we can only know a finite piece of it, a finite segment, however long —then it is impossible to determine, from the properties of this segment, whether the sequence of which it is a part is a mathematical or an empirical sequence. Only when a rule of construction is given— that is, an 'intensional' rule—can we decide whether a sequence is mathematical or empirical.

Since we wish to tackle our infinite sequences with the help of the concept of a limit (of relative frequencies), we must restrict our investigation to mathematical sequences, and indeed to those for which the corresponding sequence of relative frequencies is convergent. This restriction amounts to introducing an axiom of convergence. (The problems connected with this axiom will not be dealt with until sections 63 to 66, since it turns out to be convenient to discuss them along with the 'law of great numbers'.)

Thus we shall be concerned only with *mathematical sequences*. Yet we shall be concerned only with those mathematical sequences of which we expect, or conjecture, that they approximate, as regards frequencies, to *empirical sequences of a chance-like or random character*; for these are our main interest. But to expect, or to conjecture, of a mathematical sequence that it will, as regards frequencies, approximate to an empirical one is nothing else than *to frame a hypothesis*— a hypothesis about the frequencies of the empirical sequence.[1]

The fact that our estimates of the frequencies in empirical random

---

[1] Later, in sections 65 to 68, I will discuss the *problem of decidability* of frequency hypotheses, that is to say, the problem whether a conjecture or hypothesis of this kind can be tested; and if so, how; whether it can be corroborated in any way; and whether it is falsifiable. *Cf.* also appendix *ix.

sequences are hypotheses is without any influence on the way we may calculate these frequencies. Clearly, in connection with *finite* classes, it does not matter in the least how we obtain the frequencies from which we start our calculations. These frequencies may be obtained by actual counting, or from a mathematical rule, or from a hypothesis of some kind or other. Or we may simply invent them. In calculating frequencies we accept some frequencies as given, and derive other frequencies from them.

The same is true of estimates of frequencies in *infinite* sequences. Thus the question as to the 'sources' of our frequency estimates is not a problem of the calculus of probability; which, however, does not mean that it will be excluded from our discussion of the problems of probability theory.

In the case of infinite empirical sequences we can distinguish two main 'sources' of our hypothetical estimates of frequencies—that is to say, two ways in which they may suggest themselves to us. One is an estimate based upon an *'equal-chance hypothesis'* (or equi-probability hypothesis), the other is an estimate based upon an *extrapolation of statistical findings*.

By an *'equal-chance hypothesis'* I mean a hypothesis asserting that the probabilities of the various primary properties are equal: it is a hypothesis asserting *equal distribution*. Equal-chance hypotheses are usually based upon considerations of *symmetry*.[2] A highly typical example is the conjecture of equal frequencies in dicing, based upon the symmetry and geometrical equivalence of the six faces of the cube.

For frequency hypotheses based on *statistical extrapolation*, estimates of rates of mortality provide a good example. Here statistical data about mortality are empirically ascertained; and *upon the hypothesis that past trends will continue to be very nearly stable*, or that they will not change much—at least during the period immediately ahead—an extrapolation to unknown cases is made from known cases, *i.e.* from occurrences which have been empirically classified, and counted.

People with inductivist leanings may tend to overlook the hypothetical character of these estimates: they may confuse a hypothetical estimate, *i.e.* a frequency-prediction based on statistical extrapolation, with one of its empirical 'sources'—the classifying and actual counting of past occurrences and sequences of occurrences. The claim is often

[2] Keynes deals with such questions in his analysis of the *principle of indifference*. *Cf. op. cit.*, Chapter IV, pp. 41-64.

made that we 'derive' estimates of probabilities—that is, predictions of frequencies—from past occurrences which have been classified and counted (such as mortality statistics). But from a logical point of view there is no justification for this claim. We have made no logical derivation at all. What we may have done is to advance a non-verifiable hypothesis which nothing can ever justify logically: the conjecture that frequencies will remain *constant*, and so permit of extrapolation. Even *equal-chance hypotheses* are held to be 'empirically derivable' or 'empirically explicable' by some believers in inductive logic who suppose them to be based upon statistical experience, that is, upon empirically observed frequencies. For my own part I believe, however, that in making this kind of hypothetical estimate of frequency we are often guided solely by our reflections about the significance of symmetry, and by similar considerations. I do not see any reason why these conjectures should be inspired only by the accumulation of a large mass of inductive observations. However, I do not attach much importance to these questions about the origins or 'sources' of our estimates. (*Cf.* section 2.) What is more important, in my opinion, is to be quite clear about the fact that every predictive estimate of frequency, including one which we may get from statistical extra-polation—and certainly all those that refer to infinite empirical sequences—will always be pure conjecture since it will always go far beyond anything which we are entitled to affirm on the basis of observations.

My distinction between equal-chance hypotheses and statistical extrapolations corresponds fairly well to the classical distinction between '*a priori*' and '*a posteriori*' probabilities. But since these terms are used in so many different senses,[3] and since they are, moreover, heavily tainted with philosophical associations, they are better avoided.

In the following examination of the axiom of randomness, I shall attempt to find mathematical sequences which approximate to random empirical sequences; which means that I shall be examining frequency-hypotheses.[*2]

---

[3] Born and Jordan, for instance, in *Elementare Quantenmechanik* (1930), p. 308, use the first of these terms in order to denote a hypothesis of equal distribution. A. A. Tschuprow, on the other hand, uses the expression '*a priori* probability' for all frequency *hypotheses*, in order to distinguish them from their *statistical tests*, *i.e.* the results, obtained *a posteriori*, of empirical counting,

[*2] This is precisely the programme here alluded to in note *1 above, and carried out in appendices iv and *vi.

58. *An Examination of the Axiom of Randomness.*

The concept of an ordinal selection (*i.e.* of a selection according to position) and the concept of a neighbourhood-selection, have both been introduced and explained in section 55. With the help of these concepts I will now examine von Mises's axiom of randomness—the principle of the excluded gambling system—in the hope of finding a weaker requirement which is nevertheless able to take its place. In von Mises's theory this 'axiom' is part of his definition of the concept of a collective: he demands that the limits of frequencies in a collective shall be insensitive to any kind of systematic selection whatsoever. (As he points out, a gambling system can always be regarded as a systematic selection.)

Most of the criticism which has been levelled against this axiom concentrates on a relatively unimportant and superficial aspect of its formulation. It is connected with the fact that, among the possible selections, there will be the selection, say, of those throws which come up five; and within this selection, obviously, the frequency of the fives will be quite different from what it is in the original sequence. This is why von Mises in his formulation of the axiom of randomness speaks of what he calls 'selections' or 'choices' which are 'independent of the result' of the throw in question, and are thus defined without making use of the property of the element to be selected.[1] But the many attacks levelled against this formulation[2] can all be answered merely by pointing out that we can formulate von Mises's axiom of randomness without using the questionable expressions at all.[3] For we may put it, for example, as follows: The limits of the frequencies in a collective shall be insensitive both to ordinal and to neighbourhood selection, and also to all combinations of these two methods of selection.

With this formulation the above mentioned difficulties disappear. Others however remain. Thus it might be impossible to *prove* that the concept of a collective, defined by means of so strong an axiom of randomness, is not self-contradictory; or in other words, that the

[1] *Cf.* for example von Mises's *Wahrscheinlichkeit, Statistik und Wahrheit* (1928), p. 25; English translation, 1939, p. 33.
[2] *Cf.* for instance, Feigl, *Erkenntnis* I, 1930, p. 256, where that formulation is described as 'not mathematically expressible'. Reichenbach's criticism, in *Mathematische Zeitschrift* 34, 1932, p. 594 *f.*, is very similar.
[3] This has also been observed by Dörge, who did not, however, enlarge on his observation.

class of 'collectives' is not empty. (The necessity for proving this has been stressed by Kamke.[4]) At least it seems to be impossible to construct an *example* of a collective and in that way to show that collectives exist. This is because an example of an infinite sequence which is to satisfy certain conditions can only be given by a mathematical rule. But for a collective in von Mises's sense there can be, by definition, no such rule, since any rule could be used as a gambling system or as a system of selection. This criticism seems indeed unanswerable if *all possible* gambling systems are ruled out.[*1]

Against the idea of excluding all gambling systems, another objection may be raised, however: that it really demands *too much*. If we are going to axiomatize a system of statements—in this case the theorems of the calculus of probability, particularly the special theorem of multiplication or Bernoulli's theorem—then the axioms chosen should not only be sufficient for the derivation of the theorems of the system, but also (if we can make them so) *necessary*. Yet the exclusion of *all* systems of selection can be shown to be *unnecessary* for the deduction of Bernoulli's theorem and its corollaries. It is quite sufficient to postulate the exclusion of a special class of neighbourhood-selection: it suffices to demand that the sequence should be insensitive to selections according to arbitrarily chosen *n*-tuples of predecessors; that is to say, that it should be *n-free from after-effects for every n*, or more briefly, that it should be 'absolutely free'.

I therefore propose to replace von Mises's principle of the excluded gambling system by the less exacting requirement of 'absolute freedom', in the sense of *n*-freedom for every *n*, and accordingly to define chance-like *mathematical* sequences as those which fulfil this requirement. The chief advantage of this is that it does not exclude *all* gambling systems, so that it is possible to give mathematical rules for constructing sequences which are 'absolutely free' in our sense,

---

[4] *Cf.* for instance, Kamke, *Einführung in die Wahrscheinlichkeitstheorie* (1932), p. 147, and *Jahresbericht der Deutschen mathem. Vereinigung* **42**, 1932. Kamke's objection must also be raised against Reichenbach's attempt to improve the axiom of randomness by introducing *normal sequences*, since he did not succeed in proving that this concept is *non-empty*. *Cf.* Reichenbach, *Axiomatik der Wahrscheinlichkeitsrechnung*, *Mathematische Zeitschrift* **34**, 1932, p. 606.

[*1] It is, however, answerable if any given *denumerable* set of gambling systems is to be ruled out; for then an example of a sequence may be constructed (by a kind of diagonal method). See section *54 of the *Postscript* (text after note 5), on A. Wald.

and hence to construct examples. (*Cf.* section (a) of appendix iv.) Thus Kamke's objection, discussed above, is met. For we can now prove that the concept of chance-like mathematical sequences is not empty, and is therefore consistent.[*2]

It may seem odd, perhaps, that we should try to trace the highly irregular features of chance sequences by means of mathematical sequences which must conform to the strictest rules. Von Mises's axiom of randomness may seem at first to be more satisfying to our intuitions. It seems quite satisfying to learn that a chance sequence must be completely irregular, so that every conjectured regularity will be found to fail, in some later part of the sequence, if only we keep on trying hard to falsify the conjecture by continuing the sequence long enough. But this intuitive argument benefits my proposal also. For if chance sequences are irregular, then, *a fortiori*, they will not be regular sequences of one particular type. And our requirement of 'absolute freedom' does no more than exclude one particular type of regular sequence, though an important one.

That it is an important type may be seen from the fact that by our requirement we implicitly exclude the following three types of gambling systems (*cf.* the next section). First we exclude 'normal' or 'pure'[*3] neighbourhood selections, *i.e.* those in which we select according to some *constant characteristic of the neighbourhood*. Secondly we exclude 'normal' ordinal selection which picks out elements whose distance apart is constant, such as the elements numbered k, $n + k$, $2n + k$ . . . and so on. And finally, we exclude many combinations of these two types of selection (for example the selection of every nth element, provided its neighbourhood has certain specified constant characteristics). A characteristic property of all these selections is that they do not refer to an absolute first element of the sequence; they may thus yield the same selected sub-sequence if the numbering of the original sequence begins with another (appropriate) element. Thus the gambling systems which are excluded by my requirement are those which could be used without knowing the first element of the sequence: the systems excluded are invariant with respect to certain (linear) transformations: they are the *simple* gambling

---

[*2] The reference to appendix iv is of considerable importance here. Also, most of the objections which have been raised against my theory were answered in the following paragraph of my text.

[*3] *Cf.* the last paragraph of section 60, below.

systems (*cf.* section 43). Only [*4] gambling systems which refer to the absolute distances of the elements from an absolute (initial) element[5] are not excluded by my requirement.

The requirement of n-freedom for every n—of 'absolute freedom'— also seems to agree quite well with what most of us, consciously or unconsciously, believe to be true of chance sequences; for example that the result of the next throw of a die does not depend upon the results of preceding throws. (The practice of shaking the die before the throw is intended to ensure this 'independence'.)

59. *Chance-Like Sequences. Objective Probability.*

In view of what has been said I now propose the following definition.

An event-sequence or property-sequence, especially an alternative, is said to be 'chance-like' or 'random' if and only if the limits of the frequencies of its primary properties are 'absolutely free', *i.e.* insensitive to every selection based upon the properties of any n-tuple of predecessors. A frequency-limit corresponding to a sequence which is random is called the *objective probability* of the property in question, within the sequence concerned; it is symbolized by F. This may also be put as follows. Let the sequence α be a chance-like or random-like sequence with the primary property β; in this case, the following holds:

$$_\alpha F(\beta) = {_\alpha F'(\beta)}$$

We shall have to show now that our definition suffices for the derivation of the main theorems of the mathematical theory of probability, especially Bernoulli's theorem. Subsequently—in section 64—the definition here given will be modified so as to make it independent of the concept of a *limit* of frequencies.[*1]

[*4] The word 'only' is *only* correct if we speak of (*predictive*) gambling systems *cf.* note *3 to section 60, below, and note 6 to section *54 of my *Postscript*.

[5] Example: the selection of all terms whose number is a prime.

[*1] At present I should be inclined to use the concept of 'objective probability' differently—that is, in a wider sense, so as to cover all 'objective' *interpretations of the formal calculus of probabilities*, such as the frequency interpretation and, more especially, the propensity interpretation which is discussed in the *Postscript*. Here, in section 59, the concept is used merely as an auxiliary concept in the construction of a certain form of the frequency theory.

60. *Bernoulli's Problem.*

The first binomial formula which was mentioned in section 56, *viz.*

$$_{\alpha_{(n)}}F''(m) = {}^{n}C_{m}\,p^{m}q^{n-m} \tag{1}$$

holds for finite sequences of overlapping segments. It is derivable on the assumption that the *finite* sequence α is at least $n-1$-free. Upon the same assumption, we immediately obtain an exactly corresponding formula for infinite sequences; that is to say, if α is infinite and at least $n-1$-free, then

$$_{\alpha_{(n)}}F'(m) = {}^{n}C_{m}\,p^{m}q^{n-m} \tag{2}$$

Since chance-like sequences are absolutely free, *i.e.* n-free for every n, formula (2), the *second* binomial formula, must also apply to them; and it must apply to them, indeed, for whatever value of n we may choose.

In what follows, we shall be concerned *only* with chance-like sequences, or random sequences (as defined in the foregoing section). We are going to show that, for *chance-like sequences*, a third binomial formula (3) must hold in addition to formula (2); it is the formula

$$_{\alpha_{n}}F(m) = {}^{n}C_{m}\,p^{m}q^{n-m} \tag{3}$$

Formula (3) differs from formula (2) in two ways: First, it is asserted for sequences of adjoining segments α_{n} instead of for sequences of overlapping segments $\alpha_{(n)}$. Secondly, it does not contain the symbol F' but the symbol F. This means that it asserts, by implication, that the *sequences of adjoining segments* are in their turn chance-like, or random; for F, *i.e.* objective probability, is defined only for chance-like sequences.

The question, answered by (3), of the objective probability of the property m in a sequence of adjoining segments—*i.e.* the question of the value of $_{\alpha_{n}}F(m)$—I call, following von Mises, 'Bernoulli's problem'.[1] For its solution, and hence for the derivation of the third binomial formula (3), it is sufficient to assume that α is chance-like or

---

[1] The corresponding question for sequences of *overlapping* segments, *i.e.* the problem of $\alpha_{(n)}F'(m)$, answered by (2), can be called the 'quasi-Bernoulli problem'; *cf.* note 1 to section 56 as well as section 61.

random.[2] (Our task is equivalent to that of showing that the special theorem of multiplication holds for the sequence of adjoining segments of a random sequence α.)

The proof[*1] of formula (3) may be carried out in two steps. First we show that formula (2) holds not only for sequences of overlapping segments $\alpha_{(n)}$, but also for sequences of adjoining sequences α_n. Secondly, we show that the latter are 'absolutely free'. (The order of these steps cannot be reversed, because a sequence of overlapping segments α is definitely *not* 'absolutely free'; in fact, a sequence of this kind provides a typical example of what may be called 'sequences with after-effects'.[3])

First step. Sequences of adjoining segments α_n are subsequences of $\alpha_{(n)}$. They can be obtained from these by normal ordinal selection. Thus if we can show that the limits of the frequencies in overlapping sequences $_{\alpha_{(n)}}F'(m)$ are insensitive to normal ordinal selection, we have taken our first step (and even gone a little farther); for we shall have proved the formula:

$$_{\alpha_n}F'(m) \;=\; _{\alpha_{(n)}}F'(m) \qquad\qquad (4)$$

I shall first sketch this proof in the case of $n = 2$; *i.e.* I shall show that

$$_{\alpha_2}F'(m) \;=\; _{\alpha_{(2)}}F'(m) \qquad (m \leqslant 2) \qquad (4a)$$

is true; it will then be easy to generalize this formula for every n.

From the sequence of overlapping segments $\alpha_{(2)}$ we can select two and only two distinct sequences $\alpha_{(2)}$ of adjoining segments; one, which will be denoted by (A), contains the first, third, fifth, . . . , segments of $\alpha_{(2)}$, that is, the pairs of α consisting of the numbers 1,2; 3,4; 5,6; . . . The other, denoted by (B), contains the second, fourth, sixth, . . . , segements of $\alpha_{(2)}$, that is, the pairs of elements of α consisting of

---

[2] Reichenbach (*Axiomatik der Wahrscheinlichkeitsrechnung*, *Mathematische Zeitschrift* **34**, 1932, p. 603) implicitly contests this when he writes, '. . . normal sequences are also free from after-effect, *whilst the converse does not necessarily hold*'. But Reichenbach's normal sequences are those for which (3) holds. (My proof is made possible by the fact that I have departed from previous procedure, by defining the concept 'freedom from after-effect' not directly, but with the help of '*n*-freedom from after-effect', thus making it accessible to the procedure of mathematical induction.)

[*1] Only a sketch of the proof is here given. Readers not interested in the proof may turn to the last paragraph of the present section.

[3] Von Smoluchowski based his theory of the Brownian movement on after-effect sequences, *i.e.* on sequences of overlapping segments.

the numbers 2,3; 4,5; 6,7; . . . , *etc.* Now assume that formula (4a) does *not* hold for *one* of the *two* sequences, (A) or (B), so that the segment (*i.e.* the pair) 0,0 occurs *too often* in, say, the sequence (A); then in sequence (B) a complementary deviation must occur; that is, the segment 0,0 will occur *not often enough* ('too often', or 'not often enough', as compared with the binomial formula). But this contradicts the assumed 'absolute freedom' of α. For if the pair 0,0 occurs in (A) more often than in (B), then in sufficiently long segments of α the pair 0,0 must appear more often at certain *characteristic distances* apart than at other distances. The more frequent distances would be those which would obtain if the 0,0 pairs belonged to *one* of the two α_2-sequences. The less frequent sequences would be those which would obtain if they belonged to *both* α_2-sequences. But this would contradict the assumed 'absolute freedom' of α; for according to the second binomial formula, the 'absolute freedom' of α entails that the frequency with which a particular sequence of the length n occurs in any $\alpha_{(n)}$-sequence depends *only* on the number of ones and zeros occurring in it, and not on their *arrangement* in the sequence.[*2]

This proves (4a); and since this proof can easily be generalized for any n, the validity of (4) follows; which completes the first step of the proof.

Second step. The fact that the α_n-sequences are 'absolutely free' can be shown by a very similar argument. Again, we first consider α_2-sequences only; and with respect to these it will only be shown, to start with, that they are 1-free. Assume that one of the two α_2-sequences, *e.g.* the sequence (A), is *not* 1-free. Then in (A) after at least *one* of the segments consisting of two elements (a particular α-pair), say after the segment 0,0, another segment, say 1,1, must follow more often than would be the case if (A) were 'absolutely free'; this means that the segment 1,1 would appear with greater frequency in the sub-sequence selected from (A) according to the predecessor-segment 0,0 than the binomial formula would lead us to expect.

This assumption, however, contradicts the 'absolute freedom' of the sequence α. For if the segment 1,1 follows in (A) the segment 0,0

[*2] The following formulation may be intuitively helpful: if the 0,0 pairs are more frequent in certain characteristic distances than in others, then this fact may be easily used as the basis of a simple system which would somewhat improve the chances of a gambler. But gambling systems of this type are incompatible with the 'absolute freedom' of the sequence. The same consideration underlies the 'second step' of the proof.

too frequently then, by way of compensation, the converse must take place in (B); for otherwise the quadruple 0,0,1,1 would, in a sufficiently long segment of α, occur too often at certain *characteristic distances* apart—namely at the distances which would obtain if the double pairs in question belonged to one and the same α_2-sequence. Moreover, at other *characteristic distances* the quadruple would occur not often enough—at those distances, namely, which would obtain if they belonged to *both* α_2-sequences. Thus we are confronted with precisely the same situation as before; and we can show, by analogous considerations, that the assumption of a preferential occurrence at characteristic distances is incompatible with the assumed 'absolute freedom' of α.

This proof can again be generalized, so that we may say of α-sequences that they are not only 1-free but n-free for every n; and hence that they are *chance-like*, or random.

This completes our sketch of the two steps. Thus we are now entitled to replace, in (4), F' by F; and this means that we may accept the claim that the third binomial formula solves Bernoulli's problem.

Incidentally we have shown that sequences $\alpha_{(n)}$ of overlapping segments are insensitive to *normal ordinal selection* whenever α is 'absolutely free'.

The same is also true for sequences α_n of adjoining segments, because every normal ordinal selection from α_n can be regarded as a normal ordinal selection from $\alpha_{(n)}$; and it must therefore apply to the sequence α itself, since α is identical with both $\alpha_{(1)}$ and α_1.

We have thus shown, among other things, that from 'absolute freedom'—which means insensitiveness to a special type of neighbourhood selection—insensitiveness to normal ordinal selection follows. A further consequence, as can easily be seen, is insensitiveness to any 'pure' neighbourhood selection (that is, selection according to a constant characterization of its neighbourhood—a characterization that does not vary with the ordinal number of the element). And it follows, finally, that 'absolute freedom' will entail insensitivity to all*3 combinations of these two types of selection.

---

*3 Here the word 'all' is, I now believe, mistaken, and should be replaced, to be a little more precise, by 'all those ... that might be used as gambling systems'. *Cf.* footnote *4 to section 58 above; and footnote 6 (referring to A. Wald) to section *54 of my *Postscript*.

61. *The Law of Great Numbers (Bernoulli's Theorem).*

Bernoulli's theorem, or the (first[1]) 'law of great numbers' can be derived from the third binomial formula by purely arithmetical reasoning, under the assumption that we can take n to the limit, $n \to \infty$. It can therefore be asserted only of infinite sequences α; for it is only in these that the n-segments of α_n- sequences can increase in length indefinitely. And it can be asserted only of such sequences α as are 'absolutely free', for it is only under the assumption of n-freedom for every n that we can take n to the limit, $n \to \infty$.

Bernoulli's theorem provides the solution of a problem which is closely akin to the problem which (following von Mises) I have called 'Bernoulli's problem', *viz.* the problem of the value of $_{\alpha_n}F(m)$. As indicated in section 56, an n-segment may be said to have the property 'm' when it contains precisely m ones; the relative frequency of ones within this (finite) segment is then, of course, m/n. We may now define: An n-segment of α has the property 'Δp' if and only if the relative frequency of its ones deviates by less than δ from the value $_{\alpha}F(1) = p$, *i.e.* the probability of ones in the sequence α; here, δ is any small fraction, chosen as near to zero as we like (but different from zero). We can express this condition by saying: an n segment has the property 'Δp' if and only if $\left| \dfrac{m}{n} - p \right| < \delta$; otherwise, the segment has the property '$\overline{\Delta p}$'. Now Bernoulli's theorem answers the question of the value of the frequency, or probability, of segments of this kind— of segments possessing the property Δp—within the α_n-sequences; it thus answers the question of the value of $_{\alpha_n}F(\Delta p)$.

Intuitively one might guess that if the value δ (with $\delta > 0$) is fixed, and if n increases, then the frequency of these segments with the property Δp, and therefore the value of $_{\alpha_n}F(\Delta p)$, will also increase (and that its increase will be monotonic). Bernoulli's proof (which can be found in any textbook on the calculus of probability) proceeds by evaluating this increase with the help of the binomial formula. He finds that if n increases without limit, the value of $_{\alpha_n}F(\Delta p)$ approaches the maximal value 1, for any fixed value of δ, however small. This may be expressed in symbols by

$$\lim_{n \to \infty} {}_{\alpha_n}F(\Delta p) = 1 \qquad \text{(for any value of } \Delta p) \quad (1)$$

---

[1] Von Mises distinguishes Bernoulli's—or Poisson's—theorem from its inverse which he calls 'Bayes's theorem' or 'the second law of great numbers'.

This formula results from transforming the *third* binomial formula for sequences of *adjoining* segments. The analogous *second* binomial formula for sequences of *overlapping* segments would immediately lead, by the same method, to the corresponding formula

$$\lim_{n \to \infty} {}_{\alpha_{(n)}} F'(\Delta p) = 1 \tag{2}$$

which is valid for sequences of overlapping segments and normal ordinal selection from them, and hence for sequences with *after-effects* (which have been studied by Smoluchowski[2]). Formula (2) itself yields (1) in case sequences are selected which do not overlap, and which are therefore *n*-free. (2) may be described as a variant of Bernoulli's theorem; and what I am going to say here about Bernoulli's theorem applies *mutatis mutandis* to this variant.

Bernoulli's theorem, *i.e.* formula (1), may be expressed in words as follows. Let us call a long finite segment of some fixed length, selected from a random sequence α, a 'fair sample' if, and only if, the frequency of the ones *within this segment* deviates from *p*, *i.e.* the value of the probability of the ones *within the random sequence* α, by no more than some small fixed fraction (which we may freely choose). We can then say that the probability of chancing upon a fair sample approaches 1 as closely as we like if only we make the segments in question sufficiently long.[*1]

In this formulation the word *'probability'* (or *'value of the probability'*) occurs twice. How is it to be interpreted or translated here? In the sense of my frequency definition it would have to be translated as follows (I italicize the two translations of the word 'probability' into the frequency language): *The overwhelming majority* of all sufficiently long finite segments will be 'fair samples'; that is to say, their relative frequency will deviate from the *frequency value p* of the random sequence in question by an arbitrarily fixed small amount; or, more briefly: The *frequency p* is realized, approximately, in *almost all* sufficiently long segments. (How we arrive at the value *p* is irrelevant to our present discussion; it may be, say, the result of a hypothetical estimate.)

Bearing in mind that the Bernoulli frequency ${}_{\alpha_n} F(\Delta p)$ increases

[2] *Cf.* note 3 to section 60, and note 5 to section 64.
[*1] This sentence has been reformulated (without altering its content) in the translation by introducing the concept of a 'fair sample': the original operates only with the definiens of this concept.

monotonically with the increasing length n of the segment and that it decreases monotonically with decreasing n, and that, therefore, the value of the relative frequency is comparatively rarely realized in short segments, we can also say:

Bernoulli's theorem states that short segments of 'absolutely free' or chance-like sequences will often show relatively great deviations from p and thus relatively great fluctuations, while the longer segments, in most cases, will show smaller and smaller deviations from p with increasing length. Consequently, most deviations in sufficiently long segments will become as small as we like; or in other words, great deviations will become as rare as we like.

Accordingly, if we take a very long segment of a random sequence, in order to find the frequencies within its sub-sequences by counting, or perhaps by the use of other empirical and statistical methods, then we shall get, in the vast majority of cases, the following result. There is a characteristic average frequency, such that the relative frequencies in the whole segment, and in almost all long sub-segments, will deviate only slightly from this average, whilst the relative frequencies of smaller sub-segments will deviate further from this average, and the more often, the shorter we choose them. This fact, this statistically ascertainable behaviour of finite segments, may be referred to as their 'quasi-convergent-behaviour'; or as the fact that random sequences are statistically stable.*2

Thus Bernoulli's theorem asserts that the smaller segments of chance-like sequences often show large fluctuations, whilst the large segments always behave in a manner suggestive of constancy or convergence; in short, that we find disorder and randomness in the small, order and constancy in the great. It is this behaviour to which the expression 'the law of great numbers' refers.

62. Bernoulli's Theorem and the Interpretation of Probability Statements.

We have just seen that in the verbal formulation of Bernoulli's theorem the word 'probability' occurs twice.

The frequency theorist has no difficulty in translating this word, in both cases, in accordance with its definition: he can give a clear interpretation of Bernoulli's formula and the law of great numbers. Can the adherent of the subjective theory in its logical form do the same?

*2 Keynes says of the 'Law of Great Numbers' that 'the "Stability of Statistical Frequencies" would be a much better name for it'. (Cf. his Treatise, p. 336.)

The subjective theorist who wants to define 'probability' as 'degree of rational belief' is perfectly consistent, and within his rights, when he interprets the words 'The probability of . . . approaches to 1 as closely as we like' as meaning, 'It is *almost certain*[1] that . . .'. But he merely obscures his difficulties when he continues '. . . that the relative frequency will deviate *from its most probable value p* by less than a given amount . . .', or in the words of Keynes,[2] 'that the proportion of the event's occurrences will diverge from *the most probable proportion p* by less than a given amount . . .'. This sounds like good sense, at least on first hearing. But if here too we translate the word '*probable*' (sometimes suppressed) in the sense of the subjective theory then the whole story runs: 'It is almost certain that the relative frequencies deviate from the value *p* of the degree of rational belief by less than a given amount . . .', which seems to me complete nonsense.*[1] For relative frequencies can be compared only with relative frequencies, and can deviate or not deviate only from relative frequencies. And clearly, it must be inadmissible to give *after* the deduction of Bernoulli's theorem a meaning to *p* different from the one which was given to it before the deduction.[3]

---

[1] Von Mises also uses the expression 'almost certain', but according to him it is of course to be regarded as *defined* by 'having a frequency close to 1'.

[2] Keynes, *A Treatise on Probability* (1921), p. 338. *The *preceding* passage in quotation marks had to be inserted here because it re-translates the passage I quoted from the German edition of Keynes on which my text relied.

*[1] It may be worth while to be more explicit on this point. Keynes writes (in a passage preceding the one quoted above): 'If the probability of an event's occurrence under certain conditions is *p*, then . . . the most probable proportion of its occurrences to the total number of occasions is *p* . . .' This ought to be translatable, according to his own theory, into: 'If the degree of rational belief in the occurrence of an event is *p*, then *p* is also a proportion of occurrences, *i.e.* a relative frequency—that, namely, in whose emergence the degree of our rational belief is greatest.' I am not objecting to the latter use of the expression 'rational belief'. (It is the use which might also be rendered by 'It is almost certain that . . .'.) What I do object to is the fact that *p* is at one time a degree of rational belief and at another a frequency; in other words, I do not see why an empirical frequency should be equal to a degree of rational belief; or that it can be proved to be so by any theorem, however deep. (*Cf.* also section 49 and appendix *ix.*)

[3] This was first pointed out by von Mises in a similar connection in *Wahrscheinlichkeit, Statistik und Wahrheit* (1928), p. 85 (2nd edition 1936, p. 136; the relevant words are missing in the English translation). It may be further remarked that relative frequencies cannot be compared with 'degrees of certainty of our knowledge' if only because the ordering of such degrees of certainty is *conventional* and need not be carried out by correlating them with fractions between 0 and 1. Only if the metric of the subjective degrees of certainty is *defined* by correlating relative frequencies with it (but *only* then) can it be permissible to derive the law of great numbers within the framework of the subjective theory (*cf.* section 73).

Thus we see that the subjective theory is incapable of interpreting Bernouilli's formula in terms of the *statistical* law of great numbers. Derivation of statistical laws is possible only within the framework of the frequency theory. If we start from a strict subjective theory, we shall never arrive at statistical statements—not even if we try to bridge the gulf with Bernoulli's theorem.[*2]

63. *Bernoulli's Theorem and the Problem of Convergence.*

From the point of view of epistemology, my deduction of the law of great numbers, outlined above, is unsatisfactory; for the part played in our analysis by the axiom of convergence is far from clear.

I have in effect tacitly introduced an axiom of this kind, by confining my investigation to mathematical sequences with frequency limits. (*Cf.* section 57.) Consequently one might even be tempted to think that our result—the derivation of the law of great numbers—is trivial; for the fact that 'absolutely free' sequences are *statistically stable* might be regarded as entailed by their convergence which has been assumed axiomatically, if not implicitly.

But this view would be mistaken, as von Mises has clearly shown. For there are sequences[1] which satisfy the axiom of convergence although Bernoulli's theorem does not hold for them, since with a frequency close to 1, segments of any length occur in them which may deviate from p to any extent. (The existence of the limit p is in these cases due to the fact that the deviations, although they may increase without limit, cancel each other.) Such sequences *look* as if they were divergent in arbitrarily large segments, even though the corresponding frequency sequences are in fact convergent. Thus the law of great numbers is anything but a trivial consequence of the axiom of convergence, and this axiom is quite insufficient for its deduction. This is why my modified axiom of randomness, the requirement of 'absolute freedom', cannot be dispensed with.

Our reconstruction of the theory, however, suggests the possi-

---

[*2] But it is possible to use Bernoulli's theorem as a bridge from the *objective* interpretation in terms of 'propensities' to statistics. *Cf.* sections *49 to *57 of my *Postscript*.

[1] As an example von Mises cites the sequence of figures occupying the last place of a six-figure table of square roots. *Cf.* for example, *Wahrscheinlichkeit, Statistik und Wahrheit* (1928), p. 86 *f.*; (2nd edition 1936, p. 137; English translation, p. 165), and *Wahrscheinlichkeitsrechnung* (1931), p. 181 *f.*

bility that the law of great numbers may be *independent* of the axiom of convergence. For we have seen that Bernoulli's theorem follows immediately from the binomial formula; moreover, I have shown that the first binomial formula can be derived for *finite sequences* and so, of course, without any axiom of convergence. All that had to be assumed was that the reference-sequence α was at least *n*—1-free; an assumption from which the validity of the special multiplication theorem followed, and with it that of the first binomial formula. In order to make the transition to the limit, and to obtain Bernoulli's theorem, it is only necessary to assume that we may make *n* as large as we like. From this it can be seen that Bernoulli's theorem is true, approximately, even for *finite* sequences, if they are *n*-free for an *n* which is sufficiently large.

It seems therefore that the deduction of Bernoulli's theorem does not depend upon an axiom postulating the existence of a frequency limit, but *only* on 'absolute freedom' or randomness. The limit concept plays only a subordinate rôle: it is used for the purpose of applying some conception of relative frequency (which, in the first instance, is only defined for finite classes, and without which the concept of *n*-freedom cannot be formulated) to sequences that can be continued indefinitely.

Moreover, it should not be forgotten that Bernoulli himself deduced his theorem within the framework of the classical theory, which contains no axiom of convergence; also, that the definition of probability as a *limit* of frequencies is only an *interpretation*—and not the only possible one—of the classical formalism.

I shall try to justify my conjecture—the independence of Bernoulli's theorem of the axiom of convergence—by deducing this theorem without assuming anything except *n*-freedom (to be appropriately defined).*1 And I shall try to show that it holds even for those mathematical sequences whose primary properties possess *no frequency limits*.

---

*1 I still consider my old doubt concerning the assumption of an axiom of convergence, and the possibility of doing without it, perfectly justified: it is justified by the developments indicated in appendix iv, note *2, and in appendix *vi, where it is shown that randomness (if defined by 'shortest random-like sequences') *entails* convergence which therefore need not be separately postulated. Moreover, my reference to the classical formalism is justified by the development of the neo-classical (or measure-theoretical) theory of probability, discussed in chapter *iii of the *Postscript*; in fact, it is justified by Borel's 'normal numbers'. But I do not agree any longer with the view implicit in the next sentence of my text, although I agree with the remaining paragraphs of this section.

183

Only if this can be shown shall I regard my deduction of the law of great numbers as satisfactory from the point of view of the epistemologist. For it is a 'fact of experience'—or so at least we are sometimes told—that chance-like empirical sequences show that peculiar behaviour which I have described as 'quasi-convergent' or 'statistically stable'. (*Cf.* section 61.) By recording statistically the behaviour of long segments one can establish that the relative frequencies approach closer and closer to a definite value, and that the intervals within which the relative frequencies fluctuate become smaller and smaller. This so-called 'empirical fact', so much discussed and analysed, which is indeed often regarded as the empirical corroboration of the law of great numbers, can be viewed from various angles. Thinkers with inductivist leanings mostly regard it as a fundamental law of nature, not reducible to any simpler statement; as a peculiarity of our world which has simply to be accepted. They believe that expressed in a suitable form—for example in the form of the axiom of convergence—this law of nature should be made the basis of the theory of probability which would thereby assume the character of a natural science.

My own attitude to this so-called 'empirical fact' is different. I am inclined to believe that it is reducible to the chance-like character of the sequences; that it may be derived from the fact that these sequences are n-free. I see the great achievement of Bernoulli and Poisson in the field of probability theory precisely in their discovery of a way to show that this alleged 'fact of experience' is a tautology, and that from disorder in the small (provided it satisfies a suitably formulated condition of n-freedom), there follows logically a kind of order of stability in the large.

If we succeed in deducing Bernoulli's theorem without *assuming* an axiom of convergence, then we shall have reduced the epistemological problem of the law of great numbers to one of axiomatic independence, and thus to a purely logical question. This deduction would also explain why the axiom of convergence works quite well in all practical applications (in attempts to calculate the approximate behaviour of empirical sequences). For even if the restriction to convergent sequences should turn out to be unnecessary, it can certainly not be inappropriate to use convergent mathematical sequences for calculating the approximate behaviour of empirical sequences which, on logical grounds, are statistically stable.

64. *Elimination of the Axiom of Convergence. Solution of the 'Fundamental Problem of the Theory of Chance'.*

So far frequency limits have had no other function in our reconstruction of the theory of probability than that of providing an unambiguous concept of relative frequency applicable to infinite sequences, so that with its help we may define the concept of 'absolute freedom' (from after-effects). For it is a *relative frequency* which is required to be insensitive to selection according to predecessors.

Earlier we restricted our inquiry to alternatives with frequency limits, thus tacitly introducing an axiom of convergence. Now, so as to free us from this axiom, I shall remove the restriction without replacing it by any other. This means that we shall have to construct a frequency concept which can take over the function of the discarded frequency limit, and which may be applied to *all* infinite reference sequences.[*1]

One frequency concept fulfilling these conditions is the concept of a *point of accumulation of the sequence of relative frequencies*. (A value *a* is said to be a point of accumulation of a sequence if after any given element there are elements deviating from *a* by less than a given amount, however small.) That this concept is applicable without restriction to all infinite reference sequences may be seen from the fact that for every finite alternative *at least one* such point of accumulation must exist for the sequence of relative frequencies which corresponds to it. Since relative frequencies can never be greater than 1 nor less than 0, a sequence of them must be bounded by 1 and 0. And as an infinite bounded sequence, it must (according to a famous theorem of Bolzano and Weierstrass) have *at least one* point of accumulation.[1]

For brevity, every point of accumulation of the sequence of relative frequencies corresponding to an alternative α will be called 'a *middle frequency* of α'. We can then say: If a sequence α has *one and only one* middle frequency, then this is at the same time its frequency limit; and conversely: if it has no frequency limit, then it has more than one[2] middle frequency.

[*1] In order not to *postulate* convergence, I appealed in the following paragraph to what can be *demonstrated*—the existence of points of accumulation. All this becomes unnecessary if we adopt the method described in note *1 to section 57, and in appendix *vi.

[1] A fact which, surprisingly enough, has not hitherto been utilized in probability theory.

[2] It can easily be shown that if more than *one* middle frequency exists in a reference sequence then the values of these middle frequencies form a continuum.

The idea of a middle frequency will be found very suitable for our purpose. Just as previously it was our *estimate*—perhaps a hypothetical estimate—that p was the frequency limit of a sequence α, so we now work with the estimate that p is a middle frequency of α. And provided we take certain necessary precautions,[3] we can make *calculations* with the help of these estimated middle frequencies, in a way analogous to that in which we calculate with frequency limits. Moreover the concept of middle frequency is applicable to all possible infinite reference sequences, without any restriction.

If we now try to interpret our symbol $_\alpha F'(\beta)$ as a middle frequency, rather than a frequency limit, and if we accordingly alter the definition of objective probability (section 59), most of our formulae will still be derivable. One difficulty arises however: middle frequencies are *not unique*. If we estimate or conjecture that a middle frequency is $_\alpha F'(\beta) = p$, then this does not exclude the possibility that there are values of $_\alpha F'(\beta)$ other than p. If we postulate that this shall not be so, we thereby introduce, by implication, the axiom of convergence. If on the other hand we define objective probability without such a postulate of uniqueness,[4] then we obtain (in the first instance, at least) a *concept of probability which is ambiguous*; for under certain circumstances a sequence may possess at the same time several middle frequencies which are 'absolutely free' (*cf.* section c of appendix iv). But this is hardly acceptable, since we are accustomed to work with *unambiguous or unique* probabilities; to assume, that is to say, that for one and the same property there can be one and only one probability p, within one and the same reference sequence.

However, the difficulty of defining a unique probability concept without the limit axiom can easily be overcome. We may introduce the requirement of uniqueness (as is, after all, the most natural procedure) as the last step, *after* having postulated that the sequence shall be 'absolutely free'. This leads us to propose, as a solution of our

[3] The concept of 'independent selection' must be interpreted more strictly than hitherto, since otherwise the validity of the special multiplication theorem cannot be proved. For details see my work mentioned in note 3 to section 51. (*This is now superseded by appendix *vi.)

[4] We can do this because it must be possible to apply the theory for finite classes (with the exception of the theorem of uniqueness) immediately to middle frequencies. If a sequence α has a middle frequency p, then it must contain—whatever the term with which the counting starts—segments of any *finite* magnitude, the frequency of which deviates from p as little as we choose. The calculation can be carried out for these. That p is free from after-effect will then mean that this middle frequency of α is also a middle frequency of any predecessor selection of α.

problem, the following modification of our definition of chance-like sequences, and of objective probability.

Let α be an alternative (with one or several middle frequencies). Let the ones of α have one and only one middle frequency p that is 'absolutely free'; then we say that α is chance-like or random, and that p is the objective probability of the ones, within α.

It will be helpful to divide this definition into two axiomatic requirements.*2

(1) Requirement of randomness: for an alternative to be chance-like, there must be at least one 'absolutely free' middle frequency, *i.e.* its objective probability p.

(2) Requirement of uniqueness: for one and the same property of one and the same chance-like alternative, there must be *one and only one probability* p.

The consistency of the new axiomatic system is ensured by the example previously constructed. It is possible to construct sequences which, whilst they have one and only one probability, yet possess no frequency limit (*cf.* section b of appendix iv). This shows that the new axiomatic demands are actually wider, or less exacting, than the old ones. This fact will become even more evident if we state (as we may) our old axioms in the following form:

(1) Requirement of randomness: as above.

(2) Requirement of uniqueness: as above.

(2′) Axiom of convergence: for one and the same property of one and the same chance-like alternative there exists no further middle frequency apart from its probability p.

From the proposed system of requirements we can deduce Bernoulli's theorem, and with it all the theorems of the classical calculus of probability. This solves our problem: it is now possible to deduce the law of great numbers within the framework of the frequency theory without using the axiom of convergence. Moreover, not only does the formula (1) of section 61 and the verbal formulation

---

*2 It is possible to combine the approach described in note *1 to section 57, and in appendices iv and *vi, with these two requirements by retaining requirement (1) and replacing requirement (2) by the following:

(+2) Requirement of finitude: the sequence must become, from its commencement, as quickly n-free as possible, and for the largest possible n; or in other words, it must be (approximately) a *shortest* random-like sequence.

of Bernoulli's theorem remain unchanged,[5] but the interpretation we have given to it also remains unchanged: in the case of a chance-like sequence *without* a frequency limit it will still be true that almost all sufficiently long sequences show only small deviations from p. In such sequences (as in chance-like sequences with frequency limits) segments of any length behaving quasi-divergently will of course occur at times, *i.e.* segments which deviate from p by any amount. But such segments will be comparatively rare, since they must be compensated for by extremely long parts of the sequence in which all (or almost all) segments behave quasi-convergently. As calculation shows, these stretches will have to be longer by several orders of magnitude, as it were, than the divergently-behaving segments for which they compensate.*[3]

This is also the place to solve the '*fundamental problem of the theory of chance*' (as it was called in section 49). The seemingly paradoxical inference from the unpredictability and irregularity of singular events to the applicability of the rules of the probability calculus to them is indeed valid. It is valid provided we can express the irregularity, with a fair degree of approximation, in terms of the hypothetical assumption that one only of the recurring frequencies—of the 'middle frequencies'—so occurs in any selection according to predecessors that no after-effects result. For upon these assumptions it is possible to prove that the law of great numbers is tautological. It is admissible and not self-contradictory (as has sometimes been asserted[6]) to uphold the conclusion that in an irregular sequence in which, as it were, anything may happen at one time or another—though some things only rarely—a certain regularity or stability will appear in very large sub-sequences. Nor is this conclusion trivial, since we need for it specific mathematical tools (the Bolzano and Weierstrass theorem, the concept of n-freedom, and Bernoulli's theorem). The apparent

---

[5] The quasi-Bernoulli formulae (symbol: F') also remain unambiguous for chance-like sequences (according to the new definition), although 'F'' now symbolizes only a middle frequency.

*[3] I am in full agreement with what follows here, even though any reference to 'middle frequencies' becomes redundant if we adopt the method described in section 57, note *1, and appendix iv.

[6] *Cf.*, for instance, Feigl, *Erkenntnis* I, 1930, p. 254: 'In the law of great numbers an attempt is made to reconcile two claims which prove on closer analysis to be in fact mutually contradictory. On the one hand . . . every arrangement and distribution is supposed to be able to occur once. On the other hand, these occurrences . . . are to appear with a corresponding frequency.' (That there is in fact no incompatibility here is proved by the construction of model sequences; *cf.* appendix iv.)

paradox of an argument from unpredictability to predictability, or from ignorance to knowledge, disappears when we realize that the assumption of irregularity can be put in the form of a *frequency hypothesis* (that of freedom from after-effects), and that it must be put in this form if we want to show the validity of that argument.

It now also becomes clear why the older theories have been unable to do justice to what I call the 'fundamental problem'. The subjective theory, admittedly, can deduce Bernoulli's theorem; but it can never consistently interpret it in terms of frequencies, after the fashion of the law of great numbers (*cf.* section 62). Thus it can never explain the statistical success of probability predictions. On the other hand, the older frequency theory, by its axiom of convergence, explicitly postulates regularity in the large. Thus within this theory the problem of inference from irregularity in the small to stability in the large does not arise, since it merely involves inference from stability in the large (axiom of convergence), coupled with irregularity in the small (axiom of randomness) to a special form of stability in the large (Bernoulli's theorem, law of great numbers).*4

The axiom of convergence is not a necessary part of the foundations of the calculus of probability. With this result I conclude my analysis of the mathematical calculus.[7]

We now return to the consideration of more distinctively methodological problems, especially the problem of how to decide probability statements.

65. *The Problem of Decidability.*

In whatever way we may define the concept of probability, or whatever axiomatic formulations we choose: so long as the binomial formula is derivable within the system, *probability statements will not*

---

*4 What is said in this paragraph implicitly enhances the significance, for the solution of the 'fundamental problem', of an *objectively* interpreted neo-classical theory. A theory of this kind is described in chapter *iii of my *Postscript*.

[7] *Cf.* note 3 to section 51. In retrospect I wish to make it clear that I have taken a conservative attitude to von Mises's four points (*cf.* end of section 50). I too define probability only with reference to *random sequences* (which von Mises calls 'collectives'). I too set up a (modified) axiom of randomness, and in determining the *task of the calculus of probability* I follow von Mises without reservation. Thus our differences concern only the limit axiom which I have shown to be superfluous and which I have replaced by the demand for uniqueness, and the axiom of randomness which I have so modified that model sequences can be constructed. (Appendix iv.) As a result, Kamke's objection (*cf.* note 3 to section 53) ceases to be valid.

be falsifiable. Probability hypotheses *do not rule out anything observable*; probability estimates cannot contradict, or be contradicted by, a basic statement; nor can they be contradicted by a conjunction of any finite number of basic statements; and accordingly not by any finite number of observations either.

Let us assume that we have proposed an equal-chance hypothesis for some alternative α; for example, that we have estimated that tosses with a certain coin will come up '1' and '0' with equal frequency, so that $_{\alpha}F(1) = {}_{\alpha}F(0) = \frac{1}{2}$; and let us assume that we find, empirically, that '1' comes up over and over again without exception: then we shall, no doubt, abandon our estimate in practice, and regard it as falsified. But there can be no question of falsification in a logical sense. For we can surely observe only a finite sequence of tosses. And although, according to the binomial formula, the probability of chancing upon a very long finite segment with great deviations from $\frac{1}{2}$ is exceedingly small, it must yet always remain greater than zero. A sufficiently rare occurrence of a finite segment with even the greatest deviation can thus never contradict the estimate. In fact, we must expect it to occur: this is a consequence of our estimate. The hope that the calculable *rarity* of any such segment will be a means of falsifying the probability estimate proves illusory, since even a frequent occurrence of a long and greatly deviating segment may always be said to be nothing but one occurrence of an even longer and more greatly deviating segment. Thus there are no sequences of events, given to us extensionally, and therefore no finite *n*-tuple of basic statements, which could falsify a probability statement.

Only an infinite sequence of events—defined intensionally by a rule—could contradict a probability estimate. But this means, in view of the considerations set forth in section 38 (*cf.* section 43), that probability hypotheses are unfalsifiable because their dimension is infinite. We should therefore really describe them as empirically uninformative, as void of empirical content.[1]

Yet any such view is clearly unacceptable in face of the *successes* which physics has achieved with predictions obtained from hypothetical estimates of probabilities. (This is the same argument as has been used here much earlier against the interpretation of probability statements as tautologies by the subjective theory.) Many of

[1] But not as void of 'logical content' (*cf.* section 35); for clearly, not every frequency hypothesis holds tautologically for every sequence.

these estimates are not inferior in scientific significance to any other physical hypothesis (for example, to one of a determinist character). And a physicist is usually quite well able to decide whether he may for the time being accept some particular probability hypothesis as 'empirically confirmed', or whether he ought to reject it as 'practically falsified', *i.e.*, as useless for purposes of prediction. It is fairly clear that this 'practical falsification' can be obtained only through a methodological decision to regard highly improbable events as ruled out—as prohibited. But with what right can they be so regarded? Where are we to draw the line? Where does this 'high improbability' begin?

Since there can be no doubt, from a purely logical point of view, about the fact that probability statements cannot be falsified, the equally indubitable fact that we use them empirically must appear as a fatal blow to my basic ideas on method which depend crucially upon my criterion of demarcation. Nevertheless I shall try to answer the questions I have raised—which constitute the problem of decidability—by a resolute application of these very ideas. But to do this, I shall first have to analyse the logical form of probability statements, taking account both of the logical inter-relations between them and of the logical relations in which they stand to basic statements.*1

66. *The Logical Form of Probability Statements.*

Probability estimates are *not* falsifiable. Neither, of course, are they verifiable, and this for the same reasons as hold for other hypotheses, seeing that no experimental results, however numerous and favourable, can ever finally establish that the relative frequency of 'heads' is $\frac{1}{2}$, and will *always* be $\frac{1}{2}$.

---

*1 I believe that my emphasis upon the irrefutability of probabilistic hypotheses—which culminates in section 67—was healthy: it laid bare a problem which had not been discussed previously (owing to the general emphasis on verifiability rather than falsifiability, and the fact that probability statements are, as explained in the next section, in some sense verifiable or 'confirmable'). Yet my reform, proposed in note *1 to section 57 (see also note *2 to section 64), changes the situation entirely. For this reform, apart from achieving other things, amounts to the adoption of a methodological rule, like the one proposed below in section 68, which makes probability hypotheses falsifiable. The problem of decidability is thereby transformed into the following problem: since empirical sequences can only be expected to *approximate* to shortest random-like sequences, what is acceptable and what is unacceptable as an approximation? The answer to this is clearly that closeness of approximation is a matter of degree, and that the determination of this degree is one of the main problems of mathematical statistics, and of the theory of corroboration. See also appendix *ix, especially my 'Third Note'.

Probability statements and basic statements can thus neither contradict one another nor entail one another. And yet, it would be a mistake to conclude from this that no kind of logical relations hold between probability statements and basic statements. And it would be equally wide off the mark to believe that while logical relations do obtain between statements of these two kinds (since sequences of observations may obviously agree more or less closely with a frequency statement), the analysis of these relations compels us to introduce a special probabilistic logic[1] which breaks the fetters of classical logic. In opposition to such views I believe that the relations in question can be fully analysed in terms of the 'classical' logical relations of *deducibility* and *contradiction*.[*1]

From the non-falsifiability and non-verifiability of probability statements it can be inferred that they have no falsifiable consequences, and that they cannot themselves be consequences of verifiable statements. But the converse possibilities are not excluded. For it may be (a) that they have unilaterally verifiable consequences (purely existential consequences, or there-is-consequences) or (b) that they are themselves consequences of unilaterally falsifiable universal statements (all-statements).

Possibility (b) will scarcely help to clarify the logical relation between probability statements and basic statements: it is only too obvious that a non-falsifiable statement, *i.e.* one which says very little, can belong to the consequence class of one which is falsifiable, and which thus says more.

What is of greater interest for us is possibility (a) which is by no means trivial, and which in fact turns out to be fundamental for our analysis of the relation between probability statements and basic statements. For we find that from every probability statement, an infinite class of existential statements can be deduced, but not *vice versa*. (Thus the probability statement asserts more than does any of these existential statements.) For example, let p be a probability which has been estimated, hypothetically, for a certain alternative (and let $0 \neq p \neq 1$); then we can deduce from this estimate, for instance, the existential consequence that both ones and zeros will occur in the

---

[1] *Cf.* Section 80, especially notes 3 and 6.
[*1] Although I do not disagree with this, I now believe that the probabilistic concepts 'almost deducible' and 'almost contradictory' are extremely useful in connection with our problem; see appendix *ix, and chapter *iii of the *Postscript*.

sequence. (Of course many far less simple consequences also follow—for example, that segments will occur which deviate from p only by a very small amount.)

But we can deduce much more from this estimate; for example that there will 'over and over again' be an element with the property '1' and another element with the property '0'; that is to say, that after *any* element x there will occur in the sequence an element y with the property '1', and also an element z with the property '0'. A statement of this form ('for *every* x there is a y with the observable, or extensionally testable, property β') is both non-falsifiable—because it has no falsifiable consequences—and non-verifiable—because of the 'all' or 'for every' which made it hypothetical.*² Nevertheless, it can be better, or less well 'confirmed'—in the sense that we may succeed in verifying many, few, or none of its existential consequences; thus it stands to the basic statement in the relation which appears to be characteristic of probability statements. Statements of the above form may be called 'universalized existential statements' or (universalized) *'existential hypotheses'*.

My contention is that the relation of probability estimates to basic statements, and the possibility of their being more, or less, well 'confirmed', can be understood by considering the fact that from all probability estimates, existential hypotheses are *logically deducible*. This suggests the question whether the probability statements themselves may not, perhaps, have the form of existential hypotheses.

Every (hypothetical) probability estimate entails the conjecture that the empirical sequence in question is, approximately, chance-like

---

*² Of course, I never intended to suggest that *every* statement of the form 'for every x, there is a y with the observable property β' is non-falsifiable and thus non-testable: obviously, the statement 'for every toss with a penny resulting in 1, there is an immediate successor resulting in 0' is both falsifiable and in fact falsified. What creates non-falsifiability is not just the form 'for every x there is a y such that ...' but the fact that the 'there is' is *unbounded*—that the occurrence of the y may be delayed beyond all bounds: in the probabilistic case, y may, *as it were, occur as late as it pleases*. An element '0' may occur at once, or after a thousand tosses, or after any number of tosses: it is this fact that is responsible for non-falsifiability. If, on the other hand, the distance of the place of occurrence of y from the place of occurrence of x is *bounded*, then the statement 'for every x there is a y such that ...' may be falsifiable.

My somewhat unguarded statement in the text (which tacitly presupposed section 15) has led, to my surprise, in some quarters to the belief that *all* statements—or 'most' statements, whatever this may mean—of the form 'for every x there is a y such that ...' are non-falsifiable; and this has then been repeatedly used as a criticism of the falsifiability criterion. See, for example, *Mind* **54**, 1945, pp. 119 f. The whole problem of these 'all-and-some statements' (this term is due to J. W. N. Watkins) is discussed more fully in my *Postscript*; see especially sections *24 f.

or random. That is to say, it entails the (approximate) applicability, and the truth, of the axioms of the calculus of probability. Our question is, therefore, equivalent to the question whether these axioms represent what I have called 'existential hypotheses'.

If we examine the two requirements proposed in section 64 then we find that the requirement of randomness has in fact the form of an existential hypothesis.[2] The requirement of uniqueness, on the other hand, has not this form; it cannot have it, since a statement of the form 'There is *only one* . . .' must have the form of a universal statement. (It can be translated as 'There are not more than one . . .' or 'All . . . are identical'.)

Now it is my thesis here that it is only the 'existential constituent', as it might be called, of probability estimates, and therefore the requirement of randomness, which establishes a logical relation between them and basic statements. Accordingly, the requirement of uniqueness, as a universal statement, would have no extensional consequences whatever. That a value p with the required properties exists, can indeed be extensionally 'confirmed'—though of course only provisionally; but not that *only one* such value exists. This latter statement, which is universal, could be extensionally significant only if basic statements could *contradict* it; that is to say, if basic statements could establish the existence of more than one such value. Since they cannot (for we remember that non-falsifiability is bound up with the binomial formula), the requirement of uniqueness must be extensionally without significance.[*3]

This is the reason why the logical relations holding between a probability estimate and basic statements, and the graded 'confirmability' of the former, are unaffected if we eliminate the requirement of uniqueness from the system. By doing this we could give the system the form of a pure existential hypothesis.[3] But we should

---

[2] It can be put in the following form: For every positive ϵ, for every predecessor n-tuple, and every element with the ordinal number x there is an element, selected according to predecessor selection, with the ordinal number $y>x$ such that the frequency up to the term y deviates from a fixed value p by an amount less than ϵ.

[*3] The situation is totally different if the requirement $(+2)$ of note *2 to section 64 is adopted: this is empirically significant, and renders the probability hypotheses falsifiable (as asserted in note *1 to section 65).

[3] The formulae of the probability calculus are also derivable in this axiomatization, only the formulae must be interpreted as existential formulae. The theorem of Bernoulli, for example, would no longer assert that the single probability value for a particular n of $_{a_n}F(\Delta p)$ lies near to 1, but only that (for a particular n) among the various probability values of $_{a_n}F(\Delta p)$ there is at least one which lies near to 1.

then have to give up the uniqueness of probability estimates,*⁴ and thereby (so far as uniqueness is concerned) obtain something different from the usual calculus of probability.

Therefore the requirement of uniqueness is obviously not super-fluous. What, then, is its logical function?

Whilst the requirement of randomness helps to establish a relation between probability statements and basic statements, the requirement of uniqueness regulates the relations between the various probability statements themselves. Without the requirement of uniqueness some of these might, as existential hypotheses, be derivable from others, but they could never contradict one another. Only the requirement of uniqueness ensures that probability statements can contradict one another; for by this requirement they acquire the form of a con-junction whose components are a universal statement and an existential hypothesis; and statements of this form can stand to one another in exactly the same fundamental logical relations (equivalence, derivability, compatibility, and incompatibility) as can 'normal' universal statements of any theory—for example, a falsifiable theory.

If we now consider the axiom of convergence, then we find that it is like the requirement of uniqueness in that it has the form of a non-falsifiable universal statement. But it demands more than our requirement does. This additional demand, however, cannot have any extensional significance either; moreover, it has no logical or formal but *only an intensional* significance: it is a demand for the exclusion of all intensionally defined (*i.e.* mathematical) sequences without frequency limits. But from the point of view of applications, this exclusion proves to be without significance even intensionally, since in applied probability theory we do not of course deal with the mathematical sequences themselves but only with hypothetical estimates about empirical sequences. The exclusion of sequences without frequency limits could therefore only serve to warn us against treating those empirical sequences as chance-like or random of which we hypothetically assume that they have no frequency limit. But what possible action could we take in response to this warning?⁴

---

*⁴ As has been shown in the new footnote *2 to section 64, any special *requirement* of uniqueness can be eliminated, without sacrificing uniqueness.

⁴ Both the axiom of randomness and the axiom of uniqueness can properly be regarded as such (intensional) warnings. For example, the axiom of randomness cautions us not to treat sequences as random if we suppose (no matter on what grounds) that

What sort of considerations or conjectures about the possible convergence or divergence of empirical sequences should we indulge in or abstain from, in view of this warning, seeing that criteria of convergence are no more applicable to them than are criteria of divergence? All these embarrassing questions[5] disappear once the axiom of convergence has been got rid of.

Our logical analysis thus makes transparent both the form and the function of the various partial requirements of the system, and shows what reasons tell against the axiom of randomness and in favour of the requirement of uniqueness. Meanwhile the problem of decidability seems to be growing ever more menacing. And although we are not obliged to call our requirements (or axioms) 'meaningless',[6] it looks as if we were compelled to describe them as non-empirical. But does not this description of probability statements—no matter what words we use to express it—contradict the main idea of our approach?

67. *A Probabilistic System of Speculative Metaphysics.*

The most important use of probability statements in physics is this: certain physical regularities or observable physical effects are interpreted as 'macro laws'; that is to say, they are interpreted, or explained, as mass phenomena, or as the observable results of hypothetical and not directly observable 'micro events'. The macro laws are deduced from probability estimates by the following method: we show that observations which agree with the observed regularity in question are to be expected with a probability very close to 1, *i.e.* with a probability which deviates from 1 by an amount which can be made as small as we choose. When we have shown this, then we say that

---

[5] Similar misgivings make Schlick object to the limit axiom (*Die Naturwissenschaften* 19, 1931, p. 158).

[6] Here the positivist would have to recognize a whole hierarchy of 'meaninglessnesses'. To him, non-verifiable natural laws appear 'meaningless' (*cf.* section 6, and quotations in notes 1 and 2), and thus still more so probability hypotheses which are neither verifiable nor falsifiable. Of our axioms, the axiom of uniqueness, which is not extensionally significant, would be more meaningless than the meaningless axiom of irregularity, which at least has extensional consequences. Still more meaningless would be the limit axiom, since it is not even intensionally significant.

---

certain gambling systems will be successful for them. The axiom of uniqueness cautions us not to attribute a probability q (with $q \neq p$) to a sequence which we suppose can be approximately described by means of the hypothesis that its probability equals p.

by our probability estimate we have 'explained' the observable effect in question as a macro effect.

But if we use probability estimates in this way for the 'explanation' of observable regularities *without introducing special precautions*, then we may immediately become involved in speculations which in accordance with general usage can well be described as typical of *Speculative metaphysics*.

For since probability statements are not falsifiable, it must always be possible in this way to 'explain', by probability estimates, *any regularity we please*. Take, for example, the law of gravity. We may contrive hypothetical probability estimates to 'explain' this law in the following way. We select events of some kind to serve as elementary or atomic events; for instance the movement of a small particle. We select also what is to be a primary property of these events; for instance the direction and velocity of the movement of a particle. We then assume that these events show a chance-like distribution. Finally we calculate the probability that all the particles within a certain finite spatial region, and during a certain finite period of time—a certain 'cosmic period'—will with a specified accuracy move, accidentally, in the way required by the law of gravity. The probability calculated will, of course, be very small; negligibly small, in fact, but still not equal to zero. Thus we can raise the question how long an *n*-segment of the sequence would have to be, or in other words, how long a duration must be assumed for the whole process, in order that we may expect, with a probability close to 1 (or deviating from 1 by not more than an arbitrarily small value ε) the occurrence of one such cosmic period in which, as the result of an accumulation of accidents, our observations will all agree with the law of gravity. For any value as close to 1 as we choose, we obtain a definite, though extremely large, finite number. We can then say: if we assume that the segment of the sequence has this very great length—or in other words, that the 'world' lasts long enough—then our assumption of randomness entitles us to expect the occurrence of a cosmic period in which the law of gravity will seem to hold good, although 'in reality' nothing ever occurs but random scattering. This type of 'explanation' by means of an assumption of randomness is applicable to any regularity we choose. In fact we can in this way 'explain' our whole world, with all its observed regularities, as a phase in a random chaos—*as an accumulation of purely accidental coincidences*.

It seems clear to me that speculations of this kind are 'metaphysical', and that they are without any significance for science. And it seems equally clear that this fact is connected with their non-falsifiability—with the fact that we can always and in all circumstances indulge in them. My criterion of demarcation thus seems to agree here quite well with the general use of the word 'metaphysical'.

Theories involving probability, therefore, if they are applied without special precautions, are not to be regarded as scientific. We must rule out their metaphysical use if they are to have any use in the practice of empirical science.[*1]

68. Probability in Physics.

The problem of decidability troubles only the methodologist, not the physicist.[*1] If asked to produce a practically applicable concept of probability, the physicist might perhaps offer something like a *physical definition of probability*, on lines such as the following: There are certain experiments which, even if carried out under controlled conditions, lead to varying results. In the case of some of these experiments—those which are 'chance-like', such as tosses of a coin—frequent repetition leads to results with relative frequencies which, upon further repetition, approximate more and more to some fixed

[*1] When writing this, I thought that speculations of the kind described would be easily recognizable as useless, just because of their unlimited applicability. But they seem to be more tempting than I imagined. For it has been argued, for example by J. B. S. Haldane (in *Nature* 122, 1928, p. 808; *cf.* also his *Inequality of Man*, pp. 163 f.) that if we accept the probability theory of entropy, we must regard it as certain, or as almost certain, that the world will wind itself up again accidentally if only we wait long enough. This argument has of course been frequently repeated since by others. Yet it is, I think, a perfect example of the kind of argument here criticized, and one which would allow us to expect, with near certainty, anything we liked. Which all goes to show the dangers inherent in the existential form shared by probability statements with most of the statements of metaphysics. (*Cf.* section 15.)

[*1] The problem here discussed has been treated in a clear and thorough way long ago by the physicists P. and T. Ehrenfest, *Encycl. d. Math, Wiss.* 4th Teilband, Heft 6 (12.12.1911) section 30. They treated it as a *conceptual and epistemological* problem. They introduced the idea of 'probability hypotheses of first, second, . . . k th order': a probability hypothesis of second order, for example, is an estimate of the frequency with which certain frequencies occur in an aggregate of aggregates. However, P. and T. Ehrenfest do not operate with anything corresponding to the idea of a *reproducible effect* which is here used in a crucial way in order to solve the problem which they expounded so well. See especially the opposition between Boltzmann and Planck to which they refer in notes 247 f., and which can, I believe, be resolved by using the idea of a reproducible effect. For under appropriate experimental conditions, fluctuations may lead to reproducible effects, as Einstein's theory of Brownian movement showed so impressively. See also note *1 to section 65, and appendices *vi and *ix.

value which we may call the *probability* of the event in question. This value is '. . . empirically determinable through long series of experiments to any degree of approximation';[1] which explains, incidentally, why it is possible to falsify a hypothetical estimate of probability.

Against definitions on these lines both mathematicians and logicians will raise objections; in particular the following:

(1) The definition does not agree with the calculus of probability since, according to Bernoulli's theorem, only *almost* all very long segments are statistically stable, *i.e.* behave as if convergent. For that reason, probability cannot be defined by this stability, *i.e.* by quasi-convergent behaviour. For the expression 'almost all'—which ought to occur in the *definiens*—is itself only a synonym for 'very probable'. The definition is thus circular; a fact which can be easily concealed (but not removed) by dropping the word 'almost'. This is what the physicist's definition did; and it is therefore unacceptable.

(2) When is a series of experiments to be called '*long*'? Without being given a criterion of what is to be called 'long', we cannot know when, or whether, we have reached an approximation to the probability.

(3) How can we know that the desired *approximation* has in fact been reached?

Although I believe that these objections are justified, I nevertheless believe that we can retain the physicist's definition. I shall support this belief by the arguments outlined in the previous section. These showed that probability hypotheses lose all informative content when they are allowed unlimited application. The physicist would never use them in this way. Following his example I shall disallow the unlimited application of probability hypotheses: I propose that we take *the methodological decision never to explain physical effects, i.e. reproducible regularities, as accumulations of accidents.* This decision naturally modifies the concept of probability: it narrows it.*[2] Thus objection (1) does not

---

[1] The quotation is from Born-Jordan *Elementare Quantenmechanik* (1930), p. 306, *cf.* also the beginning of Dirac's *Quantum Mechanics*, 1930, p. 10 of the 1st edition, 1930. A parallel passage (slightly abbreviated) is to be found on p. 14 of the 3rd edition, 1947. See also Weyl, *Gruppentheorie und Quantenmechanik* (2nd edition, 1931, p. 66); English translation by H. P. Robertson: *The Theory of Groups and Quantum Mechanics* (1931), p. 74 *f.*

*[2] The methodological decision or rule here formulated narrows the concept of probability—just as it is narrowed by the decision to adopt *shortest* random-like sequences as mathematical models of empirical sequences, *cf.* note *1 to section 65.

affect my position, for I do not assert the identity of the physical and the mathematical concepts of probability at all; on the contrary, I deny it. But in place of (1), a new objection arises.

(1′) When can we speak of 'accumulated accidents'? Presumably in the case of a small probability. But when is a probability '*small*'? We may take it that the proposal which I have just submitted rules out the use of the method (discussed in the preceding section) of manufacturing an arbitrarily large probability out of a small one by changing the formulation of the mathematical problem. But in order to carry out the proposed decision, we have to know what we are to regard as *small*.

In the following pages it will be shown that the proposed methodological rule agrees with the physicist's definition, and that the objections raised by questions (1′), (2), and (3) can be answered with its help. To begin with, I have in mind only *one* typical case of the application of the calculus of probability: I have in mind the case of certain reproducible macro effects which can be described with the help of precise (macro) laws—such as gas pressure—and which we interpret, or explain, as due to a very large accumulation of micro processes, such as molecular collisions. Other typical cases (such as statistical fluctuations or the statistics of chance-like individual processes) can be reduced without much difficulty to this case.[*3]

Let us take a macro effect of this type, described by a well-corroborated law, which is to be reduced to random sequences of micro events. Let the law assert that under certain conditions a physical magnitude has the value p. We assume the effect to be 'precise', so that no measurable fluctuations occur, *i.e.* no deviations from p beyond that interval, $\pm\varphi$ (the interval of imprecision; *cf.* section 37) within which our measurements will in any case fluctuate, owing to the imprecision inherent in the prevailing technique of measurement. We now propose the hypothesis that p is a probability within a sequence α of micro events; and further, that n micro events contribute towards producing the effect. Then (*cf.* section 61) we can calculate for every chosen value δ, the probability $_{\alpha_n}F(\Delta p)$, i.e. the probability that the value measured will fall within the interval Δp. The complementary probability may be denoted by 'ε'. Thus we have $_{\alpha_n}F(\overline{\Delta p}) = \varepsilon$.

[*3] I am now a little dubious about the words 'without much difficulty'; in fact, in all cases, except those of the extreme macro effects discussed in this section, very subtle statistical methods have to be used. See also appendix *ix, especially my 'Third Note'.

According to Bernoulli's theorem, ε tends to zero as n increases without limit.

We assume that ε is so 'small' that it can be neglected. (Question (1′) which concerns what 'small' means, in this assumption, will be dealt with soon.) The Δp is to be interpreted, clearly, as the interval within which the measurements approach the value p. From this we see that the three quantities: ε, n, and Δp correspond to the three questions (1′), (2), and (3). Δp or δ can be chosen arbitrarily, which restricts the arbitrariness of our choice of ε and n. Since it is our task to deduce the exact macro effect p ($\pm\varphi$) we shall not assume δ to be greater than φ. As far as the reproducible effect p is concerned, the deduction will be satisfactory if we can carry it out for some value δ ⩽ φ. (Here φ is given, since it is determined by the measuring technique.) Now let us choose δ so that it is (approximately) equal to φ. Then we have reduced question (3) to the two other questions, (1′) and (2).

By the choice of δ (*i.e.* of Δp) we have established a relation between n and ε, since to every n there now corresponds uniquely a value of ε. Thus (2), *i.e.* the question When is n sufficiently long? has been reduced to (1′), *i.e.* the question When is ε small? (and *vice versa*).

But this means that *all three questions* could be answered if only we could decide *what particular value of* ε is to be neglected as 'negligibly small'. Now our methodological rule amounts to the decision to neglect *small* values of ε; but we shall hardly be prepared to commit ourselves for ever to a definite value of ε.

If we put our question to a physicist, that is, if we ask him what ε he is prepared to neglect—0.001, or 0.000001, or . . . ? he will presumably answer that ε does not interest him at all; that he has chosen not ε but n; and that he has chosen n in such a way as to make the correlation between n and Δp largely *independent of any changes* of the value ε which we might choose to make.

The physicists's answer is justified, because of the mathematical peculiarities of the Bernoullian distribution: it is possible to determine for every n the functional dependence between ε and Δp.*4 An

---

*4 The remarks that follow in this paragraph (and some of the discussions later in this section) are, I now believe, clarified and superseded by the considerations in appendix *ix; see especially points 8 *ff* of my Third Note. With the help of the methods there used, it can be shown that almost all possible statistical samples of large size n will strongly undermine a given probabilistic hypothesis, that is to say, give it a high

examination of this function shows that for *every* ('large') n there exists a characteristic value of Δp such that in the neighbourhood of this value Δp is highly insensitive to changes of ε. This insensitiveness increases with increasing n. If we take an n of an order of magnitude which we should expect in the case of extreme mass-phenomena, then, in the neighbourhood of its characteristic value, Δp is so highly insensitive to changes of ε that Δp hardly changes at all even if the order of magnitude of ε changes. Now the physicist will attach little value to more sharply defined boundaries of Δp. And in the case of typical mass phenomena, to which this investigation is restricted, Δp can, we remember, be taken to correspond to the interval of precision $\pm\varphi$ which depends upon our technique of measurement; and this has no sharp bounds but only what I called in section 37 'condensation bounds'. We shall therefore call n large when the insensitivity of Δp in the neighbourhood of its characteristic value, which we can determine, is at least so great that even changes in order of magnitude of ε cause the value of Δp to fluctuate only within the condensation bounds of $\pm\varphi$. (If $n \to \infty$, then Δp becomes completely insensitive.) But if this is so, then we need no longer concern ourselves with the exact determination of ε: *the decision to neglect a small ε suffices*, even if we have not exactly stated what has to be regarded as 'small'. It amounts to the decision to work with the characteristic values of Δp above mentioned, which are insensitive to changes of ε.

The rule that extreme improbabilities have to be neglected (a rule which becomes sufficiently explicit only in the light of the above) agrees with the demand for *scientific objectivity*. For the obvious objection to our rule is, clearly, that even the greatest improbability always remains a probability, however small, and that consequently even the most improbable processes—*i.e.* those which we propose to neglect—will some day happen. But this objection can be disposed

---

negative degree of corroboration; and we may decide to interpret this as refutation or falsification. Of the remaining samples, most will support the hypothesis, that is to say, give it a high *positive* degree of corroboration. Comparatively few samples of large size n will give a probabilistic hypothesis an undecisive degree of corroboration (whether positive or negative). Thus we can expect to be able to refute a probabilistic hypothesis, in the sense here indicated; and we can expect this perhaps even more confidently than in the case of a non-probabilistic hypothesis. The methodological rule or decision to regard (for a large n) a negative degree of corroboration as a falsification is, of course, a specific case of the methodological rule or decision discussed in the present section—that of neglecting certain extreme improbabilities.

of by recalling *the idea of a reproducible physical effect*—an idea which is closely connected with that of objectivity (*cf.* section 8). I do not deny the possibility that improbable events might occur. I do not, for example, assert that the molecules in a small volume of gas may not, perhaps, for a short time spontaneously withdraw into a part of the volume, or that in a greater volume of gas spontaneous fluctuations of pressure will never occur. What I do assert is that such occurrences would not be physical effects, because, on account of their immense improbability, *they are not reproducible at will*. Even if a physicist happened to observe such a process, he would be quite unable to reproduce it, and therefore would never be able to decide what had really happened in this case, and whether he may not have made an observational mistake. If, however, we find *reproducible* deviations from a macro effect which has been deduced from a probability estimate in the manner indicated, then we must assume that the probability estimate is *falsified*.

Such considerations may help us to understand pronouncements like the following of Eddington's in which he distinguishes two kinds of physical laws: 'Some things never happen in the physical world because they are *impossible*; others because they are too *improbable*. The laws which forbid the first are primary laws; the laws which forbid the second are secondary laws.'[2] Although this formulation is perhaps not beyond criticism (I should prefer to abstain from non-testable assertions about whether or not extremely improbable things occur), it agrees well with the physicist's application of probability theory.

Other cases to which probability theory may be applied, such as statistical fluctuations, or the statistics of chance-like individual events, are reducible to the case we have been discussing, that of the precisely measurable macro effect. By statistical fluctuations I understand phenomena such as the Brownian movement. Here the interval of precision of measurement ($\pm\varphi$) is smaller than the interval Δp characteristic of the number n of micro events contributing to the effect; hence measureable deviations from p are to be expected as highly probable. The fact that such deviations occur will be testable, since the fluctuation itself becomes a reproducible effect; and to this effect my earlier arguments apply: fluctuations beyond a certain magnitude (beyond some interval Δp) must not be reproducible,

---

[2] Eddington, *The Nature of the Physical World* (C.U.P. 1928, p. 75).

according to my methodological requirements, nor long sequences of fluctuations in one and the same direction, etc. Corresponding arguments would hold for the statistics of chance-like individual events.

I may now summarize my arguments regarding the problem of decidability.

Our question was: How can probability hypotheses—which, we have seen, are non-falsifiable—play the part of natural laws in empirical science? Our answer is this: Probability statements, in so far as they are not falsifiable, are metaphysical and without empirical significance; and in so far as they are used as empirical statements they are used as falsifiable statements.

But this answer raises another question: *How is it possible* that probability statements—which are not falsifiable—can be *used* as falsifiable statements? (The fact that they can be so used is not in doubt: the physicist knows well enough when to regard a probability assumption as falsified.) This question, we find, has two aspects. On the one hand, we must make the possibility of using probability statements understandable in terms of their logical form. On the other hand, we must analyse the rules governing their use as falsifiable statements.

According to section 66, accepted basic statements may agree more or less well with some proposed probability estimate; they may represent better, or less well, a typical segment of a probability sequence. This provides the opportunity for the application of some kind of *methodological rule*; a rule, for instance, which might demand that the agreement between basic statements and the probability estimate should conform to some minimum standard. Thus the rule might draw some arbitrary line and decree that only reasonably representative segments (or reasonably 'fair samples') are 'permitted', while a-typical or non-representative segments are 'forbidden'.

A closer analysis of this suggestion showed us that the dividing line between what is permitted and what is forbidden need not be drawn quite as arbitrarily as might have been thought at first. And in particular, that there is no need to draw it 'tolerantly'. For it is possible to frame the rule in such a way that the dividing line between what is permitted and what is forbidden is determined, just as in the case of other laws, by the attainable precision of our measurements.

Our methodological rule, proposed in accordance with the criterion of demarcation, does not forbid the occurrence of a-typical segments; neither does it forbid the repeated occurrence of deviations (which, of course, are typical for probability sequences). What this rule forbids is the predictable and reproducible occurrence of systematic deviations; such as deviations in a particular direction, or the occurrence of segments which are a-typical in a definite way. Thus it requires not a mere rough agreement, but the best possible one *for everything that is reproducible and testable*; in short, for all *reproducible effects*.

69. Law and Chance.

One sometimes hears it said that the movements of the planets obey strict laws, whilst the fall of a die is fortuitous, or subject to chance. In my view the difference lies in the fact that we have so far been able to predict the movement of the planets successfully, but not the individual results of throwing dice.

In order to deduce predictions one needs laws and initial conditions; if no suitable laws are available or if the initial conditions cannot be ascertained, the scientific way of predicting breaks down. In throwing dice, what we lack is, clearly, sufficient knowledge of initial conditions. With sufficiently precise measurements of initial conditions it would be possible to make predictions in this case also; but the rules for correct dicing (shaking the dice-box) are so chosen as to prevent us from measuring initial conditions. The rules of play and other rules determining the conditions under which the various events of a random sequence are to take place I shall call the '*frame conditions*'. They consist of such requirements as that the dice shall be 'true' (made from homogeneous material), that they shall be well shaken, etc.

There are other cases in which prediction may be unsuccessful. Perhaps it has not so far been possible to formulate suitable laws; perhaps all attempts to find a law have failed, and all predictions have been falsified. In such cases we may despair of ever finding a satisfactory law. (But it is not likely that we shall give up trying unless the problem does not interest us much—which may be the case, for example, if we are satisfied with frequency predictions.) In no case, however, can we say with finality that there are no laws in a particular field. (This is a consequence of the impossibility of verification.) This means that my view makes the concept of chance *subjective*.*¹ I speak of 'chance' when our knowledge does not suffice for prediction; as in the case of dicing, where we speak of 'chance' because we have no knowledge of the initial conditions. (Conceivably a physicist equipped with good instruments could predict a throw which other people could not predict.)

In opposition to this subjective view, an objective view has sometimes been advocated. In so far as this uses the metaphysical idea that events are, or are not, determined in themselves, I shall not examine it further here. (*Cf.* sections 71 and 78.) If we are successful with our prediction, we may speak of 'laws'; otherwise we can know nothing about the existence or non-existence of laws or of irregularities.*²

Perhaps more worth considering than this metaphysical idea is the following view. We encounter 'chance' in the objective sense, it may be said, when our probability estimates are corroborated; just as we encounter causal regularities when our predictions deduced from laws are corroborated.

The definition of chance implicit in this view may not be altogether useless, but it should be strongly emphasized that the concept so defined is not opposed to the concept of law: it was for this reason that I called probability sequences chance-*like*. In general, a sequence of experimental results will be chance-like if the frame conditions which define the sequence differ from the initial conditions; when the individual experiments, carried out under identical frame conditions, will proceed under different initial conditions, and so yield different results. Whether there are chance-like sequences whose elements are in no way predictable, I do not know. From the fact that a sequence is chance-like we may not even infer that its elements are not predictable, or that they are 'due to chance' in the subjective

---

*¹ This does not mean that I made any concession here to a subjective interpretation of *probability*, or of *disorder* or *randomness*.

*² In this paragraph, I dismissed (because of its metaphysical character) a metaphysical theory which I am now, in my *Postscript*, anxious to recommend because it seems to me to open new vistas, to suggest the resolution of serious difficulties, and to be, perhaps, true. Although when writing this book I was aware of holding metaphysical beliefs, and although I even pointed out the suggestive value of metaphysical ideas for science, I was not alive to the fact that some metaphysical doctrines were rationally arguable and, in spite of being irrefutable, criticizable. See especially the last section of my *Postscript*.

sense of insufficient knowledge; and least of all may we infer from this fact the 'objective' fact that there are no laws.*³

Not only is it impossible to infer from the chance-like character of the sequence anything about the conformity to law, or otherwise, of the *individual events*: it is not even possible to infer from the corroboration of probability estimates that the *sequence itself* is completely irregular. For we know that chance-like sequences exist which are constructed according to a mathematical rule. (*Cf.* appendix iv.) The fact that a sequence has a Bernoullian distribution is not a symptom of the absence of law, and much less identical with the absence of law 'by definition'.[1] In the success of probability predictions we must see no more than a symptom of the absence of *simple* laws in the structure of the *sequence* (*cf.* sections 43 and 58)—as opposed to the events constituting it. The assumption of freedom from after-effect, which is equivalent to the hypothesis that such *simple* laws are not discoverable, is corroborated, but that is all.

70. *The Deducibility of Macro Laws from Micro Laws.*

There is a doctrine which has almost become a prejudice, although it has recently been criticized severely—the doctrine that *all* observable events must be explained as macro events; that is to say, as averages or accumulations or summations of certain micro events. (The doctrine is somewhat similar to certain forms of materialism.) Like other doctrines of its kind, this seems to be a metaphysical hypostatization of a methodological rule which in itself is quite unobjectionable. I mean the rule that we should see whether we can simplify or generalize or unify our theories by employing explanatory hypotheses of the type mentioned (that is to say, hypotheses explaining observable effects as summations or integrations of micro events). In evaluating the success of such attempts, it would be a mistake to think that *non-statistical*

---

*³ It would have been clearer, I think, had I argued as follows. We can never repeat an experiment precisely—all we can do is to keep *certain* conditions constant, within certain limits. It is therefore no argument for objective fortuity, or chance, or absence of law, if certain aspects of our results repeat themselves, while others vary irregularly; especially if the conditions of the experiment (as in the case of spinning a penny) are designed with a view to making conditions vary. So far, I still agree with what I have said. But there may be *other* arguments for objective fortuity; and one of these, due to Alfred Landé ('Landé's blade') is highly relevant in this context. It is now discussed at length in my *Postscript*, sections *90, *f*.

[1] As Schlick says in *Die Kausalität in der gegenwärtigen Physik*, *Naturwissenschaften* **19**, 1931, p. 157.

hypotheses about the micro events and their laws of interaction could ever be sufficient to explain macro events. For we should need, in addition, hypothetical *frequency estimates*, since statistical conclusions can only be derived from statistical premises. These frequency estimates are always independent hypotheses which at times may indeed occur to us whilst we are engaged in studying the laws pertaining to micro events, but which can never be derived from these laws. Frequency estimates form a special class of hypotheses: they are prohibitions which, as it were, concern regularities in the large.[1] Von Mises has stated this very clearly: 'Not even the tiniest little theorem in the kinetic theory of gases follows from classical physics alone, without additional assumptions of a statistical kind.'[2]

Statistical estimates, or frequency statements, can never be derived simply from laws of a 'deterministic' kind, for the reason that in order to deduce any prediction from such laws, initial conditions are needed. In their place, assumptions about the statistical *distribution* of initial conditions—that is to say specific statistical assumptions—enter into every deduction in which statistical laws are obtained from micro assumptions of a deterministic or 'precise' character.[*1]

It is a striking fact that the frequency assumptions of theoretical physics are to a great extent *equal-chance hypotheses*, but this by no means implies that they are 'self-evident' or *a priori* valid. That they are far from being so may be seen from the wide differences between classical statistics, Bose-Einstein statistics, and Fermi-Dirac statistics. These show how special assumptions may be combined with an equal-

---

[1] A. March well says (*Die Grundlagen der Quantenmechanik* 1931, p. 250) that the particles of a gas cannot behave '... as they choose; each one must behave in accordance with the behaviour of the others. It can be regarded as one of the most fundamental principles of quantum theory that the whole is more than the mere sum of the parts'.

[2] Von Mises, *Über kausale und statistische Gesetzmässigkeiten in der Physik, Erkenntnis* I, 1930, p. 207 (cf. *Naturwissenschaften* 18, 1930).

[*1] The thesis here advanced by von Mises and taken over by myself has been contested by various physicists, among them P. Jordan (see *Anschauliche Quantentheorie*, 1936, p. 282, where Jordan uses as argument against my thesis the fact that certain forms of the ergodic hypothesis have recently been proved). But in the form that *probabilistic conclusions need probabilistic premises*—for example, measure-theoretical premises into which certain equiprobabilistic assumptions enter—my thesis seems to me supported rather than invalidated by Jordan's examples. Another critic of this thesis was Albert Einstein who attacked it in the last paragraph of an interesting letter which is here reprinted in appendix *xii. I believe that, at that time, Einstein had in mind a subjective interpretation of probability, and a principle of indifference (which looks in the subjective theory as if it were not an assumption about equiprobabilities). Much later Einstein adopted, at least tentatively, a frequency interpretation (of the quantum theory).

chance hypothesis, leading in each case to different definitions of the reference sequences and the primary properties for which equal distribution is assumed.

The following example may perhaps illustrate the fact that frequency assumptions are indispensable even when we may be inclined to do without them.

Imagine a waterfall. We may discern some odd kind of regularity: the size of the currents composing the fall varies; and from time to time a splash is thrown off from the main stream; yet throughout all such variations a certain regularity is apparent which strongly suggests a statistical effect. Disregarding some unsolved problems of hydrodynamics (concerning the formation of vortices, etc.) we can, in principle, predict the path of any volume of water—say a group of molecules—with any desired degree of precision, if sufficiently precise initial conditions are given. Thus we may assume that it would be possible to foretell of any molecule, far above the waterfall, at which point it will pass over the edge, where it will reach bottom, etc. In this way the path of any number of particles may, in principle, be calculated; and given sufficient initial conditions we should be able, in principle, to deduce any one of the individual statistical fluctuations of the waterfall. But only this or that *individual* fluctuation could be so obtained, not the recurring statistical regularities we have described, still less the general statistical distribution as such. In order to explain these we need statistical estimates—at least the assumption that certain initial conditions will again and again recur for many different groups of particles (which amounts to a universal statement). We obtain a statistical result if, and only if, we make such specific statistical assumptions—for example, assumptions concerning the frequency distribution of recurring initial conditions.

71. *Formally Singular Probability Statements.*

I call a probability statement 'formally singular' when it ascribes a probability to a single occurrence, or to a single element of a certain class of occurrences;[*1] for example, 'the probability of throwing

---

[*1] The term *'formalistic'* in the German text was intended to convey the idea of a statement which is singular in form (or 'formally singular') although its intended meaning is in fact defined by statistical statements.

five with the next throw of this die is 1/6' or 'the probability of throwing five with any single throw (of this die) is 1/6'. From the standpoint of the frequency theory such statements are as a rule regarded as not quite correct in their formulation, since probabilities cannot be ascribed to single occurrences, but only to infinite sequences of occurrences or events. It is easy, however, to interpret these statements as correct, by appropriately defining formally singular probabilities with the help of the concept of objective probability or relative frequency. I use '$_\alpha P_k(\beta)$' to denote the formally singular probability that a certain occurrence k has the property β, in its capacity as an element of a sequence α—in symbols:[1] $k \, \varepsilon \, \alpha$—and I then define the formally singular probability as follows:

$$_\alpha P_k(\beta) = {}_\alpha F(\beta) \qquad (k \, \varepsilon \, \alpha) \quad \text{(Definition)}$$

This can be expressed in words as: The formally singular probability that the event k has the property β—given that k is an element of the sequence α—is, by definition, equal to the probability of the property β within the reference sequence α.

This simple, almost obvious, definition proves to be surprisingly useful. It can even help us to clarify some intricate problems of modern quantum theory. (*Cf.* sections 75–76.)

As the definition shows, a formally singular probability statement would be incomplete if it did not explicitly state a reference-class. But although α is often not explicitly mentioned, we usually know in such cases which α is meant. Thus the first example given above does not specify any reference sequence α, but it is nevertheless fairly clear that it relates to all sequences of throws with true dice.

In many cases there may be several different reference sequences for an event k. In these cases it may be only too obvious that different formally singular probability statements can be made about the same event. Thus the probability that an individual man k will die within a given period of time may assume very different values according to whether we regard him as a member of his age-group, or of his occupational group, etc. It is not possible to lay down a general rule as to which out of several possible reference-classes should be chosen. (The narrowest reference-class may often be the most suitable, provided that it is numerous enough to allow the probability estimate

---

[1] The sign '... ε ...', called the copula, means '... is an element of the class (or sequence) ...'

to be based upon reasonable statistical extrapolation, and to be supported by a sufficient amount of corroborating evidence.)

Not a few of the so-called paradoxes of probability disappear once we realize that different probabilities may be ascribed to one and the same occurrence or event, as an element of different reference-classes. For example, it is sometimes said that the probability $_\alpha P_k(\beta)$ of an event *before its occurrence* is different from the probability of the same event after it has occurred: before, it may equal 1/6, while afterwards it can only be equal to 1 or 0. This view is, of course, quite mistaken. $_\alpha P_k(\beta)$ is always the same, both before and after the occurrence. Nothing has changed except that, on the basis of the information $k \ \varepsilon \ \beta$ (or $k \ \varepsilon \ \bar{\beta}$)—information which may be supplied to us upon observing the occurrence—we may choose a new reference-class, namely β (or $\bar{\beta}$), and then ask what is the value of $_\beta P_k(\beta)$. The value of this probability is of course 1; just as $_{\bar{\beta}} P_k(\beta) = 0$. Statements informing us about the actual outcome of single occurrences—statements which are not about some frequency but rather of the form '$k \ \varepsilon \ \varphi$'—cannot change the probability of these occurrences; they may, however, suggest to us the choice of another reference-class.

The concept of a formally singular probability statement provides a kind of bridge to the *subjective* theory, and thereby also, as will be shown in the next section, to the theory of range. For we might agree to interpret formally singular probability as 'degree of rational belief' (following Keynes)—provided we allow our 'rational beliefs' to be guided by an objective *frequency statement*. This then is the information upon which our beliefs depend. In other words, it may happen that we know nothing about an event except that it belongs to a certain reference-class in which some probability estimate has been successfully tested. This information does not enable us to predict what the property of the event in question will be; but it enables us to express all we know about it by means of a formally singular probability statement which looks like *an indefinite prediction about the particular event in question*.[*2]

[*2] At present I think that the question of the relation between the various interpretations of probability theory can be tackled in a much simpler way—by giving a formal system of axioms or postulates and proving that it is satisfied by the various interpretations. Thus I regard most of the considerations advanced in the rest of this chapter (sections 71 and 72) as being superseded. See appendix *iv, and chapters *ii, *iii, and *v of my *Postscript*. But I still agree with most of what I have written, provided my '*reference classes*' are determined by the conditions defining an experiment, so that the 'frequencies' may be considered as the result of propensities.

Thus I do not object to the subjective interpretation of probability statements about single events, *i.e.* to their interpretation as indefinite predictions—as confessions, so to speak, of our deficient knowledge about the particular event in question (concerning which, indeed, nothing follows from a frequency statement). I do not object, that is to say, so long as we clearly recognize that the *objective frequency statements are fundamental, since they alone are empirically testable.* I reject, however, any interpretation of these formally singular probability statements—these indefinite predictions—as statements about an *objective state of affairs,* other than the objective statistical state of affairs. What I have in mind is the view that a statement about the probability 1/6 in dicing is not a mere confession that we know nothing definite (subjective theory), but rather an assertion about the next throw—an assertion that its result is objectively both indeterminate and undetermined—something which as yet hangs in the balance.*³ I regard all attempts at this kind of objective interpretation (discussed at length by Jeans, among others) as mistaken. Whatever indeterministic airs these interpretations may give themselves, they all involve the metaphysical idea that not only can we deduce and test predictions, but that, in addition, nature is more or less 'determined' (or 'undetermined'); so that the success (or failure) of predictions is to be explained not by the laws from which they are deduced, but over and above this by the fact that nature is actually constituted (or not constituted) according to these laws.*⁴

72. The Theory of Range.

In section 34 I said that a statement which is falsifiable to a higher degree than another statement can be described as the one which is logically more *improbable*; and the less falsifiable statement as the one which is logically more *probable*. The logically less probable statement entails¹ the logically more probable one. Between this

---

*³ I do not now object to the view that an event may hang in the balance, and I even believe that probability theory can best be interpreted as *a theory of the propensities of events* to turn out one way or another. (See my *Postscript.*) But I should still object to the view that probability theory *must* be so interpreted. That is to say, I regard the propensity interpretation as a conjecture about the structure of the world.

*⁴ This somewhat disparaging characterization fits perfectly my own views which I now submit to discussion in the 'Metaphysical Epilogue' of my *Postscript,* under the name of 'the propensity interpretation of probability'.

¹ Usually (*cf.* section 35).

concept of *logical probability* and that of objective or formally singular *numerical probability* there are affinities. Some of the philosophers of probability (Bolzano, von Kries, Waismann) have tried to base the calculus of probability upon the concept of logical range, and thus upon a concept which (*cf.* section 37) coincides with that of logical probability; and in doing so, they also tried to work out the affinities between logical and numerical probability.

Waismann[2] has proposed to measure the degree of interrelatedness between the logical ranges of various statements (their ratios, as it were) by means of the relative frequencies corresponding to them, and thus to treat the frequencies as determining a *system of measurement for ranges*. I think it is feasible to erect a theory of probability on this foundation. Indeed we may say that this plan amounts to the same thing as correlating relative frequencies with certain 'indefinite predictions' —as we did in the foregoing section, when defining formally singular probability statements.

It must be said, however, that this method of defining probability is only practicable when a frequency theory has already been constructed. Otherwise one would have to ask how the frequencies used in defining the system of measurement were defined in their turn. If, however, a frequency theory is at our disposal already, then the introduction of the theory of range becomes really superfluous. But in spite of this objection I regard the practicability of Waismann's proposal as significant. It is satisfactory to find that a more comprehensive theory can bridge the gaps—which at first appeared unbridgeable—between the various attempts to tackle the problem, especially between the subjective and the objective interpretations. Yet Waismann's proposal calls for some slight modification. His concept of a ratio of ranges (*cf.* note 2 to section 48) not only presupposes that ranges can be compared with the help of their subclass relations (or their entailment relations); but it also presupposes, more generally, that even ranges which only partially overlap (ranges of non-comparable statements) can be made comparable. This latter assumption, however, which involves considerable difficulties, is superfluous. It is possible to show that in the cases concerned (such as cases of randomness) the comparison of subclasses and that of frequencies must lead to analogous results. This justifies the procedure of correlating

[2] Waismann, *Logische Analyse des Wahrscheinlichkeitsbegriffes* (*Erkenntnis* I, 1930, p. 128 *f.*).

frequencies to ranges in order to measure the latter. In doing so, we make the statements in question (non-comparable by the subclass method) comparable. I will indicate roughly how the procedure described might be justified.

If between two property classes γ and β the subclass relation

$$\gamma \subset \beta$$

holds, then we have:

$$(k)[Fsb(k \, \varepsilon \, \gamma) \geqslant Fsb(k \, \varepsilon \, \beta)] \qquad \text{(cf. section 33)}$$

so that the logical probability or the range of the statement $(k \, \varepsilon \, \gamma)$ must be smaller than, or equal to, that of $(k \, \varepsilon \, \beta)$. It will be equal only if there is a reference class α (which may be the universal class) with respect to which the following rule holds which may be said to have the form of a 'law of nature':

$$(x)\{[x \, \varepsilon \, (\alpha.\beta)] \rightarrow (x \, \varepsilon \, \gamma)\}.$$

If this 'law of nature' does not hold, so that we may assume randomness in this respect, then the inequality holds. But in this case we obtain, provided α is denumerable, and acceptable as a reference sequence:

$$_{\alpha}F(\gamma) < {}_{\alpha}F(\beta).$$

This means that, in the case of randomness, a comparison of ranges must lead to the same inequality as a comparison of relative frequencies. Accordingly, if we have randomness, we may correlate relative frequencies with the ranges in order to make the ranges measurable. But this is just what we did, although indirectly, in section 71, when we defined the formally singular probability statement. Indeed, from the assumptions made, we might have inferred immediately that

$$_{\alpha}P_k(\gamma) < {}_{\alpha}P_k(\beta).$$

So we have come back to our starting point, the problem of the interpretation of probability. And we now find that the conflict between objective and subjective theories, which at first seemed so obdurate, may be eliminated altogether by the somewhat obvious definition of formally singular probability.

CHAPTER IX

SOME OBSERVATIONS ON QUANTUM THEORY

OUR analysis of the problem of probability has placed instruments at our disposal which we may now put to the test, by applying them to one of the topical problems of modern science; and I will try, with their help, to analyse, and to clarify, some of the more obscure points of modern quantum theory.

My somewhat audacious attempt to tackle, by philosophical or logical methods, one of the central problems of physics, is bound to arouse the suspicion of the physicist. I admit that his scepticism is healthy and his suspicions well-founded; yet I have some hope that I may be able to overcome them. Meanwhile it is worth remembering that in every branch of science, questions may crop up which are mainly logical. It is a fact that quantum physicists have been eagerly participating in epistemological discussions. This may suggest that they themselves feel that the solution of some of the still unsolved problems in quantum theory has to be sought in the no-man's-land that lies between logic and physics.

I will begin by setting down in advance the main conclusions which will emerge from my analysis.

(1) There are some mathematical formulae in quantum theory which have been interpreted by Heisenberg in terms of his uncertainty principle; that is, as statements about ranges of uncertainty due to the limits of precision which we may attain in our measurements. These formulae, as I shall try to show, are to be interpreted as formally singular probability statements (*cf.* section 71); which means that they in their turn must be interpreted statistically. So interpreted the formulae in question assert that *certain relations hold between certain ranges of statistical 'dispersion' or 'variance' or 'scatter'*. (They will be here called 'statistical scatter relations'.)

(2) Measurements of a higher degree of precision than is permitted by the uncertainty principle are not, I shall try to show, incompatible with the system of formulae of quantum theory, or with its statistical interpretation. Thus quantum theory would not necessarily be refuted if measurements of such a degree of precision should ever become possible.

(3) The existence of limits of attainable precision which was asserted by Heisenberg would therefore not be a logical consequence deducible from the formulae of the theory. It would be, rather, a separate or an additional assumption.

(4) Moreover this additional assumption of Heisenberg's actually *contradicts*, as I shall try to show, the formulae of quantum theory if they are statistically interpreted. For not only are more precise measurements compatible with the quantum theory, but it is even possible to describe imaginary experiments which show the possibility of more exact measurements. In my view it is this contradiction which creates all those difficulties by which the admirable structure of modern quantum physics is beset; so much so that Thirring could say of quantum theory that it 'has remained an impenetrable mystery to its creators, on their own admission'.[1]

What follows here might be described, perhaps, as an inquiry into the foundations of quantum theory.[2] In this, I shall avoid all mathematical arguments and, with one single exception, all mathematical formulae. This is possible because I shall not question the correctness of the system of the mathematical formulae of quantum theory. I shall only be concerned with the logical consequences of its physical interpretation which is due to Born.

As to the controversy over 'causality', I propose to dissent from the indeterminist metaphysic so popular at present. What distinguishes it from the determinist metaphysic until recently in vogue among physicists is not so much its greater lucidity as its greater sterility.

In the interests of clarity, my criticism is often severe. It may

---

[1] H. Thirring, *Die Wandlung des Begriffssystems der Physik* (essay in *Krise und Neuaufbau in den exakten Wissenschaften, Fünf Wiener Vorträge*, by Mark, Thirring, Hahn, Nobeling, Menger; Verlag Deuticke, Wien und Leipzig, 1933, p. 30).

[2] In what follows I confine myself to discussing the interpretation of quantum physics, but I omit problems concerning wave-fields (Dirac's theory of emission and absorbtion; 'second quantization' of the Maxwell-Dirac field-equations). I mention this restriction because there are problems here, such as the interpretation of the equivalence between a quantized wave-field and a corpuscular gas, to which my arguments apply (if at all) only if they are adapted to these problems with great care.

therefore be just as well to say here that I regard the achievement of the creators of modern quantum theory as one of the greatest in the whole history of science.[*1]

73. *Heisenberg's Programme and the Uncertainty Relations.*

When he attempted to establish atomic theory on a new basis, Heisenberg started with an epistemological programme:[1] to rid the theory of 'unobservables', that is, of magnitudes inaccessible to experimental observation; to rid it, one might say, of metaphysical elements. Such unobservable magnitudes did occur in Bohr's theory, which preceded Heisenberg's own: nothing observable by experiment corresponded to the orbits of the electrons or even to the frequencies of their revolutions (for the emitted frequencies which could be observed as spectral lines could not be identified with the frequencies of the electron's revolutions). Heisenberg hoped that by eliminating these unobservable magnitudes, he might manage to cure Bohr's theory of its shortcomings.

There is a certain similarity between this situation and the one with which Einstein was confronted when trying to re-interpret the Lorentz-Fitzgerald hypothesis of contraction. This hypothesis tried to explain the negative result of the experiments of Michelson and Morley by making use of unobservable magnitudes such as the movements relative to Lorentz's immobile ether; *i.e.* of magnitudes inaccessible to experimental testing. Both in this case and in that of Bohr's theory, the theories needing reform explained certain observable natural processes; but both made use of the unsatisfactory assumption that physical events and physically defined magnitudes exist which nature succeeds in hiding from us by making them for ever inaccessible to observational tests.

Einstein showed how the unobservable events involved in Lorentz's

---

[*1] I have not changed my mind on this point, nor on the main points of my criticism. But I have changed my interpretation of quantum theory together with my interpretation of probability theory. My present views are to be found in my *Postscript* where I argue, independently of the quantum theory, in favour of *indeterminism*. Yet with the exception of section 77 (which is based upon a mistake) I still regard the present chapter as important—especially section 76.

[1] W. Heisenberg, *Zeitschrift für Physik* **33**, 1925, p. 879; in what follows I mainly refer to Heisenberg's work *Die physikalischen Prinzipien der Quantentheorie* (1930). English translation by C. Eckart and F. C. Hoyt: *The Physical Principles of the Quantum Theory*, Chicago, 1930.

theory could be eliminated. One might be inclined to say the same of Heisenberg's theory, or at least of its mathematical content. However there still seems to be room for improvement. Even from the point of view of Heisenberg's own interpretation of his theory, it does not seem that his programme has been fully carried out. Nature still succeeds in hiding from us most cunningly some of the magnitudes embodied in the theory.

This state of affairs is connected with the so-called *uncertainty principle* enunciated by Heisenberg. It may, perhaps, be explained as follows. Every physical measurement involves an exchange of energy between the object measured and the measuring apparatus (which might be the observer himself). A ray of light, for example, might be directed upon the object, and part of the dispersed light reflected by the object might be absorbed by the measuring apparatus. Any such exchange of energy will alter the state of the object which, after being measured, will be in a state different from before. Thus the measurement yields, as it were, knowledge of a state which has just been destroyed by the measuring process itself. This interference by the measuring process with the object measured can be neglected in the case of macroscopic objects, but not in the case of atomic objects; for these may be very strongly affected, for example by irradiation with light. It is thus impossible to infer from the result of the measurement the precise state of an atomic object immediately *after* it has been measured. *Therefore the measurement cannot serve as basis for predictions.* Admittedly, it is always possible to ascertain, by means of new measurements, the state of the object after the previous measurement, but the system is thereby again interfered with in an incalculable way. And admittedly, it is always possible to arrange our measurements in such a way that certain of the characteristics of the state to be measured—for example the momentum of the particle—are not disturbed. But this can only be done at the price of interfering the more severely with certain other characteristic magnitudes of the state to be measured (in this case the position of the particle). If two magnitudes are mutually correlated in this way then the theorem holds for them that they cannot simultaneously be measured with precision, although each may be separately so measured. Thus if we increase the precision of one of the two measurements—say the momentum p_x, thereby reducing the range or interval of error Δp_x—then we are bound to decrease the precision of the measurement of the position co-ordinate x, *i.e.* to expand the interval Δx. In this way, the greatest precision attainable is, according to Heisenberg, limited by the uncertainty relation,[2]

$$\Delta x \cdot \Delta p_x \geqslant \frac{h}{4\pi}.$$

Similar relations hold for the other co-ordinates. The formula tells us that the product of the two ranges of error is at least of the order of magnitude of h, where h is Planck's quantum of action. It follows from this formula that a completely precise measurement of one of the two magnitudes will have to be purchased at the price of complete indeterminacy in the other.

According to Heisenberg's uncertainty relations, every measurement of the position interferes with the measurement of the corresponding component of the momentum. Thus it is in principle impossible to predict *the path of a particle*. 'In the new mechanics, the concept "path" has no definite meaning whatever. . . .'[3]

But here the first difficulty arises. The uncertainty relations apply only to the magnitudes (characteristic of physical states) which belong to the particle after the measurement has been made. The position and momentum of an electron *up to the instant of measuring* can be ascertained in principle with unlimited precision. This follows from the very fact that it is after all possible to carry out several measuring operations in succession. Accordingly, by combining the results of (a) two measurements of position, (b) measurement of position preceded by measurement of momentum, and (c) measurement of position followed by measurement of momentum, it would be possible to calculate, with the help of the data obtained, the precise position and momentum co-ordinates for the whole period of time *between* the two measurements. (To start with, we may confine our considerations only to this period.[4]) But these precise calculations are, according to Heisenberg, useless for prediction: it is therefore impossible to test them. This is so because the calculations are valid for the path between the two experiments only if the second is the

---

[2] For the derivation of this formula *cf.* note 2 to section 75.

[3] March, *Die Grundlagen der Quantenmechanik* (1931), p. 55.

[4] I shall show in detail in section 77 and in appendix vi that the case (b) will in certain circumstances also enable us to calculate the past of the electron *before* the first measurement was taken. (The next quotation from Heisenberg seems to allude to this fact.) * I now regard this footnote, like section 77, as mistaken.

immediate successor of the first in the sense that no interference has occurred between them. Any test that might be arranged for the purpose of checking the path between the two experiments is bound to disturb it so much that our calculations of the exact path become invalid. Heisenberg says about these exact calculations: '... whether one should attribute any physical reality to the calculated past history of the electron is a pure matter of taste'.[5] By this he clearly wishes to say that such untestable calculations of paths are from the physicist's point of view without any significance. Schlick comments on this passage of Heisenberg's as follows: 'I would have expressed myself even more strongly, in complete agreement with the fundamental views of both Bohr and Heisenberg themselves, which I believe to be incontestable. If a statement concerning the position of an electron in atomic dimensions is not verifiable then we cannot attribute any sense to it; it becomes impossible to speak of the "path" of a particle between two points at which it has been observed.'[6] (Similar remarks are to be found in March[7], Weyl,[8] and others.)

Yet as we have just heard, *it is possible to calculate* such a 'senseless' or metaphysical path in terms of the new formalism. And this shows that Heisenberg has failed to carry through his programme. For this state of affairs only allows of two interpretations. The first would be that the particle has an exact position and an exact momentum (and therefore also an exact path) but that it is impossible for us to measure them both simultaneously. If this is so then nature is still bent on hiding certain physical magnitudes from our eyes; not indeed the position, nor yet the momentum, of the particle, but the combination of these two magnitudes, the *'position-cum-momentum'*, or the 'path'. This interpretation regards the uncertainty principle as a limitation of our knowledge; thus it is *subjective*. The other possible interpretation, which is an *objective* one, asserts that it is inadmissible or incorrect or metaphysical to attribute to the particle anything

---

[5] Heisenberg. *Die Physikalischen Prinzipien der Quantentheorie* (1930), p. 15. (The English translation, p. 20, puts it very well: 'is a matter of personal belief'.)

[6] Schlick, *Die Kausalität in der gegenwärtigen Physik*, *Die Naturwissenschaften* 19, 1931, p. 159.

[7] March, *op. cit. passim* (e.g. p. 1 f. and p. 57).

[8] Weyl, *Gruppentheorie und Quantenmechanik* (2nd edition, 1931), p. 68 (*cf.* the last quotation in section 75, below:'... the meaning of these concepts....'). *The paragraph referred to seems to have been omitted in the English translation, *The Theory of Groups and Quantum Mechanics*, 1931.

like a sharply defined 'position-cum-momentum' or 'path': it simply *has* no 'path', but only either an exact position combined with an inexact momentum, or an exact momentum combined with an inexact position. But if we accept this interpretation then, again, the formalism of the theory contains metaphysical elements; for a 'path' or 'position-cum-momentum' of the particle, as we have seen, is exactly calculable—for those periods of time during which it is in principle impossible to test it by observation.

It is illuminating to see how the champions of the uncertainty relation vacillate between a subjective and an objective approach. Schlick for instance writes, immediately after upholding the objective view, as we have seen: 'Of natural events themselves it is impossible to assert meaningfully any such thing as "haziness" or "inaccuracy". It is only to our own thoughts that anything of this sort can apply (more especially, if we do not know which statements . . . are true)': a remark which is obviously directed *against* that very same objective interpretation which assumes that it is not our knowledge, but the momentum of the particle, which gets 'blurred' or 'smeared', as it were, by having its position precisely measured.*1 Similar vacillations are shown by many other authors. But whether one decides in favour of the objective or the subjective view, the fact remains that Heisenberg's programme has not been carried out and that he has not succeeded in his self-imposed task of expelling all metaphysical elements from atomic theory. Nothing whatever is therefore to be gained by attempting, with Heisenberg, to fuse the two opposing interpretations by a remark such as '. . . an "objective" physics in this sense, *i.e.* a sharp division of the world into object and subject has indeed ceased to be possible'.[9] Heisenberg has not so far accomplished his self-imposed task: he has not yet purged quantum theory of its metaphysical elements.

74. *A Brief Outline of the Statistical Interpretation of Quantum Theory.*

In his derivation of the uncertainty relations, Heisenberg follows Bohr in making use of the idea that atomic processes can be just as

---

*1 The expression 'smeared' is due to Schrödinger. The problem of the objective existence or non-existence of a 'path'—whether the path is 'smeared', or whether it is merely not fully known—is, I believe, fundamental. Its importance has been enhanced by the experiment of Einstein, Podolski and Rosen, discussed in appendices *xi and *xii.

[9] Heisenberg, *Physikalische Prinzipien*, p. 49.

well represented by the 'quantum-theoretical image of a particle' as by the 'quantum-theoretical image of a wave'.

This idea is connected with the fact that modern quantum theory has advanced along two different roads. Heisenberg started from the classical particle theory of the electron which he re-interpreted according to quantum theory; whilst Schrödinger started from the (likewise 'classical') wave-theory of de Broglie: he co-ordinated with each electron a 'wave-packet', *i.e.* a group of oscillations which by interference strengthen each other within a small region and extinguish each other outside it. Schrödinger later showed that his wave-mechanics led to results mathematically equivalent to those of Heisenberg's particle mechanics.

The paradox of the equivalence of two so fundamentally different images as those of particle and wave was resolved by Born's statistical interpretation of the two theories. He showed that the wave theory too can be taken as a particle theory; for Schrödinger's wave equation can be interpreted in such a way that it gives us the *probability of finding the particle* within any given region of space. (The probability is determined by the square of the amplitude of the wave; it is great within the wave-packet where the waves reinforce each other, and vanishes outside it.)

That the quantum theory should be interpreted *statistically* was suggested by various aspects of the problem situation. Its most important task—the deduction of the atomic spectra—had to be regarded as a *statistical* task ever since Einstein's hypothesis of photons (or light-quanta). For this hypothesis interpreted the observed light-effects as mass-phenomena, as due to the incidence of many photons. 'The experimental methods of atomic physics have, . . . under the guidance of experience, become concerned, exclusively, with statistical questions. Quantum mechanics, which furnishes the systematic theory of the observed regularities, corresponds in every way to the present state of experimental physics; for it confines itself, from the outset, to statistical questions and to statistical answers.'[1]

It is only in its application to problems of atomic physics that quantum theory obtains results which differ from those of classical mechanics. In its application to macroscopic processes its formulae yield with close approximation those of classical mechanics. 'According to quantum theory, the laws of classical mechanics are

[1] Born-Jordan, *Elementare Quantenmechanik* (1930), p. 322 *f.*

valid if they are regarded as statements about the relations between statistical averages', says March.[2] In other words, the classical formulae can be deduced as macro-laws.

In some expositions the attempt is made to *explain* the statistical interpretation of the quantum theory by the fact that the precision attainable in measuring physical magnitudes is limited by Heisenberg's uncertainty relations. It is argued that, *owing to this uncertainty* of measurements in any atomic experiments, '. . . the result will not in general be determinate, *i.e.* if the experiment is repeated several times under identical conditions several different results may be obtained. If the experiment is repeated a large number of times it will be found that each particular result will be obtained in a definite fraction of the total number of times, so that one can say there is a definite probability of its being obtained any time the experiment is performed' (Dirac).[3] March too writes with reference to the uncertainty relation: 'Between the present and the future there hold . . . only probability relations; from which it becomes clear that the character of the new mechanics must be that of a statistical theory.'[4]

I do not think that this analysis of the relations between the uncertainty formulae and the statistical interpretation of the quantum theory is acceptable. It seems to me that the logical relation is just the other way round. For we can derive the uncertainty formulae from Schrödinger's wave equation (which is to be interpreted statistically), but not this latter from the uncertainty formulae. If we are to take due account of these relations of derivability, then the interpretation of the uncertainty formulae will have to be revised.

75. *A Statistical Re-Interpretation of the Uncertainty Formulae.*

Since Heisenberg it is accepted as an established fact that any simultaneous measurements of position and momentum with a precision exceeding that permitted by his uncertainty relations would contradict

---

[2] March, *Die Grundlagen der Quantenmechanik* (1931), p. 170.

[3] Dirac, *Quantum Mechanics* (1930), p. 10. *(From the 1st edition.) A parallel passage, slightly more emphatic, occurs on p. 14 of the 3rd edition. '. . . in general the result will not be determinate, *i.e.*, if the experiment is repeated several times under identical conditions several different results may be obtained. It is a law of nature, though, that if the experiment is repeated a large number of times, each particular result will be obtained in a definite fraction of the total number of times, so that there is a definite *probability* of its being obtained.'

[4] March, *Die Grundlagen der Quantenmechanik*, p. 3.

quantum theory. The 'prohibition' of exact measurements, it is believed, can be logically derived from quantum theory, or from wave mechanics. On this view, the theory would have to be regarded as falsified if experiments resulting in measurements of 'forbidden accuracy' could be carried out.[1]

I believe this view to be false. Admittedly, it is true that Heisenberg's formulae ($\Delta x \Delta p_x \geqslant \dfrac{h}{4\pi}$ etc.) result as logical conclusions from the theory;[2] but the *interpretation* of these formulae as rules limiting attainable precision of measurement, in Heisenberg's sense, does not follow from the theory. Therefore measurements more exact than those permissible according to Heisenberg cannot logically contradict the quantum theory, or wave mechanics. I shall accordingly draw a sharp distinction between the *formulae*, which I shall call the 'Heisenberg formulae' for short, and their *interpretation*—also due to Heisenberg—as uncertainty relations; that is, as statements imposing *limitations upon the attainable precision of measurement*.

When working out the mathematical deduction of the Heisenberg formulae one has to employ the wave-equation or some equivalent assumption, *i.e.* an assumption which can be *statistically interpreted* (as we saw in the preceding section). But if this interpretation is adopted, then the description of a single particle by a wave-packet is undoubtedly nothing else but *a formally singular probability statement* (*cf.* section 71). The wave-amplitude determines, as we have seen, the probability of detecting the particle at a certain place; and it is just this kind of probability statement—the kind that refers to a single particle (or event)—which we have called 'formally singular'. If one accepts the statistical interpretation of quantum theory, then one is bound to interpret those statements—such as the Heisenberg formulae—which can be derived from the formally singular probability statements of the theory, as probability statements in their turn,

---

[1] I refrain from criticizing here the very widespread and rather naïve view that Heisenberg's arguments furnish conclusive proof of the impossibility of all such measurements; *cf.* for instance, Jeans, *The New Background of Science*, 1933, p. 233; 2nd edition, 1934, p. 237: 'Science has found no way out of this dilemma. On the contrary, it has proved that there is no way out.' It is clear, of course, that no such proof can ever be furnished, and that the principle of uncertainty could, at best, be deducible from the hypotheses of quantum and wave mechanics and could be empirically refuted together with them. In a question like this, we may easily be misled by plausible assertions such as the one made by Jeans.

[2] Weyl supplies a strict logical deduction: *Gruppentheorie und Quantenmechanik*, 2nd edition 1931, pp. 68 and 345; English translation, pp. 77 and 393 *f.*

and again as formally singular, if they apply to a single particle. They too must therefore be interpreted, ultimately, as *statistical assertions*.

As against the subjective interpretation, 'The more precisely we measure the position of a particle the less we can know about its momentum', I propose that an objective and statistical interpretation of the uncertainty relations should be accepted as being the fundamental one; it may be phrased somewhat as follows. Given an aggregate of particles and a selection (in the sense of a physical separation) of those which, at a certain instant, and with a certain given degree of precision, have a certain position x, we shall find that their momenta p_x will show random scattering; and the range of scatter, Δp_x, will thereby be the greater, the smaller we have made Δx, *i.e.* the range of scatter or imprecision allowed to the positions. And *vice versa*: if we select, or separate, those particles whose momenta p_x all fall within a prescribed range Δp_x, then we shall find that their positions will scatter in a random manner, within a range Δx which will be the greater, the smaller we have made Δp_x, *i.e.* the range of scatter or imprecision allowed to the momenta. And finally: if we try to select those particles which have both the properties Δx and Δp_x, then we can physically carry out such a selection—that is, physically separate the particles—only if both ranges are made sufficiently great to satisfy the equation $\Delta x \cdot \Delta p_x \geqslant \frac{h}{4\pi}$. This objective interpretation of the Heisenberg formulae takes them as asserting that certain relations hold between certain ranges of scatter; and I shall refer to them, if they are interpreted in this way, as the '*statistical scatter relations*'.[*1]

In my statistical interpretation I have so far made no mention of *measurement*; I have referred only to *physical selection*.[3] It is now necessary to clarify the relation between these two concepts.

I speak of physical selection or physical separation if, for example, we screen off, from a stream of particles, all except those which pass

---

[*1] I still uphold the objective interpretation here explained, with one important change, however. Where, in this paragraph, I speak of 'an aggregate of particles' I should now speak of 'an aggregate—or of a sequence—of repetitions of an experiment undertaken with *one* particle (or *one* system of particles)'. Similarly, in the following paragraphs; for example, the 'ray' of particles should be re-interpreted as consisting of repeated experiments with (one or a few) particles—selected by screening off, or by shutting out, particles which are not wanted.

[3] Weyl too, among others, writes of 'selections'; see *Gruppentheorie und Quantenmechanik*, p. 67 *ff.*, English translation p. 76 *ff.*; but unlike me he does not contrast measurement and selection.

through a narrow aperture Δx, that is, through a range Δx allowed to their position. And I shall say of the particles belonging to the ray thus isolated that they have been selected physically, or technically, according to their property Δx. It is only this process, or its result, the physically or technically isolated ray of particles, which I describe as a 'physical selection'—in contradistinction to a merely 'mental' or 'imagined' selection, such as we make when speaking of the class of all those particles which have passed, or will pass, through the range Δp; that is, of a class within a wider class of particles from which it has not been physically screened off.

Now every physical selection can of course be regarded as a *measurement*, and can actually be used as such.[4] If, say, a ray of particles is selected by screening off or shutting out all those which do not pass through a certain positional range ('place-selection') and if later the momentum of one of these particles is measured, then we can regard the place-selection as a measurement of position, because we learn from it that the particle has passed through a certain position (though *when* it was there we may sometimes not know, or may only learn from another measurement). On the other hand, we must not regard every measurement as a physical selection. Imagine, for example, a monochromatic ray of electrons flying in the direction x. By using a Geiger counter, we can then record those electrons that arrive at a certain position. By the time-intervals between the impacts upon the counter, we may also measure spatial intervals; that is to say, we measure their positions in the x direction up to the moment of impact. But in taking these measurements we do not make a physical selection of the particles according to their positions in the x direction. (And indeed, these measurements will generally yield a completely random distribution of the positions in the x direction.)

Thus in their physical application, our statistical scatter relations come to this. If one tries, by whatever physical means, to obtain *as homogeneous an aggregate of particles as possible*, then this attempt will encounter a definite barrier in these scatter-relations. For example, we can obtain by means of physical selection a plane monochromatic ray—say, a ray of electrons of equal momentum. But if we attempt to make this aggregate of electrons still more homogeneous—perhaps

---

[4] By a 'measurement' I mean, in conformity with linguistic usage accepted by physicists, not only direct measuring operations but also measurements obtained indirectly by calculation (in physics these are practically the only measurements that occur).

by screening off part of it—so as to obtain electrons which not only have the same momentum but have also passed through some narrow slit determining a positional range Δx, then we are bound to fail. We fail because any selection according to the position of the particles amounts to an interference with the system which will result in increased scattering of the momentum components p_x, so that the scattering will increase (in accordance with the law expressed by the Heisenberg formula) with the narrowing of the slit. And conversely: if we are given a ray selected according to position by being passed through a slit, and if we try to make it 'parallel' (or 'plane') and monochromatic, then we have to destroy the selection according to position since we cannot avoid increasing the width of the ray. (In the ideal case—for example, if the p_x components of the particles are all to become equal to o—the width would have to become infinite.) If the homogeneity of a selection has been increased as far as possible (*i.e.* as far as the Heisenberg formulae permit, so that the sign of equality in these formulae becomes valid) then this selection may be called *a pure case*.[5]

Using this terminology, we can formulate the statistical scatter relations thus: There is no aggregate of particles more homogeneous than a pure case.[*2]

It has not till now been taken sufficiently into account that to the mathematical derivation of the Heisenberg formulae from the fundamental equations of quantum theory there must correspond, precisely, a derivation of the *interpretation* of the Heisenberg formulae from the *interpretation* of these fundamental equations. March for instance has described the situation just the other way round (as indicated in the previous section): the statistical interpretation of quantum theory appears in his presentation as a consequence of the Heisenberg limitation upon attainable precision. Weyl on the other hand gives

[5] The term is due to Weyl (*Zeitschrift für Physik* **46**, 1927, p. 1) and J. von Neumann (*Göttinger Nachrichten*, 1927, p. 245). If, following Weyl (*Gruppentheorie und Quantenmechanik*, p. 70; English translation p. 79; *cf.* also Born-Jordan, *Elementare Quantenmechanik*, p. 315), we characterize the pure case as one '. . . which it is impossible to produce by a combination of two statistical collections different from it', then pure cases satisfying this description need not be pure momentum or place selections. They could be produced, for example, if a place-selection were effected with some chosen degree of precision, and the momentum with the greatest precision still attainable.

[*2] In the sense of note *1, this should, of course, be re-formulated: 'There is no experimental arrangement capable of producing an aggregate or sequence of experiments with results more homogeneous than a pure case.'

a strict derivation of the Heisenberg formulae from the wave equation —an equation which he interprets in statistical terms. Yet he interprets the Heisenberg formulae—which he has just derived from a statistically interpreted premise—as limitations upon attainable precision. And he does so in spite of the fact that he notices that this interpretation of the formulae runs counter in some respects to the statistical interpretation of Born. For according to Weyl, Born's interpretation is subject to 'a correction' in the light of the uncertainty relations. 'It is not merely the case that position and velocity of a particle are just subject to statistical laws, while being precisely determined in every single case. Rather, the very meaning of these concepts depends on the measurements needed to ascertain them; and an exact measurement of the position robs us of the possibility of ascertaining the velocity.'[6]

The conflict perceived by Weyl between Born's statistical interpretation of quantum theory and Heisenberg's limitations upon attainable precision does indeed exist; but it is sharper than Weyl thinks. Not only is it impossible to derive the limitations of attainable precision from the statistically interpreted wave-equation, but the fact (which I have still to demonstrate) that neither the possible experiments nor the actual experimental results agree with Heisenberg's interpretation can be regarded as a decisive argument, as a kind of *experimentum crucis*, in favour of the statistical interpretation of the quantum theory.

76. An Attempt to Eliminate Metaphysical Elements by Inverting Heisenberg's Programme; with Applications.

If we start from the assumption that the formulae which are peculiar to quantum theory are probability hypotheses, and thus statistical statements, then it is difficult to see how prohibitions of single events could be deduced from a statistical theory of this character (except perhaps in the cases of probabilities equal to one or to zero). The belief that single measurements can contradict the formulae of quantum physics seems logically untenable; just as untenable as the belief that a contradiction might one day be detected between a formally singular probability statement $_\alpha P_k(\beta) = p$ (say, 'the probability that the throw k will be a five equals $1/6$') and one of

[6] Weyl, *Gruppentheorie und Quantenmechanik*, p. 68. *The paragraph here cited seems to be omitted in the English translation.

the following two statements: $k \, \varepsilon \, \beta$ ('the throw is in fact a five') or $k \, \varepsilon \, \bar{\beta}$ ('the throw is in fact not a five').

These simple considerations provide us with the means to refute any of the alleged proofs which have been designed to show that exact measurements of position and momentum would contradict the quantum theory; or which have been designed, perhaps, to show that the mere assumption that any such measurements are physically possible must lead to contradictions within the theory. For any such proof must make use of quantum-theoretical considerations applied to *single* particles; which means that it has to make use of formally singular probability statements, and further, that it must be possible to translate the proof—word for word, as it were—into the statistical language. If we do this then we find that there is no contradiction between the single measurements which are assumed to be precise, and the quantum theory in its statistical interpretation. There is only an apparent contradiction between these precise measurements and certain formally singular probability statements of the theory. (In appendix v an example of this type of proof will be examined.)

But whilst it is wrong to say that the quantum theory *rules out* exact measurements, it is yet correct to say that from formulae which are peculiar to the quantum theory—provided they are interpreted statistically—*no precise singular predictions can be derived*. (I do not count either the law of conservation of energy nor the law of conservation of momentum among the formulae peculiar to quantum theory.)

This is so because in view of the scatter relations, we must fail, more especially, to produce precise initial conditions, by experimentally manipulating the system (*i.e.* by what we have called physical selection). Now it is indeed true that the normal technique of the experimenter is to *produce* or to *construct* initial conditions; and this allows us to derive from our statistical scatter relations the theorem—which, however, only holds for this '*constructive*' experimental technique—that from quantum theory we cannot obtain any singular predictions, but only frequency predictions.[1]

This theorem sums up my attitude to all those imaginary experiments discussed by Heisenberg (who here largely follows Bohr) with the object of proving that it is impossible to make measurements of a precision forbidden by his uncertainty principle. The point is in

[1] The term 'constructive experimental technique' is used by Weyl, *Gruppentherie und Quantenmechanik*, p. 67; English translation p. 76.

every case the same: the statistical scatter makes it impossible to *predict* what the path of the particle will be after the measuring operation.

It might well seem that not much has been gained by our re-interpretation of the uncertainty principle. For even Heisenberg asserts in the main (as I have tried to show) no more than that our *predictions* are subject to this principle; and as in this matter I agree with him up to a point, it might be thought that I am only quarrelling about words rather than debating any substantial issue. But this would hardly do justice to my argument. Indeed I think that Heisenberg's view and mine are diametrically opposed. This will be shown at length in my next section. Meanwhile I shall attempt to resolve the typical difficulties inherent in Heisenberg's interpretation; and I shall try to make clear how, and why, these difficulties arise.

First we must examine the difficulty over which, as we have seen, Heisenberg's programme comes to grief. It is the occurrence, in the formalism, of precise statements of position–cum–momentum; or in other words, of exact calculations of a path (*cf.* section 73) whose physical reality Heisenberg is obliged to leave in doubt, while others, such as Schlick, deny it outright. But the experiments in question, (a), (b), and (c)—see section 73—can all be interpreted in statistical terms. For example, combination (c), *i.e.* a measurement of position followed by a measurement of momentum, may be realized by an experiment such as the following. We select a ray according to position with the help of a diaphragm with a narrow slit (position-measurement). We then measure the momentum of those particles which were travelling from the slit in a definite direction. (This second measurement will of course produce a new scatter of positions.) The two experiments together will then determine precisely the path of all those particles which belong to the second selection, in so far as this path lies between the two measurements: both position and momentum between the two measurements can be precisely calculated.

Now these measurements and calculations, which correspond precisely to the elements regarded as superfluous in Heisenberg's interpretation, are on my interpretation of the theory anything but superfluous. Admittedly, they do not serve as initial conditions or as a basis for the derivation of predictions; but they are indispensable nevertheless:

they are needed for testing our predictions, which are *statistical predictions*. For what our statistical scatter relations assert is that the momenta must scatter when positions are more exactly determined, and *vice versa*. This is a prediction which would not be testable, or falsifiable, if we were not in a position to measure and calculate, with the help of experiments of the kind described, the various scattered momenta which occur immediately after any selection according to position has been made.*[1]

The statistically interpreted theory, therefore, not only does not rule out the possibility of exact single measurements, but would be untestable, and thus 'metaphysical', if these were impossible. So the fulfilment of Heisenberg's programme, the elimination of metaphysical elements, is here achieved, but by a method the very opposite of his. For while he tried to exclude magnitudes which he regarded as inadmissible (though without entirely succeeding), I invert the attempt, so to speak, by showing that the formalism which contains these magnitudes is correct just because *the magnitudes are not metaphysical*. Once we have given up the dogma embodied in Heisenberg's limitation upon attainable precision, there is no longer any reason why we should doubt the physical significance of these magnitudes. The scatter relations are frequency predictions about paths; and therefore these paths must be measureable—in precisely the same way as, say, throws of five must be empirically ascertainable—if we are to be able to test our frequency predictions about these paths, or about these throws.

*[1] I consider this paragraph (and also the first sentence of the next paragraph) as one of the most important in this discussion, and as one with which I can still agree completely. Since misunderstandings continue, I will explain the matter more fully. The *scatter relations* assert that, if we arrange for a sharp selection of the position (by a slit in a screen, say), the momenta will scatter as a consequence. (Rather than becoming 'indeterminate', the single momenta become 'unpredictable' in a sense which allows us to predict that they will scatter.) This is a prediction which we must test by *measuring the single momenta*, so as to determine their statistical distribution. These measurements of the single momenta (which will lead to a new scatter—but this we need not discuss) will give in each single case results as precise as we like, and at any rate very much more precise than Δp, i.e. the mean width of the region of the scatter. Now these measurements of the various single momenta allow us to calculate their values back to the place where the position was selected, and measured, by the slit. And this 'calculation of the past history' of the particle (*cf.* note 3 to section 73) is essential; without it, we could not assert that we were measuring the momenta immediately after the positions were selected; and thus we could not assert that we were testing the scatter relations—which we do in fact with any experiment which shows an increase of scatter as a consequence of a decrease of the width of a slit. So it is only the precision of the *prediction* which becomes 'blurred' or 'smeared' in consequence of the scatter relations, but never the precision of a *measurement*.

Heisenberg's rejection of the concept of path, and his talk of 'non-observable magnitudes', clearly show the influence of philosophical and especially of positivistic ideas. Under the same influence, March writes: 'One may say perhaps without fear of being misunderstood . . . that for the physicist a body has reality only in the instant in which he observes it. Naturally nobody is so mad as to assert that a body ceases to exist the moment we turn our backs to it; but it does cease, in that moment, to be an object of inquiry for the physicist, because there exists no possibility of saying anything about it which is based on experiment.'[2] In other words, the hypothesis that a body moves in this or that path whilst it is not being observed is *non-verifiable*. This, of course, is obvious, but uninteresting. What is important, however, is that this and similar hypotheses are *falsifiable*: on the basis of the hypothesis that it moves along a certain path we are able to predict that the body will be observable in this or that position; and this is a prediction which can be refuted. That the quantum theory does *not* exclude this kind of procedure will be seen in the next section. But in fact what we have said here is quite sufficient;[*2] for it disposes of all the difficulties connected with the 'meaninglessness' of the concept of path. How much this helps to clear the air will best be realized if we remember the drastic conclusions which were drawn from the alleged failure of the concept of path. Schlick formulates them thus: 'Perhaps the most concise way of describing the situation under review is to say (as the most eminent investigators of quantum problems do) that the validity of the ordinary spatio-temporal concepts is confined to the sphere of the macroscopically observable, and that they are not applicable to atomic dimensions'.[3] Here Schlick is probably alluding to Bohr who writes: 'Therefore one may assume that where the general problem of quantum theory is concerned, it is not a mere question of a change of mechanical and electro-dynamic theories, a change which may be

---

[2] March, *Die Grundlagen der Quantenmechanik*, p. 1. * Reichenbach's position is similar; it is criticized in my *Postscript*, section *13.

[*2] The beginning of this sentence (from 'But in fact' to 'sufficient') was not in the original text. I have inserted it because I do no longer believe in the argument of 'the next section' (77), referred to in the previous sentence, and because what follows is in fact completely independent of the next section: it is based upon the argument just given according to which calculations of the past path of the electron are needed for testing the statistical predictions of the theory, so that these calculations are far from 'meaningless'.

[3] Schlick, *Die Kausalität in der gegenwärtigen Physik*, Die Naturwissenschaften **19** (1931), p. 159.

described in terms of ordinary physical concepts, but the deep-seated failure of our spatio-temporal images which till now have been used in the description of natural phenomena.'[4] Heisenberg adopted this idea of Bohr's, namely the renunciation of spatio-temporal descriptions, as the basis of his programme of research. His success seemed to show that this was a fruitful renunciation. But in fact, the programme was never carried through. The frequent and unavoidable, if surreptitious, use of spatio-temporal concepts now seems justifiable in the light of our analysis. For this has shown that the statistical scatter relations are statements about the scatter of position-cum-momentum, and therefore statements about paths.

Now that we have shown that the uncertainty relations are formally singular probability statements, we can also unravel the tangled web of their objective and subjective interpretations. We learned in section 71 that every formally singular probability statement can also be interpreted subjectively, as an indefinite prediction, a statement concerning the uncertainty of our knowledge. We have also seen under what assumptions the justified and necessary attempt to interpret a statement of this kind objectively is bound to fail. It is bound to fail if one tries to substitute for the statistical objective interpretation a singular objective interpretation, by attributing the uncertainty directly to the single event.[*2] Yet if one interprets the Heisenberg formulae (directly) in a subjective sense, then the position of physics as an objective science is imperilled; for to be consistent one would also have to interpret Schrödinger's probability waves subjectively. This conclusion is drawn by Jeans[5] who says: 'In brief, the particle picture tells us that our knowledge of an electron is indeterminate; the wave picture that the electron itself is indeterminate, regardless of whether experiments are performed upon it or not. Yet the content of the uncertainty principle must be exactly the same in the two cases. There is only one way of making it so: we must suppose that the wave picture provides a representation not

---

[4] Bohr, *Die Naturwissenschaften* 14 (1926), p. 1.

[*2] This is one of the points on which I have since changed my mind. *Cf.* my *Postscript*, chapter *v. But my main argument in favour of an objective interpretation remains unaffected. According to my present view, Schrödinger's theory may and should be interpreted not only as objective and singular but, at the same time, as probabilistic.

[5] Jeans, *The New Background of Science* (1933, p. 236; 2nd edition 1934, p. 240). In Jean's text, a new paragraph begins with the second sentence, *i.e.* with the words, 'Yet the content'. For the quotation that follows at the end of this paragraph, see *op. cit.*, p. 237 (2nd edition, p. 241).

of objective nature, but only of our knowledge of nature. . . .'
Schrödinger's waves are thus for Jeans *subjective probability waves*,
waves of our knowledge. And with this, the whole subjectivist
probability theory invades the realm of physics. The arguments I
have rejected—the use of Bernoulli's theorem as a 'bridge' from
nescience to statistical knowledge, and similar arguments (*cf.* section
62)—become inescapable. Jeans formulates the subjectivist attitude of
modern physics as follows: 'Heisenberg attacked the enigma of the
physical universe by giving up the main enigma—the nature of the
objective universe—as insoluble, and concentrating on the minor
puzzle of co-ordinating our observations of the universe. Thus it is
not surprising that the wave picture which finally emerged should
prove to be concerned solely with our knowledge of the universe
as obtained through our observations.'

Such conclusions will no doubt appear highly acceptable to the
positivists. Yet my own views concerning objectivity remain un-
touched. The statistical statements of quantum theory must be
inter-subjectively testable in the same way as any other statements of
physics. And my simple analysis preserves not only the possibility of
spatio-temporal descriptions, but also the objective character of physics.

It is interesting that there exists a counterpart to this subjective
interpretation of the Schrödinger waves: a non-statistical and thus a
directly (*i.e.* singular) objective interpretation. Schrödinger himself
in his famous *Collected Papers on Wave-Mechanics* has proposed some
such interpretation of his wave equation (which as we have seen is a
formally singular probability statement). He tried to identify the
particle immediately with the wave-packet itself. But his attempt
led straight to those difficulties which are so characteristic of this
kind of interpretation: I mean the ascription of uncertainty to the
physical objects themselves (objectivized uncertainties). Schrödinger
was forced to assume that the charge of the electron was 'blurred' or
'smeared' in space (with a charge density determined by the wave
amplitude); an assumption which turned out to be incompatible with
the atomic structure of electricity.[6] Born's statistical interpretation solved
the problem; but the logical connection between the statistical and the
non-statistical interpretations remained obscure. Thus it happened that
the peculiar character of other formally singular probability state-

[6] *Cf.* for instance Weyl, *Gruppentheorie und Quantenmechanik*, p. 193; English
translation p. 216 *f.*

ments—such as the uncertainty relations—remained unrecognized and that they could continue to undermine the physical basis of the theory.

I may conclude perhaps with an application of what has been said in this section to an imaginary experiment proposed by Einstein[7] and called by Jeans[8] 'one of the most difficult parts of the new quantum theory'; though I think that our interpretation makes it perfectly clear, if not trivial.[*4]

Imagine a semi-translucent mirror, *i.e.* a mirror which reflects part of the light, and lets part of it through. The formally singular probability that one given photon (or light quantum) passes through the mirror, $_\alpha P_k(\beta)$, may be taken to be equal to the probability that it will be reflected; we therefore have

$$_\alpha P_k(\beta) = {_\alpha P_k(\bar{\beta})} = \tfrac{1}{2}.$$

This probability estimate, as we know, is defined by objective statistical probabilities; that is to say, it is equivalent to the hypothesis that one half of a given class α of light quanta will pass through the mirror whilst the other half will be reflected. Now let a photon k fall upon the mirror; and let it next be experimentally ascertained that this photon has been reflected: then the probabilities seem to change suddenly, as it were, and discontinuously. It is as though before the experiment they had both been equal to $\tfrac{1}{2}$, while after the fact of the reflection became known, they had suddenly turned into 0 and to 1, respectively. It is plain that this example is really the same as that given in section 71.[*5] And it hardly helps to clarify the situation if this experiment is described, as by Heisenberg,[9] in such terms as

[7] *Cf.* Heisenberg, *Physikalische Prinzipien*, p. 29 (English translation by C. Eckart and F. C. Hoyt: *The Physical Principles of the Quantum Theory*, Chicago, 1930, p. 39).

[8] Jeans, *The New Background of Science* (1933, p. 242; 2nd edition, p. 246).

[*4] The problem following here has since become famous under the name 'The problem of the (discontinuous) *reduction of the wave packet*'. Some leading physicists told me in 1934 that they agreed with my trivial solution, yet the problem still plays a most bewildering role in the discussion of the quantum theory, after more than twenty years. I have discussed it again at length in sections *100 and *115 of the *Postscript*.

[*5] That is to say, the probabilities 'change' only in so far as α is replaced by $\bar{\beta}$. Thus $_\alpha P(\beta)$ remains unchanged $\tfrac{1}{2}$; but $_{\bar{\beta}} P(\beta)$, of course, equals 0, just as $_{\bar{\beta}} P(\bar{\beta})$ equals 1.

[9] Heisenberg, *Physikalische Prinzipien*, p. 29 (English translation: *The Physical Principles of the Quantum Theory*, Chicago, 1930, p. 39). Von Laue, on the other hand, in *Korpuskular- und Wellentheorie, Handbuch d. Radiologie* **6** (2nd edition, p. 79 of the offprint) says quite rightly: 'But perhaps is it altogether quite mistaken to correlate a wave with one single corpuscle. If we assume that the wave is, as a matter of principle, related to an aggregate of equal but mutually independent bodies, the paradoxical conclusion vanishes.' * Einstein adopted in one of his last papers a similar interpretation: *cf.* note *1 to the next section.

the following: 'By the experiment [*i.e.* the measurement by which we find the reflected photon], a kind of physical action (a reduction of wave packets) is exerted from the place where the reflected half of the wave packet is found upon another place—as distant as we choose—where the other half of the packet just happens to be'; a description to which he adds: 'this physical action is one which spreads with super-luminal velocity.' This is unhelpful since our original probabilities, $_\alpha P_k(\beta)$ and $_\alpha P_k(\bar\beta)$, remain equal to $\frac{1}{2}$. All that has happened is the choice of a new reference class—β or $\bar\beta$, instead of α—a choice strongly suggested to us by the result of the experiment, *i.e.* by the information $k \in \beta$ or $k \in \bar\beta$, respectively. Saying of the logical consequences of this choice (or, perhaps, of the logical consequences of this information) that they 'spread with super-luminal velocity' is about as helpful as saying that twice two turns with super-luminal velocity into four. A further remark of Heisenberg's, to the effect that this kind of propagation of a physical action cannot be used to transmit signals, though true, hardly improves matters.

The fate of this imaginary experiment is a reminder of the urgent need to distinguish and to define the statistical and the formally singular probability concepts. It also shows that the problem of interpretation to which quantum theory has given rise can only be approached by way of a logical analysis of the interpretation of probability statements.

77. *Decisive Experiments.*

I have now carried out the first two parts of my programme outlined in the introduction preceding section 73. I have shown (1) that the Heisenberg formulae can be statistically interpreted, and therefore (2) that their interpretation as limitations upon attainable precision does not follow logically from the quantum theory, which therefore could not be contradicted merely by our attaining a higher degree of precision in our measurements.[*1]

[*1] Point (3) of my programme has, in fact, been covered also. While I believe that, up to this point, my criticism is fully justified (even though I should now favour a different approach to the problem of interpretation, as explained in my *Postscript*), I do not wish to uphold the views expressed in the present section, 77. I no longer think that my imaginary experiment is valid. But neither do I think any longer that it is as 'decisive' as I believed when I first described it. Moreover, it can be largely replaced by the famous imaginary experiment of Einstein, Podolski, and Rosen. (*Cf.* footnote *4.) Thus while I am ready to admit the force of some of the criticism which has been levelled against the present section, I do not think that the arguments of the

'So far, so good,' someone might retort. 'I won't deny that it may be possible to view quantum mechanics in this way. But it still does not seem to me that the real physical core of Heisenberg's theory, the impossibility of making exact singular *predictions*, has even been touched by your arguments'.

If asked to elaborate his thesis by means of a physical example, my opponent might proceed as follows: 'Imagine a beam of electrons, like one in a cathode tube. Assume the direction of this beam to be the *x*-direction. We can obtain various physical selections from this beam. For example, we may select or separate a group of electrons according to their position in the *x*-direction (*i.e.* according to their *x*-co-ordinates at a certain instant); this could be done, perhaps, by means of a shutter which we open for a very short time. In this way we should obtain a group of electrons whose extension in the *x*-direction is very small. According to the scatter relations, the momenta of the various electrons of this group would differ widely in the *x*-direction (and therefore also their energies). As you rightly stated, we can test such statements about scattering. We can do this by measuring the momenta or the energies of single electrons; and as we know the position, we shall thus obtain both position and momentum. A measurement of this kind may be carried out, for example, by letting the electrons impinge upon a plate whose atoms they would excite: we shall then find, among other things, some excited atoms whose excitation requires energy in excess of the average energy of these electrons. Thus I admit that you were quite right in stressing that such measurements are both possible and significant. But—and now comes my objection—in making any such measurement we must disturb the system we are examining, *i.e.* either the single electrons, or if we measure many (as in our example), the whole electron beam. Admittedly, the theory would not be logically contradicted if we could know the momenta of the various electrons of the group before disturbing it (so long, of course, as this would not enable us to use our knowledge so as to effect a forbidden selection). But there is no way of obtaining any such knowledge concerning the single electrons without disturbing them. To conclude, it remains true that precise single predictions are impossible.'

---

preceding or succeeding sections are affected. For a criticism of the present section, see especially note *1 to my old appendix vi, and appendices *xi (9), and *xii.

To this objection I should first reply that it would not be surprising if it were correct. It is after all obvious that from a statistical theory exact singular predictions can never be derived, but only 'indefinite' (*i.e.* formally singular) single predictions. But what I assert at this stage is that although the theory does not supply any such predictions, it *does not rule them out* either. One could speak of the impossibility of singular predictions only if it could be asserted that disturbing the system or interfering with it must prevent every kind of predictive measurement.

'But that is just what I assert', my opponent will say. 'I assert, precisely, the impossibility of any such measurement. You assume that it is possible to measure the energy of one of these moving electrons without forcing it out of its path and out of the electron group. *This* is the assumption which I regard as untenable. For assuming that I possessed any apparatus with which I could make such measurements, then I should with this or some similar apparatus be able to *produce* aggregates of electrons which all (a) were limited as to their position, and (b) had the same momentum. That the existence of such aggregates would contradict the quantum theory is, of course, your view too, since it is ruled out by your own 'scatter relations', as you call them. Thus you could only reply that it is possible to conceive of an apparatus which would allow us to take measurements but not to make selections. I admit that this answer is logically permissible; but as a physicist I can only say that my instincts revolt against the idea that we could measure the momenta of electrons while being unable to eliminate, for instance, all those whose momentum exceeds (or falls short of) some given amount.'

My first answer to this would be that it all sounds quite convincing. But a strict *proof* of the contention that, if a predictive measurement is possible, the corresponding physical selection or separation would also be possible, has not been given (and it cannot be given, as will be seen soon). None of these arguments prove that the precise predictions would contradict the quantum theory. They all introduce an *additional hypothesis*. For the statement (which corresponds to Heisenberg's view) that exact single predictions are impossible, turns out to be equivalent to the hypothesis that *predictive measurements and physical selections are inseparably linked*. With this new theoretical system—the conjunction of the quantum theory with this auxiliary *'hypothesis of linkage'*—my conception must indeed clash.[1]

With this, point (3) of my programme has been carried out. But point (4) has still to be established; that is, we have still to show that the system which combines the statistically interpreted quantum theory (including, we assume, the conservation laws for momentum and energy) with the 'hypothesis of linkage', is self-contradictory. There is, I suppose, a deep-seated presumption that predictive measurement and physical selection are always linked. The prevalence of this presumption may explain why the simple arguments which would establish the opposite have never been worked out.

I wish to stress that the mainly physical considerations now to be presented do not form part of the assumptions or premises of my logical analysis of the uncertainty relations although they might be described as its fruit. In fact, the analysis so far carried out is *quite independent* of what follows; especially of the imaginary physical experiment described below,[*2] which is intended to establish the possibility of arbitrarily precise predictions of the path of single particles.

By way of introduction to this imaginary experiment I will first discuss a few simpler experiments. These are intended to show that we can without difficulty make arbitrarily precise path predictions, and also test them. At this stage I only consider predictions which do not refer to definite single particles, but refer to (all) particles within a definite small space-time region ($\Delta x.\Delta y.\Delta z.\Delta t$). In each case there is only a certain *probability* that particles are present in that region.

We again imagine a beam (an electron or light beam) of particles travelling in the x-direction. But this time we assume it to be monochromatic, so that all the particles are travelling along parallel paths in the x-direction with the same known momentum. The components in the other directions of the momentum will then also be known, that is, known to be equal to zero. Now instead of determining the position in the x-direction of a group of particles by means of a

---

[1] The auxiliary hypothesis here discussed can of course appear in a different form. My reason for choosing this particular form for critical analysis and discussion is that the objection which asserts the linkage of measurement and physical selection was actually (in conversations as well as in letters) raised against the view here advanced.

[*2] Those of my critics who rightly rejected the idea of this imaginary experiment appear to have believed that they had thereby also refuted the preceding analysis, in spite of the warning here given.

physical selection—instead, that is, of isolating the group of particles from the rest of the beam by technical means (as we did above)—we shall be content to differentiate this group from the rest merely by focusing our attention upon it. For example, we may focus our attention upon all those particles which have (with a given precision) in a given instant the place co-ordinate x, and which therefore do not spread beyond an arbitrarily small range Δx. Of each of these particles we know the momentum precisely. We therefore know for each future instant precisely where this group of particles is going to be. (It is clear that the mere existence of such a group of particles does not contradict quantum theory; only its separate existence, that is, the possibility of selecting it physically, would contradict the theory.) We can carry out the same kind of imaginary selection in connection with the other space co-ordinates. The physically selected monochromatic beam would have to be very wide in the y and z-directions (infinitely wide in the case of an ideal monochromatic beam) because in these directions the momentum is supposed to be selected with precision, *i.e.* to be equal to o; so that positions in these directions must be widely spread. Nevertheless we can again focus our attention upon a very narrow partial ray. Again, we shall not only know the position but also the momentum of every particle of this ray. We shall therefore be able to predict for every particle of this narrow ray (which we have, as it were, selected in imagination) at which point, and with what momentum, it will impinge upon a photographic plate set in its path, and of course we can test this prediction empirically (as with the former experiment).

Imaginary selections, analogous to the one just made from a 'pure case' of a particular type, can be made from other types of aggregates. For example, we may take a monochromatic beam from which a physical selection has been made by means of a very small slit Δy (thus taking as our physical starting point a physical selection corresponding to the merely imagined selection of the preceding example). We do not know of any of the particles in which direction it will turn after passing through the slit; but if we consider one definite direction we can calculate precisely the momentum component of all particles that did turn in this particular direction. Thus the particles which after having passed through the slit travel in one definite direction again form an imagined selection. We are able to predict their position and their momentum, or in short, their paths;

and again, by putting a photographic plate in their path, we can test our predictions.

The situation is in principle the same (even though empirical tests are somewhat more difficult) in the case of the first example we considered, namely the selection of particles according to their position in the direction of travel. If we produce a physical selection corresponding to this case, then different particles will travel with different velocities, because of the spread of the momenta. The group of particles will thus spread over an increasing range in the x-direction as it proceeds. (The packet will get wider.) We can then work out the momentum of a partial group of these particles (selected in imagination) which, in a given moment, will be at a given position in the x-direction: the momentum will be the greater the farther ahead is the selected partial group (and *vice versa*). The empirical test of the prediction made in this way could be carried out by substituting for the photographic plate a moving strip of photographic film. As we could know of each point in the band the time of its exposure to the impact of the electrons we could also *predict* for each point on the band with what momentum the impacts would occur. These predictions we could *test*, for example by inserting a filter in front of the moving band or perhaps in front of the Geiger-counter (a filter in the case of light rays; in the case of electrons an electric field at right angles to the direction of the ray) followed by a selection according to direction, allowing only those particles to pass which possess a given minimum momentum. We could then ascertain whether these particles really did arrive at the predicted time or not.

The precision of the measurements involved in these tests is not limited by the uncertainty relations. These are meant to apply, as we have seen, mainly to those measurements which are used for the deduction of predictions, and not for testing them. They are meant to apply, that is to say, to '*predictive measurements*' rather than to '*non-predictive measurements*'. In sections 73 and 76 I examined three cases of such 'non-predictive' measurements, namely (a) measurement of two positions, (b) measurement of position preceded or (c) succeeded by a measurement of momentum. The above discussed measurement by means of a filter in front of a film strip of a Geiger-counter exemplifies (b), *i.e.* a selection according to momentum followed by a measurement of position. This is presumably just that case which, according to Heisenberg (*cf.* section 73), permits 'a calculation about the past of the electron'. For while in cases (a) and (c) only calculations for the time *between* the two measurements are possible, it is possible in case (b) to calculate the path *prior* to the first mesaurement, provided this measurement was a selection according to a given momentum; for such a selection does not disturb the position of the particle.*3 Heisenberg, as we know, questions the 'physical reality' of this measurement, because it permits us to calculate the momentum of the particle only upon its arrival at a precisely measured position and at a precisely measured time: the measurement seems to lack predictive content because no testable conclusion can be derived from it. Yet I shall base my imaginary experiment, intended to establish the possibility of precisely predicting the position and momentum of a definite particle, upon this particular measuring arrangement which at first sight is apparently non-predictive.

As I am about to derive such far-reaching consequences from the assumption that precise 'non-predictive' measurements of this type are possible, it seems proper to discuss the admissibility of this assumption. This is done in appendix vi.

With the imaginary experiment that follows here, I directly challenge the method of arguing which Bohr and Heisenberg have used in order to justify the interpretation of the Heisenberg formulae as limitations upon attainable precision. For they tried to justify this interpretation by showing that no imaginary experiment can be devised which will produce more exact predictive measurements. But this method of arguing can clearly not exclude the possibility that an imaginary experiment might some day be devised which (using known physical effects and laws) would show that such measurements are possible after all. It was taken for granted that any such experiment would contradict the formalism of the quantum theory and it appears that this idea determined the direction of the search for such experiments. My analysis—the carrying out of the points of my programme (1) and (2)—has however cleared the way for

---

*3 This statement (which I tried to base upon my discussion in appendix vi) was effectively criticized by Einstein (*cf.* appendix *xii), and with it, my imaginary experiment collapses. The main point is that non-predictive measurements determine the path of a particle only *between* two measurements, such as a measurement of momentum followed by one of position (or *vice versa*); it is not possible, according to quantum theory, to project the path further back, *i.e.* to the region of time before the first of these measurements. Thus the last paragraph of appendix vi is mistaken; and we cannot know, of the particle arriving at *x* (see below) whether it did come from *P*, or from somewhere else.

an imaginary experiment to be devised which shows, in *full agreement* with quantum theory, that the precise measurements in question are possible.

To carry out this experiment, I shall make use of 'imaginary selection', as before; but shall choose an arrangement such that, if a particle which is characterized by the selection really exists, we shall be able to ascertain the fact.

My experiment, in a way, forms a kind of idealization of the experiments of Compton-Simon and Bothe-Geiger.[2] Since we wish to obtain singular predictions, we cannot operate with statistical assumptions only. The non-statistical laws of the conservation of energy and momentum will have to be used. We can exploit the fact that these laws permit us to calculate what occurs when the particles collide, provided we are given two of the four magnitudes which describe the collision (*i.e.* of the moments $\mathbf{a}_1$ and $\mathbf{b}_1$ before, and $\mathbf{a}_2$ and $\mathbf{b}_2$ after the collision) and one component[3] of a third one. (The method of calculation is well known as part of the theory of the Compton-effect.[4])

Let us now imagine the following experimental arrangement. (See figure 2.) We cross two particle beams (of which one at most may be a light-ray, and one at most may be electrically non-neutral[5]) which are both '*pure cases*' in the sense that the beam A is monochromatic, that is, a selection according to the momentum $\mathbf{a}_1$, whilst the beam B passes through a narrow slit Sl and is thereby subjected to a physical selection according to position. The B-particles may be supposed to have the (absolute) momentum $\mathbf{b}_1$. Some of the particles of these two beams will collide. We now *imagine* two narrow partial rays $[A]$ and $[B]$ which intersect at the place P. The momentum of $[A]$ is known; it is $\mathbf{a}_1$. The momentum of the partial ray $[B]$ becomes calculable as soon as we have decided upon a definite direction for it; let it be $\mathbf{b}_1$. We now choose a direction PX. Attending to those

[2] Compton and Simon, *Physical Review* 25, 1924, p. 439; Bothe und Geiger, *Zeitschrift für Physik* 32, 1925, p. 639; *cf.* also Compton, *X-Rays and Electrons* (New York, 1927); *Ergebnisse der exakten Naturwissenschaft* 5, 1926, p. 267 *ff.*; Haas, *Atomtheorie* (1929), p. 229 *ff.*
[3] 'Component' to be understood here in the widest sense (either as the direction or as the absolute magnitude).
[4] *Cf.* Haas, *op. cit.*
[5] I am thinking of a light ray and any kind of corpuscular ray (negaton, positon, or neutron); in principle, however, two corpuscular rays could be used of which at least *one* is a neutron ray. (Incidentally, the words 'negatron' and 'positron', now becoming current usage, seem to me linguistic monstrosities—after all, we neither say 'positrive' nor 'protron'.)

particles of the partial ray [A] which after the collision travel in the direction PX, we can calculate their momentum $\mathbf{a_2}$, and also $\mathbf{b_2}$, i.e. the momentum after collision of the particles with which they collided. To every particle of [A] which was deflected at the point P

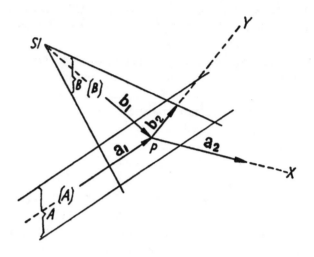

Figure 2

with the momentum $\mathbf{a_2}$, in the direction X, there must correspond a second particle, of [B], which was deflected at P with the momentum $\mathbf{b_2}$, in the calculable direction PY. We now place an apparatus at X—for instance a Geiger-counter or a moving film strip—which records the impacts of particles arriving from P at the arbitrarily restricted region X. Then we can say: as we note any such recording of a particle, we learn at the same time that a second particle must be travelling from P with the momentum $\mathbf{b_2}$ towards Y. And we also learn from the recording where this second particle was at any given moment; for we can calculate from the time of the impact of the first particle at X, and from its known velocity, the moment of its collision at P. By using another Geiger-counter at Y (or the moving film band), we can test our predictions for the second particle.[*4]

*4 Einstein, Podolski, and Rosen use a *weaker* but *valid* argument: let Heisenberg's interpretation be correct, so that we can only measure at will *either* the position *or* the momentum of the first particle at X. Then if we *measure* the position of the first particle, we can *calculate* the position of the second particle; and if we *measure* the momentum of the first particle, we can *calculate* the momentum of the second particle. But since we can make our choice—as to whether we measure position or momentum—at any time, even after the collision of the two particles has taken place,

The precision of these predictions as well as that of the measurements undertaken to test them is *in principle not subject to any of the limitations due to the uncertainty principle*, as regards both the position co-ordinate and the component of the momentum in the direction *PY*. For my imaginary experiment reduces the question of the precision with which predictions can be made about a *B*-particle deflected in *P* to the question of the precision attainable in taking measurements at *X*. These, at first, seemed to be non-predictive measurements of the time, position and momentum of the corresponding first particle [*A*]. The momentum of this particle in the *PX* direction as well as the time of its impact at *X*, *i.e.* of its position in the *PS* direction, can be measured with any desirable degree of precision (*cf.* appendix vi) if we make a momentum selection by interposing, for instance, an electrical field or a filter in front of the Geiger-counter, before we measure the position. But in consequence of this (as will be shown more fully in appendix vii) we can make predictions with any degree of precision about the *B*-particle travelling in the *PY* direction.

This imaginary experiment allows us to see not only that precise single predictions can be made, but also under what conditions they can be made, or better, under what conditions they are compatible with the quantum theory. They can be made only if we can obtain knowledge about the state of the particle without being able to create this state at will. Thus we really obtain our knowledge after the event, as it were, since at the time when we obtain it the particle will already have assumed its state of motion. Yet we can still make use of this knowledge to deduce from it testable predictions. (If the *B*-particle in question is a photon, for instance, we might be able to calculate the time of its arrival on Sirius.) The impacts of particles arriving at *X* will succeed each other at irregular time-intervals; which means that the particles of the partial ray *B* about which we are making predictions will also succeed each other after irregular time-intervals. It would contradict the quantum theory

---

it is unreasonable to assume that the second particle was in any way affected, or interfered with, by the change in the experimental arrangements resulting from our choice. Accordingly, we can calculate, with any precision we like, *either* the position *or* the momentum of the second particle *without interfering with it*; a fact which may be expressed by saying that the second particle 'has' both a precise position and a precise momentum. (Einstein expressed this by saying that both position and momentum are 'real'; whereupon he was attacked as 'reactionary'.) See also appendices *xi and *xii.

if we could alter this state of things by, for example, making these time-intervals equal. Thus we are able, as it were, to take aim and to predetermine the force of the bullet; we can also (and this *before* the bullet hits the target *Y*) calculate the exact time at which the shot was fired at *P*. Yet we cannot freely choose the moment of firing, but have to wait till the gun goes off. Nor can we prevent uncontrolled shots being fired in the direction of our target (from the neighbourhood of *P*).

It is clear that our experiment and Heisenberg's interpretation are incompatible. But since the possibility of carrying out this experiment can be deduced from the statistical interpretation of quantum physics (with the addition of the laws of energy and momentum), it appears that Heisenberg's interpretation, in contradicting it, must also contradict the statistical interpretation of quantum theory. In view of the experiments of Compton-Simon and Bothe-Geiger, it would seem that it is possible to carry out our experiment. It can be regarded as a kind of *experimentum crucis* to decide between Heisenberg's conception and a consistently statistical interpretation of quantum theory.

78. *Indeterminist Metaphysics.*

It is the task of the natural scientist to search for laws which will enable him to deduce predictions. This task may be divided into two parts. On the one hand, he must try to discover such laws as will enable him to deduce single predictions ('causal' or 'deterministic' laws or 'precision statements'). On the other hand, he must try to advance hypotheses about frequencies, that is, laws asserting probabilities, in order to deduce frequency predictions. There is nothing in these two tasks to make them in any way mutually incompatible. It is clearly not the case that whenever we make precision statements we shall make no frequency hypotheses; for some precision statements are, as we have seen, macro laws which are derivable from frequency assumptions. Nor is it the case that whenever in a particular field frequency statements are well confirmed, we are entitled to conclude that in this field *no precision statements can be made*. This situation seems plain enough. Yet the second of the two conclusions we have just rejected has been drawn again and again. Again and again we meet with the belief that where fortuity rules, regularity is ruled out. I have critically examined this belief in section 69.

The dualism of macro and micro laws—I mean the fact that we operate with both—will not be easily overcome, to judge by the present state of scientific development. What might be logically possible, however, is a reduction of all known precision statements—by interpreting them as macro laws—to frequency statements. The converse reduction is not possible. Frequency statements can never be deduced from precision statements, as we have seen in section 70. They need their own assumptions which must be specifically statistical. Only from probability estimates can probabilities be calculated.[*1]

This is the logical situation. It encourages neither a deterministic nor an indeterministic view. And should it ever become possible to work in physics with nothing but frequency statements, then we should still not be entitled to draw indeterminist conclusions; which is to say that we should still not be entitled to assert that 'there are no precise laws in nature, no laws from which predictions about the course of single or elementary processes can be deduced'. The scientist will never let anything stop him searching for laws, including laws of this kind. And however successfully we might operate with probability estimates, we must not conclude that the search for precision laws is vain.

These reflections are not by any means the outcome of the imaginary experiment described in section 77; quite the contrary. Let us assume that the uncertainty relations are not refuted by this experiment (for whatever reason—say, because the *experimentum crucis* described in appendix vi would decide against the quantum theory): even then they could only be tested as frequency statements and could only be corroborated as frequency statements. Thus in no case should we be entitled to draw indeterministic conclusions from the fact that they are well corroborated.[*2]

Is the world ruled by strict laws or not? This question I regard as metaphysical. The laws we find are always hypotheses; which means that they may always be superseded, and that they may possibly be deduced from probability estimates. Yet denying causality would

---

[*1] This view is opposed by Einstein at the end of his letter here printed in appendix *xii. But I still think that it is true.

[*2] I still believe that this analysis is essentially correct: we cannot conclude from the success of frequency predictions about penny tosses that the single penny tosses are undetermined. But we may argue in favour of, say, an indeministic metaphysical view by pointing out difficulties and contradictions which this view might be able to dissolve.

be the same as attempting to persuade the theorist to give up his search; and that such an attempt cannot be backed by anything like a proof has just been shown. The so-called 'causal principle' or 'causal law', however it may be formulated, is very different in character from a natural law; and I cannot agree with Schlick when he says, '. . . the causal law can be tested as to its truth, in *precisely the same sense* as any other natural law'.[1]

The belief in causality is metaphysical.[*3] It is nothing but a typical metaphysical hypostatization of a well justified methodological rule—the scientist's decision never to abandon his search for laws. The metaphysical belief in causality seems thus more fertile in its various manifestations than any indeterminist metaphysics of the kind advocated by Heisenberg. Indeed we can see that Heisenberg's comments have had a crippling effect on research. Connections which are not far to seek may easily be overlooked if it is continually repeated that the search for any such connections is 'meaningless'.

Heisenberg's formulae—like similar statements which can only be corroborated by their statistical consequences—do not necessarily lead to indeterminist conclusions. But this in itself does not prove that there can be no other empirical statement which justifies these or similar conclusions: for example the conclusion that the methodological rule mentioned—the decision never to abandon the search for laws—cannot fulfil its purpose, perhaps because it is futile or meaningless or 'impossible' (*cf.* note 2 to section 12) to search for laws and for singular predictions. But there could not be an empirical statement having methodological consequences which could compel us to abandon the search for laws. For a statement supposed to be free from metaphysical elements can have indeterminist conclusions only

---

[1] Schlick, *Die Kausalität in der gegenwärtigen Physik, Die Naturwissenschaften* 19, 1931, p. 155, writes as follows: (I quote the passage in full; *cf.* also my notes 7 and 8 to Section 4) 'Our attempts to find a testable statement equivalent to the principle of causality have failed; our attempts to formulate one have only led to pseudo-statements. This result, however, does not after all come as a surprise, for we have already remarked that the truth of the causal law can be tested *in the same sense* as that of any other natural law; but we have also indicated that these natural laws in their turn, when strictly analysed, do not seem to have the character of statements that are true or false, but turn out to be nothing but rules for the (trans-)formation of such statements.' Schlick had already earlier held that the causal principle should be placed on a par with natural laws. But as at that time he regarded natural laws as genuine statements he also regarded 'the causal principle . . . as an empirically testable hypothesis'. *Cf. Algemeine Erkenntnislehre* (2nd edition 1925), p. 374.

[*3] Compare with the views expressed here, and in the rest of this section, chapter *iv of the *Postscript*.

if these are falsifiable.*⁴ But they can be shown to be false only if we succeed in formulating laws, and in deducing predictions from them which are corroborated. Accordingly, if we assume that these indeterminist conclusions are *empirical hypotheses*, we ought to try hard to test them, *i.e.* to falsify them. And this means that we ought to *search* for laws and predictions. Thus we cannot obey an exhortation to abandon this search without repudiating the empirical character of these hypotheses. This shows that it would be self-contradictory to think that any empirical hypothesis could exist which might compel us to abandon the search for laws.

I do not intend to show here in detail how so many attempts to establish indeterminism reveal a mode of thought which can only be described as determinist, in the metaphysical sense. (Heisenberg for instance tries to give a causal explanation why causal explanations are impossible.*⁵) I may just remind the reader of the attempts to demonstrate that the uncertainty relations close some avenues of possible research, as does the principle of the constancy of light velocity: the analogy between the two constants c and h, the velocity of light and Planck's constant, was interpreted by saying that both set a limit, in principle, to the possibilities of research. Questions raised in the attempt to grope beyond these barriers were dismissed by the well-known method of dismissing unpalatable problems as 'pseudo'. In my view there is indeed an analogy between the two constants c and h; one which, incidentally, ensures that the constant h is no more a barrier to research than the constant c. The principle of the constancy of light velocity (and of the impossibility of exceeding this velocity) does not forbid us to search for velocities which are greater than that of light; for it only asserts that we shall not find any; that is to say, that we shall be unable to produce signals that travel faster than light. And similarly, the Heisenberg formulae ought not to be interpreted as forbidding the search for 'super-pure' cases; for they only assert that we shall not find any; and, in particular, that we cannot produce any. The laws forbidding velocities greater than that of light and 'super-pure' cases challenge the investigator, as do

---

*⁴ This, though valid *as a reply to a positivist*, is misleading as it stands; for a falsifiable statement may have all kinds of logically weak consequences, including non-falsifiable ones. (*Cf.* the fourth paragraph of section 66.)

*⁵ His argument is, in brief, that causality breaks down owing to our interference with the observed object, *i.e.* owing to a certain causal interaction.

other empirical statements, to search for the forbidden. For he can test empirical statements only by trying to falsify them.

From an historical point of view, the emergence of indeterminist metaphysics is understandable enough. For a long time, physicists believed in determinist metaphysics. And because the logical situation was not fully understood, the failure of the various attempts to deduce the light spectra—which are statistical effects—from a mechanical model of the atom was bound to produce a crisis for determinism. Today we see clearly that this failure was inevitable, since it is impossible to deduce statistical laws from a non-statistical (mechanical) model of the atom. But at that time (about 1924, the time of the theory of Bohr, Kramers, and Slater) it could not but seem as if in the mechanism of each single atom, probabilities were taking the place of strict laws. The determinist edifice was wrecked—mainly because probability statements were expressed as formally singular statements. On the ruins of determinism, indeterminism rose, supported by Heisenberg's uncertainty principle. But it sprang, as we now see, from that same misunderstanding of the meaning of formally-singular probability statements.

The lesson of all this is that we should try to find strict laws —prohibitions—that can founder upon experience. Yet we should abstain from issuing prohibitions that draw limits to the possibilities of research.

CHAPTER X

CORROBORATION,
OR HOW A THEORY STANDS UP TO TESTS

THEORIES are not verifiable, but they can be 'corroborated'. The attempt has often been made to describe theories as being neither *true* nor *false*, but instead more or less *probable*. Inductive logic, more especially, has been developed as a logic which may ascribe not only the two values 'true' and 'false' to statements, but also degrees of probability; a type of logic which will here be called *'probability logic'*. According to those who believe in probability logic, induction should determine the degree of probability of a statement. And a principle of induction should either *make it sure* that the induced statement is 'probably valid' or else it should *make it probable*, in its turn—for the principle of induction might itself be only 'probably valid'. Yet in my view, the whole problem of the probability of hypotheses is misconceived. Instead of discussing the 'probability' of a hypothesis we should try to assess what tests, what trials, it has withstood; that is, we should try to assess how far it has been able to prove its fitness to survive by standing up to tests. In brief, we should try to assess how far it has been 'corroborated'.[*1]

[*1] I introduced the terms *'corroboration'* (*'Bewährung'*) and especially *'degree of corroboration'* (*'Grad der Bewährung'*, *'Bewährungsgrad'*) in my book because I wanted a *neutral* term to describe the degree to which a hypothesis has stood up to severe tests, and thus 'proved its mettle'. By 'neutral' I mean a term not prejudging the issue whether, by standing up to tests, the hypothesis becomes 'more probable', in the sense of the probability calculus. In other words, I introduced the term 'degree of corroboration' mainly in order to be able to discuss the problem whether or not 'degree of corroboration' could be indentified with 'probability' (either in a frequency sense or in the sense of Keynes, for example).

Carnap translated my term 'degree of corroboration' (*'Grad der Bewährung'*), which I had first introduced into the discussions of the Vienna Circle, as 'degree of confirmation'. (See his 'Testability and Meaning', in *Philosophy of Science* 3, 1936; especially p. 427); and so the term 'degree of confirmation' soon became widely accepted. I did not like this term, because of some of its associations ('make firm'; 'establish firmly';

79. *Concerning the So-Called Verification of Hypotheses.*

The fact that theories are not verifiable has often been overlooked. People often say of a theory that it is verified when some of the predictions derived from it have been verified. They may perhaps admit that the verification is not completely impeccable from a logical point of view, or that a statement can never be finally established by establishing some of its consequences. But they are apt to look upon such objections as due to somewhat unnecessary scruples. It is quite true, they say, and even trivial, that we cannot know for certain whether the sun will rise tomorrow; but this uncertainty may be neglected: the fact that theories may not only be improved but that they can also be *falsified by new experiments* presents to the scientist a serious possibility which may at any moment become actual; but never yet has a theory had to be regarded as falsified owing to the sudden breakdown of a well-confirmed law. It never happens that old experiments one day yield new results. What happens is only that new experiments decide against an old theory. The old theory, even when it is superseded, often retains its validity as a kind of limiting case of the new theory; it still applies, at least with a high degree of approximation, in those cases in which it was successful before. In short, regularities which are directly testable by experiment do not change. Admittedly it is conceivable, or logically possible, that they might change; but this possibility is disregarded by empirical science and does not affect its methods. On the contrary, scientific method presupposes *the immutability of natural processes*, or the 'principle of the uniformity of nature'.

There is something to be said for the above argument, but it does not affect my thesis. It expresses the metaphysical faith in the existence of regularities in our world (a faith which I share, and without which practical action is hardly conceivable).*[1] Yet the question before us—

*[1] Cf. appendix *x, and also section *15 of my *Postscript*.

'put beyond doubt'; 'prove'; 'verify': 'to confirm' corresponds more closely to '*erhärten*' or '*bestätigen*' than to '*bewähren*'). I therefore proposed in a letter to Carnap (written, I think, about 1939) to use the term 'corroboration'. (This term had been suggested to me by Professor H. N. Parton.) But as Carnap declined my proposal, I fell in with his usage, thinking that words do not matter. This is why I myself used the term 'confirmation' for a time in a number of my publications.

Yet it turned out that I was mistaken: the associations of the word 'confirmation' did matter, unfortunately, and made themselves felt: 'degree of confirmation' was soon used—by Carnap himself—as a synonym (or 'explicans') of 'probability'. I have therefore now abandoned it in favour of 'degree of corroboration'. See also appendix *ix, and section *29 of my *Postscript*.

the question which makes the non-verifiability of theories significant in the present context—is on an altogether different plane. Consistently with my attitude towards other metaphysical questions, I abstain from arguing for or against faith in the existence of regularities in our world. But I shall try to show that *the non-verifiability of theories is methodologically important*. It is on this plane that I oppose the argument just advanced.

I shall therefore take up as relevant only one of the points of this argument—the reference to the so-called 'principle of the uniformity of nature'. This principle, it seems to me, expresses in a very superficial way an important methodological rule, and one which might be derived, with advantage, precisely from a consideration of the non-verifiability of theories.*2

Let us suppose that the sun will not rise tomorrow (and that we shall nevertheless continue to live, and also to pursue our scientific interests). Should such a thing occur, science would have to try to *explain* it, *i.e.* to derive it from laws. Existing theories would presumably require to be drastically revised. But the revised theories would not merely have to account for the new state of affairs: *our older experiences would also have to be derivable from them*. From the methodological point of view one sees that the principle of the uniformity of nature is here replaced by the postulate of *the invariance of natural laws*, with respect to both space and time. I think, therefore, that it would be a mistake to assert that natural regularities do not change. (This would be a kind of statement that can neither be argued against nor argued for.) What we should say is, rather, that it is part of our *definition* of natural laws if we postulate that they are to be invariant with respect to space and time; and also if we postulate that they are to have no exceptions. Thus from a methodological point of view, the possibility of falsifying a corroborated law is by no means without significance. It helps us to find out what we demand and expect from natural laws. And the 'principle of the uniformity of nature' can again be regarded as a metaphysical interpretation of a methodological rule—like its near relative, the 'law of causality'.

One attempt to replace metaphysical statements of this kind by principles of method leads to the 'principle of induction', supposed to govern the method of induction, and hence that of the verification

---

*2 I mean the rule that any new system of hypotheses should yield, or explain, the old, corroborated, regularities. See also section *3 (third paragraph) of my *Postscript*.

of theories. But this attempt fails, for the principle of induction is itself metaphysical in character. As I have pointed out in section 1, the assumption that the principle of induction is empirical leads to an infinite regress. It could therefore only be introduced as a primitive proposition (or a postulate, or an axiom). This would perhaps not matter so much, were it not that the principle of induction would have in any case to be treated as a *non-falsifiable statement*. For if this principle—which is supposed to validate the inference of theories—were itself falsifiable, then it would be falsified with the first falsified theory, because this theory would then be a conclusion, derived with the help of the principle of induction; and this principle, as a premise, will of course be falsified by the *modus tollens* whenever a theory is falsified which was derived from it.[*3] But this means that a falsifiable principle of induction would be falsified anew with every advance made by science. It would be necessary, therefore, to introduce a principle of induction assumed not to be falsifiable. But this would amount to the misconceived notion of a synthetic statement which is *a priori* valid, *i.e.* an irrefutable statement about reality.

Thus if we try to turn our metaphysical faith in the uniformity of nature and in the verifiability of theories into a theory of knowledge based on inductive logic, we are left only with the choice between an infinite regress and *apriorism*.

80. *The Probability of a Hypothesis and the Probability of Events: Criticism of Probability Logic.*

Even if it is admitted that theories are never finally verified, may we not succeed in making them secure to a greater or lesser extent—more probable, or less so? After all, it might be possible that the question of *the probability of a hypothesis* could be reduced, say, to that of *the probability of events*, and thus be made susceptible to mathematical and logical handling.[*1]

Like inductive logic in general, the theory of the probability of

---

[*3] The premises of the derivation of the theory would (according to the inductivist view here discussed) consist of the principle of induction *and* of observation statements. But the latter are here tacitly assumed to be unshaken and reproducible, so that they cannot be made responsible for the failure of the theory.

[*1] The present section (80) contains mainly a criticism of the attempt (Reichenbach's) to interpret *the probability of hypotheses* in terms of *a frequency theory of the probability of events*. A criticism of Keynes's approach is contained in section 83.

hypotheses seems to have arisen through a confusion of psychological with logical questions. Admittedly, our subjective feelings of conviction are of different intensities, and the degree of confidence with which we await the fulfilment of a prediction and the further corroboration of a hypothesis is likely to depend, among other things, upon the way in which this hypothesis has stood up to tests so far— upon its past corroboration. But that these psychological questions do not belong to epistemology or methodology is pretty well acknowledged even by the believers in probability logic.*[2] They argue, however, that it is possible, on the basis of inductivist decisions, to ascribe degrees of probability to *the hypotheses themselves*; and further, that it is possible to reduce this concept to that of the probability of events.

The probability of a hypothesis is mostly regarded as merely a special case of the general problem of the *probability of a statement*; and this in turn is regarded as nothing but the problem of the *probability of an event*, expressed in a particular terminology. Thus we read in Reichenbach, for example: 'Whether we ascribe probability to statements or to events is only a matter of terminology. So far we have regarded it as a case of the probability of events that the probability of 1/6 has been assigned to the turning up of a certain face of a die. But we might just as well say that it is the *statement* "the face showing the 1 will turn up" which has been assigned the probability of 1/6.'[1]

This identification of the probability of events with the probability of statements may be better understood if we recall what was said in section 23. There the concept 'event' was defined as a class of singular statements. It must therefore also be permissible to speak of the *probability of statements* in place of the probability of events. So we can regard this as being merely a change of terminology: the reference-sequences are interpreted as sequences of statements. If we think of an 'alternative', or rather of its elements, as represented by statements, then we can describe the turning up of heads by the statement 'k is heads', and its failure to turn up by the negation of this statement. In this way we obtain a sequence of statements of the form p_j, p_k, $\bar{p}_l$, p_m, $\bar{p}_n$, . . ., in which a statement p_i is sometimes characterized as 'true', and sometimes (by placing a bar over its name) as 'false'. Probability within an alternative can thus be interpreted as *the relative*

---

*[2] I am alluding here to the school of Reichenbach rather than to Keynes.
[1] Reichenbach, *Erkenntnis* I, 1930, p. 171 f.

'truth-frequency'[2] *of statements within a sequence of statements* (rather than as the relative frequency of a property).

If we like, we can call the concept of probability, so transformed, the 'probability of statements' or the 'probability of propositions'. And we can show a very close connection between this concept and the concept of 'truth'. For if the sequence of statements becomes shorter and shorter and in the end contains only one element, *i.e.* only *one single* statement, then the probability, or truth-frequency, of the sequence can assume only one of the two values 1 and 0, according to whether the single statement is true or false. The truth or falsity of a statement can thus be looked upon as a limiting case of probability; and conversely, probability can be regarded as a generalization of the concept of truth, in so far as it includes the latter as a limiting case. Finally, it is possible to define operations with truth-frequencies in such a way that the usual truth-operations of classical logic become limiting cases of these operations. And the calculus of these operations can be called *'probability logic'*.[3]

But can we really identify *the probability of hypotheses* with the probability of statements, defined in this manner, and thus indirectly with the probability of events? I believe that this identification is the result of a confusion. The idea is that the probability of a hypothesis, since it is obviously a kind of probability of a statement, must come under the head of 'probability of statements' *in the sense just defined*. But this conclusion turns out to be unwarranted; and the terminology is thus highly unsuitable. Perhaps after all it would be better never to use the expression 'probability of statements' if we have the probability of events in mind.[*2]

[2] According to Keynes, *A Treatise on Probability* (1921), p. 101 *ff.*, the expression 'truth-frequency' is due to Whitehead; *cf.* the next note.

[3] I am giving here an outline of the construction of the probability logic developed by Reichenbach (*Wahrscheinlichkeitslogik, Sitzungsberichte der Preussischen Akademie der Wissenschaften*, Physik.-mathem. Klasse **29**, 1932, p. 476 *ff.*) who follows E. L. Post (*American Journal of Mathematics* **43**, 1921, p. 184), and, at the same time, the frequency theory of von Mises. Whitehead's form of the frequency theory, discussed by Keynes, *op. cit.* p. 101 *ff.* is similar.

[*2] I still think (a) that the so-called 'probability of hypotheses' cannot be interpreted by a truth-frequency: (b) that it is better to call a probability defined by a relative frequency—whether a truth-frequency or the frequency of an event—the 'probability of an event'; (c) that the so-called 'probability of a hypothesis' (in the sense of its acceptability) is *not* a special case of the 'probability of statements'. And I should now regard the 'probability of statements' as one interpretation (the logical interpretation) among several possible interpretations of the formal calculus of probability, rather than as a truth-frequency. (*Cf.* appendices *ii, *iv, and *ix, and my *Postscript*.)

However this may be, I assert that the issues arising from the concept of a *probability of hypotheses* are not even touched by considerations based on probability logic. I assert that if one says of a hypothesis that it is not true but 'probable', then this statement can under *no* circumstances be translated into a statement about the probability of events.

For if one attempts to reduce the idea of a probability of hypotheses to that of a truth-frequency which uses the concept of a sequence of statements, then one is at once confronted with the question: *with reference to what sequence* of statements can a probability value be assigned to a hypothesis? Reichenbach identifies an 'assertion of natural science'—by which he means a scientific hypothesis—itself with a reference-sequence of statements. He says, '. . . the assertions of natural science, which are never singular statements, are in fact sequences of statements to which, strictly speaking, we must assign not the degree of probability 1 but a smaller probability value. It is therefore only probability logic which provides the logical form capable of strictly representing the concept of knowledge proper to natural science.'[4] Let us now try to follow up the suggestion that the hypotheses themselves are sequences of statements. One way of interpreting it would be to take, as the elements of such a sequence, the various singular statements which can contradict, or agree with, the hypothesis. The probability of this hypothesis would then be determined by the truth-frequency of those among these statements which agree with it. But this would give the hypothesis a probability of ½ if, on the average, it is refuted by every second singular statement of this sequence! In order to escape from this devastating conclusion, we might try two more expedients.[*3] One would be to ascribe to the hypothesis a certain probability—perhaps not a very precise one—on the basis of an estimate of the ratio of all the tests passed by it to all the tests which have not yet been attempted. But this way too leads nowhere. For this estimate can, as it happens, be computed with precision, and the result is always that the probability is zero. And finally, we could try to base our estimate upon the ratio of those tests which led to a favourable result to those which

---

[4] Reichenbach, *Wahrscheinlichkeitslogik* (*op. cit.* p. 488), p. 15 of the reprint.

[*3] It is here assumed that we have by now made up our minds that whenever there is a clear-cut falsification, we will attribute to the hypothesis the probability zero, so that the discussion is now confined to those cases in which no clear-cut falsification has been obtained.

led to an indifferent result—*i.e.* one which did not produce a clear decision. (In this way one might indeed obtain something resembling a measure of the subjective feeling of confidence with which the experimenter views his results.) But this last expedient will not do either, even if we disregard the fact that with this kind of estimate we have strayed a long way from the concept of a truth-frequency, and that of a probability of events. (These concepts are based upon the ratio of the true statements to those which are false, and we must not, of course, equate an indifferent statement with one that is objectively false.) The reason why this last attempt fails too is that the suggested definition would make the probability of a hypothesis hopelessly subjective: the probability of a hypothesis would depend upon the training and skill of the experimenter rather than upon objectively reproducible and testable results.

But I think it is altogether impossible to accept the suggestion that a hypothesis can be taken to be a sequence of statements. It would be possible if universal statements had the form: 'For every value of k it is true that at the place k so-and-so occurs.' If universal statements had this form, then we could regard basic statements (those that contradict, or agree with, the universal statement) as elements of a sequence of statements—the sequence to be taken for the universal statement. But as we have seen (*cf.* sections 15 and 28), universal statements do not have this form. Basic statements are never derivable from universal statements alone.*⁴ The latter cannot therefore be regarded as sequences of basic statements. If, however, we try to take into consideration the sequence of those negations of basic statements which *are* derivable from universal statements, then the estimate for *every* self-consistent hypothesis will lead to the same probability, namely 1. For we should then have to consider the ratio of the *non-falsified* negated basic statements which can be derived (or other derivable statements) to the *falsified* ones. This means that instead of

*⁴ As explained in section 28 above, the singular statements which *can* be deduced from a theory—the 'instantial statements'—are not of the character of basic statements or of observation statements. If we nevertheless decide to take the sequence of these statements and base our probability upon the truth frequency within this sequence, then the probability will be always equal to 1, however often the theory may be falsified; for as has been shown in section 28, note *1, almost any theory is 'verified' by almost all instances (*i.e.* by almost all places k.) The discussion following here in the text contains a very similar argument—also based upon 'instantial statements' (*i.e.* negated basic statements)—designed to show that the probability of a hypothesis, if based upon these negated basic statements, would always be equal to one.

considering a truth frequency we should have to consider the complementary value of a falsity frequency. This value however would be equal to 1. For the class of derivable statements, and even the class of the derivable negations of basic statements, are both infinite; on the other hand, there cannot be more than at most a finite number of accepted falsifying basic statements. Thus even if we disregard the fact that universal statements are never sequences of statements, and even if we try to interpret them as something of the kind and to correlate with them sequences of completely decidable singular statements, even then we do not reach an acceptable result.

We have yet to examine another, quite different, possibility of explaining the probability of a hypothesis in terms of sequences of statements. It may be remembered that we have called a given singular occurrence 'probable' (in the sense of a 'formally singular probability statement') if it is an *element of a sequence* of occurrences with a certain probability. Similarly one might try to call a hypothesis 'probable' if it is an *element of a sequence of hypotheses* with a definite truth-frequency. But this attempt again fails—quite apart from the difficulty of determining the reference sequence (it can be chosen in many ways; *cf.* section 71). For we cannot speak of a truth-frequency within a sequence of hypotheses, simply because we can never know of a hypothesis whether it is true. If we *could* know this, then we should hardly need the concept of the probability of a hypothesis at all. Now we might try, as above, to take the complement of the falsity-frequency within a sequence of hypotheses as our starting point. But if, say, we define the probability of a hypothesis with the help of the ratio of the non-falsified to the falsified hypotheses of the sequence, then, as before, the probability of *every* hypothesis within *every infinite* reference sequence will be equal to 1. And even if a *finite* reference sequence were chosen we should be in no better position. For let us assume that we can ascribe to the elements of some (*finite*) sequence of hypotheses a degree of probability between 0 and 1 in accordance with this procedure—say, the value 3/4. (This can be done if we obtain the information that this or that hypothesis belonging to the sequence has been falsified.) In so far as these *falsified* hypotheses are elements of the sequence, we thus would have to ascribe to them, *just because of this information*, not the value 0, but 3/4. And in general, the probability of a hypothesis would decrease by $1/n$ in consequence of the information that it is false, where n is the number of hypotheses in the reference sequence. All this glaringly contradicts the programme of expressing, in terms of a *'probability of hypotheses'*, the degree of reliability which we have to ascribe to a hypothesis in view of supporting or under-mining evidence.

This seems to me to exhaust the possibilities of basing the concept of the probability of a hypothesis on that of the frequency of true statements (or the frequency of false ones), and thereby on the frequency theory of the probability of events.*5

I think we have to regard the attempt to identify the probability of a hypothesis with the probability of events as a complete failure. This conclusion is quite independent of whether we accept the claim (it is Reichenbach's) that *all hypotheses of physics* are 'in reality', or

---

*5 One might summarize my foregoing attempts to make sense of Reichenbach's somewhat cryptic assertion that the probability of a hypothesis is to be measured by a truth frequency, as follows. (For a similar summary, with criticism, see the penultimate paragraph of appendix *i.)

Roughly, we can try two possible ways to define the probability of a theory. One is to count the number of experimentally testable statements belonging to the theory, and to determine the relative frequency of those which turn out to be true; this relative frequency can then be taken as a measure of the probability of a theory. We may call this a *probability of the first kind*. Secondly, we can consider the theory as an element of a class of ideological entities—say, of theories proposed by other scientists—and we can then determine the relative frequencies within this class. We may call this a *probability of the second kind*.

In my text I tried, further, to show that each of these two possibilities of making sense of Reichenbach's idea of truth frequency leads to results which must be quite unacceptable to adherents of the probability theory of induction.

Reichenbach replied to my criticism, not so much by defending his views as by attacking mine. In his paper on my book (*Erkenntnis* 5, 1935, pp. 267-284), he said that 'the results of this book are completely untenable', and explained this by a failure of my 'method'—by my failure 'to think out all the consequences' of my conceptual system.

Section iv of his paper (pp. 274 f.) is devoted to our problem—the probability of hypotheses. It begins: 'In this connection, some remarks may be added about the probability of theories—remarks which should render more complete my so far all too brief communications of the subject, and which may remove a certain obscurity which still surrounds the issue.' After this follows a passage which forms the second paragraph of the present note, headed by the word 'Roughly' (the only word which I have added to Reichenbach's text).

Reichenbach remains silent about the fact that his attempt to remove 'the obscurity which still surrounds the issue' is but a summary—a rough one, admittedly—of some pages of the very book which he is attacking. Yet in spite of this silence I feel that I may take it as a great compliment from so experienced a writer on probability (who at the time of writing his reply to my book had two books and about a dozen papers on the subject to his credit) that he does accept the results of my endeavours to 'think out the consequences' of his 'all too brief communications on the subject'. This success of my endeavours was due, I believe, to a rule of 'method': that we should always try to clarify and to strengthen our opponent's position as much as possible before criticizing him, if we wish our criticism to be worth while.

'on closer examination' nothing but probability statements (about some average frequencies within sequences of observations which always show deviations from some mean value), or whether we are inclined to make a distinction between two different *types* of natural laws—between the 'deterministic' or 'precision' laws on the one hand, and the 'probability laws' or 'hypotheses of frequency' on the other. For both of these types are hypothetical assumptions which in their turn can never become 'probable': they can only be corroborated, in the sense that they can 'prove their mettle' under fire—the fire of our tests.

How are we to explain the fact that the believers in probability logic have reached an opposite view? Wherein lies the error made by Jeans when he writes—at first in a sense with which I can fully agree—that '... we can know nothing ... *for certain*', but then goes on to say: 'At best we can only deal in *probabilities*. [And] the predictions of the new quantum theory agree so well [with the observations] that the odds in favour of the scheme having some correspondence with reality are *enormous*. Indeed, we may say the scheme is *almost certain* to be quantitatively true ...'?[5]

Undoubtedly the commonest error consists in believing that hypothetical estimates of frequencies, that is to say, hypotheses regarding probabilities, can in their turn be only probable; or in other words, in ascribing to *hypotheses of probability* some degree of an alleged *probability of hypotheses*. We may be able to produce a persuasive argument in favour of this erroneous conclusion if we remember that hypotheses regarding probabilities are, as far as their logical form is concerned (and without reference to our methodological requirement of falsifiability), neither verifiable nor falsifiable. (*Cf.* sections 65 to 68.) They are not verifiable because they are universal statements, and they are not strictly falsifiable because they can never be logically contradicted by any basic statements. They are thus (as Reichenbach puts it) *completely undecidable*.[6] Now they can, as I have tried to show, be better, or less well, '*confirmed*', which is to say that they may agree more, or less, with accepted basic state-

---

[5] Jeans, *The New Background of Science* (1934), p. 58. (Only the words 'for certain' are italicized by Jeans.)

[6] Reichenbach, *Erkenntnis* **1**, 1930, p. 169 (*cf.* also Reichenbach's reply to my note in *Erkenntnis* **3**, 1933, p. 426 *f.*). Similar ideas about the degrees of probability or certainty of inductive knowledge occur very frequently (*cf.* for instance Russell, *Our Knowledge of the External World*, 1926, p. 225 *f.*, and *The Analysis of Matter*, 1927, pp. 141 and 398).

ments. This is the point where, it may appear, probability logic comes in. The symmetry between verifiability and falsifiability accepted by classical inductivist logic suggests the belief that it must be possible to correlate with these 'undecidable' probability statements some scale of degrees of validity, something like 'continuous degrees of probability whose unattainable upper and lower limits are truth and falsity',[7] to quote Reichenbach again. According to my view, however, probability statements, just because they are completely undecidable, are *metaphysical* unless we decide to make them falsifiable by accepting a methodological rule. Thus the simple result of their non-falsifiability is not that they can be better, or less well corroborated, but *that they cannot be empirically corroborated at all*. For otherwise—seeing that they rule out nothing, and are therefore compatible with every basic statement—they could be said to be 'corroborated' by *every arbitrarily chosen basic statement* (of any degree of composition) provided it describes the occurrence of some relevant instance.

I believe that physics uses probability statements only in the way which I have discussed at length in connection with the theory of probability; and more particularly that it uses probability assumptions, just like other hypotheses, as falsifiable statements. But I should decline to join in any dispute about how physicists 'in fact' proceed, since this must remain largely a matter of interpretation.

We have here quite a nice illustration of the contrast between my view and what I called, in section 10, the 'naturalistic' view. What can be shown is, first, the internal logical consistency of my view, and secondly, that it is free from those difficulties which beset other views. Admittedly it is impossible to prove that my view is correct, and a controversy with upholders of another logic of science may well be futile. All that can be shown is that my approach to this particular problem is a consequence of the conception of science for which I have been arguing.[*6]

81. *Inductive Logic and Probability Logic.*
The probability of hypotheses cannot be reduced to the probability

---

[7] Reichenbach, *Erkenntnis* I, 1930, p. 186 (*cf.* note 4 to section 1).
[*6] The last two paragraphs were provoked by the 'naturalistic' approach sometimes adopted by Reichenbach, Neurath, and others; *cf.* section 10, above.

of events. This is the conclusion which emerges from the examination carried out in the previous section. But might not a different approach lead to a satisfactory definition of the idea of a *probability of hypotheses*?

I do not believe that it is possible to construct a concept of the probability of hypotheses which may be interpreted as expressing a 'degree of validity' of the hypothesis, in analogy to the concepts 'true' and 'false' (and which, in addition, is sufficiently closely related to the concept 'objective probability', *i.e.* to relative frequency, to justify the use of the word 'probability').[1] Nevertheless, I will now, for the sake of argument, adopt the *supposition* that such a concept has in fact been successfully constructed, in order to raise the question: how would this affect the problem of induction?

Let us suppose that a certain hypothesis—say Schrödinger's theory —is recognized as 'probable' in some definite sense; either as 'probable to this or that numerical degree', or merely as 'probable', without specification of a degree. The statement that describes Schrödinger's theory as 'probable' we may call its *appraisal*.

An appraisal must, of course, be a synthetic statement—an assertion about 'reality'—in the same way as would be the statement 'Schrödinger's theory is true' or 'Schrödinger's theory is false'. All such statements obviously say something about the adequacy of the theory, and are thus certainly not tautological.*[1] They say that a theory

---

[1] (Added while the book was in proof.) It is conceivable that for estimating degrees of corroboration, one might find a formal system showing some limited formal analogies with the calculus of probability (*e.g.* with Bayes's theorem), without however haviing anything in common with the frequency theory. I am indebted to Dr. J. Hosiasson for suggesting this possibility to me. I am satisfied, however, that it is quite impossble to tackle *the problem of induction* by such methods with any hope of success. *See also note 3 to section *57 of my *Postscript*.

* Since 1938, I have upheld the view that 'to justify the use of the word probability', as my text puts it, we should have to show that the axioms of the formal calcuus are satisfied. (*Cf.* appendices *ii to *v, and especially section *28 of my *Postscript*.) This would of course include the satisfaction of Bayes's theorem. As to the formal analogies between Bayes's theorem on *probability* and certain theorems on *degree of corroboration*, see appendix *ix, point 9 (vii) of the first note, and points (12) and (13) of section *32 of my *Postscript*.

*[1] The probability statement '$p(S,e) = r$', in words, 'Schrödinger's theory, given the evidence e, has the probability r'—a statement of relative or conditional logical probability—may certainly be tautological (provided the values of e and r are chosen so as to fit each other: if e consists only of observational reports, r will have to equal zero in a sufficiently large universe). But the 'appraisal', in our sense, would have a different form (see section 84, below, especially the text to note *2)—for example, the following: $p_k(S) = r$, where k is today's date; or in words: 'Schrödinger's theory has *today* (in view of the actual total evidence now available) a probability of r.' In order to obtain this assessment, $p_k(S) = r$, from (i) the tautological statement of relative probability

is adequate or inadequate, or that it is adequate in some degree.
Further, an appraisal of Schrödinger's theory must be a *non-verifiable*
synthetic statement, just like the theory itself. For the 'probability'
of a theory—that is, the probability that the theory will remain
acceptable—cannot, it appears, be deduced from basic statements *with
finality*. Therefore we are forced to ask: How can the appraisal be
justified? How can it be tested? (Thus the problem of induction
arises again; see section 1.)

As to the appraisal itself, this may either be asserted to be 'true',
or it may, in its turn, be said to be 'probable'. If it is regarded as 'true'
then it must be a *true synthetic statement* which has not been empirically
verified—a synthetic statement which is *a priori* true. If it is regarded
as 'probable', then we need a *new* appraisal: an appraisal of the
appraisal, as it were, and therefore an appraisal on a higher level.
But this means that we are caught up in an infinite regress. The appeal
to the probability of the hypothesis is unable to improve the precarious
logical situation of inductive logic.

Most of those who believe in probability logic uphold the view
that the appraisal is arrived at by means of a 'principle of induction'
which ascribes probabilities to the induced hypotheses. But if they
ascribe a probability to this principle of induction in its turn, then the
infinite regress continues. If on the other hand they ascribe 'truth'
to it then they are left with the choice between infinite regress and
a priorism. 'Once and for all', says Heymans, 'the theory of probability
is incapable of explaining inductive arguments; for precisely the
same problem which lurks in the one also lurks in the other (in
the empirical application of probability theory). In both cases the

---

$p(S,e) = r$, and (ii) the statement '*e* is the total evidence available today', we must apply a
principle of inference (called the 'rule of absolution' in my *Postscript*, sections *43 and *51).
This principle of inference looks very much like the *modus ponens*, and it may therefore
seem that it should be taken as analytic. But if we take it to be analytic, then this amounts
to the decision to consider p_k as *defined* by (i) and (ii), or at any rate as meaning *no more*
than do (i) and (ii) together; but in this case, p_k cannot be interpreted as being of any
practical significance: it *certainly* cannot be interpreted as a practical measure of acceptability. This is best seen if we consider that in a sufficiently large universe, $p_k(t,e) \approx 0$ for
every universal theory t, provided e consists only of singular statements. (*Cf.* appendices,
*vii and *viii.) But in practice, we certainly do accept some theories and reject others.

If, on the other hand, we interpret p_k as *degree of adequacy or acceptability*, then the
principle of inference mentioned—the 'rule of absolution' (which, on this interpretation,
becomes a typical example of a 'principle of induction')—is simply *false*, and therefore
clearly non-analytic.

conclusion goes beyond what is given in the premises.'[2] Thus nothing is gained by replacing the word 'true' by the word 'probable', and the word 'false' by the word 'improbable'. Only if *the asymmetry between verification and falsification* is taken into account—that asymmetry which results from the logical relation between theories and basic statements—is it possible to avoid the pitfalls of the problem of induction.

Believers in probability logic may try to meet my criticism by asserting that it springs from a mentality which is 'tied to the framework of classical logic', and which is therefore incapable of following the methods of reasoning employed by probability logic. I freely admit that I am incapable of following these methods of reasoning.

82. *The Positive Theory of Corroboration*: *How a Hypothesis may 'Prove its Mettle'*.

Cannot the objections I have just been advancing against the probability theory of induction be turned, perhaps, against my own view? It might well seem that they can; for these objections are based on the idea of an *appraisal*. And clearly, I have to use this idea too. I speak of the '*corroboration*' of a theory; and corroboration can only be expressed as an appraisal. (In this respect there is no difference between corroboration and probability.) Moreover, I too hold that hypotheses cannot be asserted to be 'true' statements, but that they are 'provisional conjectures' (or something of the sort); and this view, too, can only be expressed by way of an appraisal of these hypotheses.

The second part of this objection can easily be answered. The appraisal of hypotheses which indeed I am compelled to make use of, and which describes them as 'provisional conjectures' (or something of the sort) has the status of a *tautology*. Thus it does not give rise

---

[2] Heymans, *Gesetze und Elemente des wissenschaftlichen Denkens* (1890, 1894), p. 290 f.; * third edition, 1915, p. 272. Heymans's argument was anticipated by Hume in his anonymous pamphlet, *An Abstract of a Book lately published entitled A Treatise of Human Nature*, 1740. I have little doubt that Heymans did not know this pamphlet which was re-discovered and attributed to Hume by J. M. Keynes and P. Sraffa, and published by them in 1938. I knew neither of Hume's nor of Heymans's anticipation of my arguments against the probabilistic theory of induction when I presented them in 1931 in an earlier book, still unpublished, which was read by several members of the Vienna Circle. The fact that Heymans's passage had been anticipated by Hume was pointed out to me by J. O. Wisdom; *cf.* his *Foundations of Inference in Natural Science*, 1952, p. 218. Hume's passage is quoted below, in appendix *vii, text to footnote 6.

to difficulties of the type to which inductive logic gives rise. For this description only paraphrases or interprets the assertion (to which it is equivalent by definition) that strictly universal statements, *i.e.* theories, cannot be derived from singular statements.

The position is similar as regards the first part of the objection which concerns appraisals stating that a theory is corroborated. The appraisal of the corroboration is not a hypothesis, but can be derived if we are given the theory as well as the accepted basic statements. It asserts the fact that these basic statements do not ontradict the theory, and it does this with due regard to the degree of testability of the theory, and to the severity of the tests to which the theory has been subjected, up to a stated period of time.

We say that a theory is 'corroborated' so long as it stands up to these tests. The appraisal which asserts corroboration (the corroborative appraisal) establishes certain fundamental relations, *viz.* compatibility and incompatibility. We regard incompatibility as falsification of the theory. But compatibility alone must not make us attribute to the theory a positive degree of corroboration: the mere fact that a theory has not yet been falsified can obviously not be regarded as sufficient. For nothing is easier than to construct any number of theoretical systems which are compatible with any given system of accepted basic statements. (This remark applies also to all 'metaphysical' systems.)

It might perhaps be suggested that a theory should be accorded some positive degree of corroboration if it is compatible with the system of accepted basic statements, and if, in addition, part of this system can be derived from the theory. Or, considering that basic statements are not derivable from a purely theoretical system (though their negations may be so derivable), one might suggest that the following rule should be adopted: a theory is to be accorded a positive degree of corroboration if it is compatible with the accepted basic statements and if, in addition, a non-empty sub-class of these basic statements is derivable from the theory in conjunction with the other accepted basic statements.*1

*1 The tentative definition of 'positively corroborated' here given (but rejected as insufficient in the next paragraph of the text because it does not explicitly refer to the results of severe tests, *i.e.* of attempted refutations) is of interest in at least two ways. First, it is closely related to my criterion of demarcation, especially to that formulation of it to which I have attached note *1 to section 21. In fact, the two agree except for the restriction to *accepted* basic statements which forms part of the present definition.

I have no serious objections to this last formulation, except that it seems to me insufficient for an adequate characterization of the positive degree of corroboration of a theory. For we wish to speak of theories as being better, or less well, corroborated. But the *degree of corroboration* of a theory can surely not be established simply by counting the number of the corroborating instances, *i.e.* the accepted basic statements which are derivable in the way indicated. For it may happen that one theory appears to be far less well corroborated than another one, even though we have derived very many basic statements with its help, and only a few with the help of the second. As an example we might compare the hypothesis 'All crows are black' with the hypothesis (mentioned in section 37) 'the electronic charge has the value determined by Millikan'. Although in the case of a hypothesis of the former kind, we have presumably encountered many more corroborative basic statements, we shall nevertheless judge Millikan's hypothesis to be the better corroborated of the two.

This shows that it is not so much the number of corroborating instances which determines the degree of corroboration as *the severity of the various tests* to which the hypothesis in question can be, and has been, subjected. But the severity of the tests, in its turn, depends upon the *degree of testability*, and thus upon the simplicity of the hypothesis: the hypothesis which is falsifiable in a higher degree, or the simpler hypothesis, is also the one which is corroborable in a higher degree.[1]

---

[1] This is another point in which there is agreement between my view of simplicity and Weyl's; *cf.* note 7 to section 42. *This agreement is a consequence of the view, due to Jeffreys, Wrinch, and Weyl (*cf.* note 7 to section 42), that the paucity of the parameters of a function can be used as a measure of its simplicity, taken in conjunction with my view (*cf.* sections 38 ff.) that the paucity of the parameters can be used as a measure of testability or improbability—a view rejected by these authors. (See also notes *1 and *2 to sections 43.)

Thus if we omit this restriction, the present definition turns into my criterion of demarcation.

Secondly, if instead of omitting this restriction we restrict the class of the *derived* accepted basic statements further, by demanding that they should be accepted as the results of sincere attempts to refute the theory, then our definition becomes an adequate definition of 'positively corroborated', though not, of course, of 'degree of corroboration'. The argument supporting this claim is implicit in the text here following. Moreover, the basic statements so accepted may be described as 'corroborating statements' of the theory.

It should be noted that 'instantial statements' (*i.e.* negated basic statements; see section 28) cannot be adequately described as corroborating or confirming statements of the theory which they instantiate, owing to the fact that we know that *every universal law is instantiated* almost everywhere, as indicated in note *1 to section 28. (See also note *4 to section 80, and text.)

Of course, the degree of corroboration actually attained does not depend *only* on the degree of falsifiability: a statement may be falsifiable to a high degree yet it may be only slightly corroborated, or it may in fact be falsified. And it may perhaps, without being falsified, be superseded by a better testable theory from which it—or a sufficiently close approximation to it—can be deduced. (In this case too its degree of corroboration is lowered.)

The degree of corroboration of two statements may not be comparable in all cases, any more than the degree of falsifiability: we cannot define a numerically calculable degree of corroboration, but can speak only roughly in terms of positive degrees of corroboration, negative degrees of corroboration, and so forth.*2 Yet we can lay down various rules; for instance the rule that we shall not continue to accord a positive degree of corroboration to a theory which has been falsified by an inter-subjectively testable experiment based upon a falsifying hypothesis (*cf.* sections 8 and 22). (We may, however, under certain circumstances accord a positive degree of corroboration to another theory, even though it follows a kindred line of thought. An example is Einstein's photon theory, with its kinship to Newton's corpuscular theory of light.) In general we regard an inter-subjectively testable falsification as final (provided it is well tested): this is the way in which the asymmetry between verification and falsification of theories makes itself felt. Each of these methodological points contributes in its own peculiar way to the historical development of science as a process of step by step approximations. A corroborative appraisal made at a later date—that is, an appraisal made after new basic statements have been added to those already accepted—can replace a positive degree of corroboration by a negative one, but not *vice versa.* And although I believe that in the history of science it is always the theory and not the experiment, always the idea and not the observation, which opens up the way to new knowledge, I also believe that it is always the experiment which saves us from following a track that leads nowhere: which helps us out of the rut, and which challenges us to find a new way.

*2 As far as practical application to existing theories goes, this seems to me still correct; but I think now that it is possible to define 'degree of corroboration' in such a way that we can *compare* degrees of corroboration (for example, those of Newton's and of Einstein's theory of gravity). Moreover, this definition makes it even possible to attribute numerical degrees of corroboration to statistical hypotheses, and perhaps even to other statements *provided* we can attribute degrees of (absolute and relative) logical probability to them. See also appendix *ix.

Thus the degree of falsifiability or of simplicity of a theory enters into the appraisal of its corroboration. And this appraisal may be regarded as one of the logical relations between the theory and the accepted basic statements: as an appraisal that takes into consideration the severity of the tests to which the theory has been subjected.

83. *Corroborability, Testability, and Logical Probability.*[*1]

In appraising the degree of corroboration of a theory we take into account its degree of falsifiability. A theory can be the better corroborated the better testable it is. Testability, however, is converse to the concept of *logical probability*, so that we can also say that an appraisal of corroboration takes into account the logical probability of the statement in question. And this, in turn, as was shown in section 72, is related to the concept of objective probability—the probability of events. Thus by taking logical probability into account the concept of corroboration is linked, even if perhaps only indirectly and loosely, with that of the probability of events. The idea may occur to us that there is perhaps a connection here with the doctrine of the probability of hypotheses criticized above.

When trying to appraise the degree of corroboration of a theory we may reason somewhat as follows. Its degree of corroboration will increase with the number of its corroborating instances. Here we usually accord to the first corroborating instances far greater importance than to later ones: once a theory is well corroborated, further instances raise its degree of corroboration only very little. This rule however does not hold good if these new instances are very different from the earlier ones, that is if they corroborate the theory in a *new field of application*. In this case, they may increase the degree of corroboration very considerably. The degree of corroboration of a theory which has a higher degree of universality can thus be greater than that of a theory which has a lower degree of universality (and therefore a lower degree of falsifiability). In a similar way, theories of a higher degree of precision can be better corroborated than less precise ones. One of the reasons why we do not accord a positive degree of corroboration to the typical prophecies of palmists and soothsayers is that their pre-

[*1] If the terminology is accepted which I first explained in my note in *Mind*, 1938, then the word 'absolute' should be inserted here throughout (as in sections 34, etc.) before 'logical probability' (in contradistinction to 'relative' or 'conditional' logical probability); *cf.* appendices *ii, *iv, and *ix.

dictions are so cautious and imprecise that the logical probability of their being correct is extremely high. And if we are told that more precise and thus logically less probable predictions of this kind have been successful, then it is not, as a rule, their success that we are inclined to doubt so much as their alleged logical improbability: since we tend to believe that such prophecies are non-corroborable, we also tend to argue in such cases from their low degree of corroborability to their low degree of testability.

If we compare these views of mine with what is implicit in (inductive) probability logic, we get a truly remarkable result. According to my view, the corroborability of a theory—and also the degree of corroboration of a theory which has in fact passed severe tests, stand both, as it were,*² in inverse ratio to its logical probability; for they both increase with its degree of testability and simplicity. *But the view implied by probability logic is the precise opposite of this.* Its upholders let the probability of a hypothesis increase in *direct proportion* to its logical probability—although there is no doubt that they *intend* their 'probability of a hypothesis' to stand for much the same thing that I try to indicate by 'degree of corroboration'.*³

*² I said in the text '*as it were*': I did so because I did not really believe in numerical (absolute) logical probabilities. In consequence of this, I wavered, when writing the text, between the view that the degree of corroborability is *complementary* to (absolute) logical probability and the view that it is inversely proportional; or in other words, between a definition of $C(g)$, *i.e.* the degree of corroborability, by $C(g) = 1 - P(g)$ which would make *corroborability equal to content*, and by $C(g) = 1/P(g)$, where $P(g)$ is the absolute logical probability of g. In fact, definitions may be adopted which lead to either of these consequences, and both ways seem fairly satisfactory on intuitive grounds; this explains, perhaps, my wavering. There are strong reasons in favour of the first method, or else of a logarithmic scale applied to the second method. See appendix *ix.

*³ The last lines of this paragraph, especially from the italicized sentence on (it was not italicized in the original) contain the crucial point of my criticism of the probability theory of induction. The point may be summarized as follows.

We want *simple* hypotheses—hypotheses of a high *content*, a high degree of *testability*. These are also the highly *corroborable* hypotheses, for the degree of corroboration of a hypothesis depends mainly upon the severity of its tests, and thus upon its testability. Now we know that testability is the same as high (absolute) logical *improbability*, or low (absolute) logical *probability*.

But if two hypotheses, h_1 and h_2, are comparable with respect to their content, and thus with respect to their (absolute) logical probability, then the following holds: let the (absolute) logical probability of h_1 be smaller than that of h_2. Then, whatever the evidence e, the (relative) logical probability of h_1 given e can never exceed that of h_2 given e. Thus the *better testable and better corroborable hypothesis can never obtain a higher probability, on the given evidence, than the less testable one*. But this entails that *degree of corroboration cannot be the same as probability*.

This is the crucial result. My later remarks in the text merely draw the conclusion from it: if you value high probability, you must say very little—or better still, nothing at all: tautologies will always retain the highest probability.

Among those who argue in this way is Keynes who uses the expression 'a priori probability' for what I call 'logical probability'. (See note 1 to section 34.) He makes the following perfectly accurate remark[1] regarding a 'generalization' g (i.e. a hypothesis) with the 'condition' or antecedent or protasis φ and the 'conclusion' or consequent or apodosis f: 'The more comprehensive the condition φ and the less comprehensive the conclusion f, the greater à priori*[4] probability do we attribute to the generalization g. With every increase in φ this probability increases, and with every increase in f it will diminish.' This, as I said, is perfectly accurate, even though Keynes does not draw a sharp distinction*[5] between what he calls the 'probability of a generalization'—corresponding to what is here called the 'probability of a hypothesis'—and its 'a priori probability'. That Keynes nevertheless intends by his 'probability' the same as I do by my 'corroboration' may be seen from the fact that his 'probability' rises with the number of corroborating instances, and also (most important) with the increase of diversity among them. (But Keynes overlooks the fact that theories whose corroborating instances belong to widely different fields of application will usually have a correspondingly high degree of universality. Hence his two requirements for obtaining a high probability—the least possible universality and the greatest possible diversity of corroborating instances—will be, as a rule, incompatible.)

Expressed in my terminology, Keynes's theory implies that corroboration (or the probability of hypotheses) *decreases* with testability.

[1] Keynes, *A Treatise on Probability* (1921), pp. 224 f. Keynes's condition φ and conclusion f correspond (cf. note 6 to section 14) to our conditioning statement function φ and our consequence statement function f; cf. also section 36. It should be noticed that Keynes called the condition or the conclusion *more comprehensive* if its *content*, or its intension, rather than its extension, is the greater. (I am alluding to the inverse relationship holding between the intension and the extension of a term.)

*[4] Keynes persistently follows some other eminent Cambridge logicians in writing 'à priori' and 'à posteriori'; one can only say, à propos de rien—unless, perhaps, apropos of 'à propos'.

*[5] Keynes does, in fact, allow for the distinction between the a priori (or 'absolute logical', as I now call it) probability of the 'generalization' g and its probability with respect to a given piece of evidence h, and to this extent, my statement in the text needs correction. (He makes the distinction by assuming, correctly though perhaps only implicitly—see p. 225 of the *Treatise*—that if $\varphi = \varphi_1 \varphi_2$, and $f = f_1 f_2$, then the *a priori* probabilities of the various g are: $g(\varphi, f_1) \geqslant g(\varphi, f) \geqslant g(\varphi_1, f)$.) And he correctly *proves* that the (*a posteriori*) probabilities of these hypotheses g (with respect to any given piece of evidence h) are related in the same way as their *a priori* probabilities. Thus he proves that the probabilities are related like the (absolute) logical probabilities; while my cardinal point was, and still is, that their degree of corroborability and of corroboration are related in the opposite way.

He is led to this view by his belief in inductive logic.*⁶ For it is the tendency of inductive logic to make scientific hypotheses as *certain* as possible. Scientific significance is assigned to the various hypotheses only to the extent to which they can be justified by experience. A theory is regarded as scientifically valuable only because of the close *logical proximity* (*cf.* note 2 to section 48 and text) between the theory and empirical statements. But this means nothing else than that the *content* of the theory must go *as little as possible* beyond what is empirically established.*⁷ This view is closely connected with a tendency to deny the value of prediction. 'The peculiar virtue of prediction' Keynes writes² '. . . is altogether imaginary. The number of instances examined and the analogy between them are the essential points, and the question as to whether a particular hypothesis happens to be propounded before or after their examination is quite irrelevant.' In reference to hypotheses which have been '*a priori* proposed'—that is, proposed before we had sufficient support for them on inductive grounds—Keynes writes: '. . . if it is a mere guess, the lucky fact of its preceding some or all of the cases which verify it adds nothing whatever to its value.' This view of prediction is certainly consistent. But it makes one wonder why we should ever have to generalize at all. What possible reason can there be for constructing all these theories and hypotheses? The standpoint of inductive logic makes these activities quite incomprehensible. If what we value most is the securest knowledge available—and if predictions as such contribute nothing towards corroboration—why then may we not rest content with our basic statements?*⁸

Another view which gives rise to very similar questions is that

---

*⁶ See my *Postscript*, chapter *ii. In my theory of corroboration—in direct opposition to Keynes's, Jeffreys's, and Carnap's theories of probability—corroboration does not *decrease* with testability, but tends to *increase* with it.

*⁷ This may also be expressed by the unacceptable rule: 'Always choose the hypothesis which is most *ad hoc*!'

² Keynes, *op. cit.*, p. 305.

*⁸ Carnap, in his *Logical Foundations of Probability* (1950) believes in the *practical* value of predictions; nevertheless, he draws part of the conclusion here mentioned—that we might be content with our basic statements. For he says that theories (he speaks of 'laws') are 'not indispensable' for science—not even for making predictions: we can manage throughout with singular statements. 'Nevertheless', he writes (p. 575) 'it is expedient, of course, to state universal laws in books on physics, biology, psychology, *etc.*' But the question is not one of expediency—it is one of scientific curiosity. *Some scientists want to explain the world*: their aim is to find satisfactory explanatory theories—well testable, *i.e.* simple theories—and to test them. (See also appendix *x and section *15 of my *Postscript*.)

of Kaila.[3] Whilst I believe that it is the simple theories, and those which make little use of auxiliary hypotheses (*cf.* section 46) which can be well corroborated, just because of their logical improbability, Kaila interprets the situation in precisely the opposite way, on grounds similar to Keynes's. He too sees that we usually ascribe a high probability (in our terminology, a high 'probability of hypotheses') to *simple* theories, and especially to those needing few auxiliary hypotheses. But his reasons are the opposite of mine. He does not, as I do, ascribe a high probability to such theories because they are severely testable, or logically improbable; that is to say because they have, *a priori* as it were, *many opportunities of clashing with basic statements*. On the contrary he ascribes this high probability to simple theories with few auxiliary hypotheses because he believes that a system consisting of *few* hypotheses will, *a priori*, have *fewer* opportunities of clashing with reality than a system consisting of many hypotheses. Here again one wonders why we should ever bother to construct these adventurous theories. If we shrink from conflict with reality, why invite it by making assertions? Our aim being security, our safest course would be to adopt a system *without* hypotheses.

My own rule which requires that auxiliary hypotheses shall be used as sparingly as possible (the 'principle of parsimony in the use of hypotheses') has nothing whatever in common with considerations such as Kaila's. I am not interested in merely keeping down the number of our statements: I am interested in their *simplicity in the sense of high testability*. It is this interest which leads, on the one hand, to my rule that auxiliary hypotheses should be used as sparingly as possible, and on the other hand, to my demand that the number of our axioms—of our most fundamental hypotheses—should be kept down. For this latter point arises out of the demand that statements of a high level of universality should be chosen, and that a system consisting of many 'axioms' should, if possible, be deduced from (and thus explained by) one with fewer 'axioms', and with axioms of a higher level of universality.

84. Remarks Concerning the Use of the Concepts 'True' and 'Corroborated'. In the logic of science here outlined it is possible to avoid using

---

[3] Kaila, *Die Principien der Wahrscheinlichkeitslogik* (*Annales Universitatis Aboensis,* Turku 1926), p. 140.

R—L S D

the concepts 'true' and 'false'.*¹ Their place may be taken by logical considerations about derivability relations. Thus we need not say: 'The prediction p is true provided the theory t and the basic statement b are true.' We may say, instead, that the statement p follows from the (non-contradictory) conjunction of t and b. The falsification of a theory may be described in a similar way. We need not say that the theory is 'false', but we may say instead that it is contradicted by a certain set of accepted basic statements. Nor need we say of basic statements that they are 'true' or 'false', for we may interpret their acceptance as the result of a conventional decision, and the accepted statements as results of this decision.

This certainly does not mean that we are forbidden to use the concepts 'true' and 'false', or that their use creates any particular difficulty. The very fact that we can avoid them shows that they cannot give rise to any new fundamental problem. The use of the concepts 'true' and 'false' is quite analogous to the use of such concepts as

---

*¹ Not long after this was written, I had the good fortune to meet Alfred Tarski who explained to me the fundamental ideas of his theory of truth. It is a great pity that this theory—one of the two great discoveries in the field of logic made since *Principia Mathematica*—is still often misunderstood and misrepresented. It cannot be too strongly emphasized that Tarski's idea of truth (for whose definition with respect to formalized languages Tarski gave a method) is the same idea which Aristotle had in mind and indeed most people (except pragmatists): the idea that *truth is correspondence with the facts* (or with reality). But what can we possibly mean if we say of a *statement* that it corresponds with the *facts* (or with reality)? Once we realize that this correspondence cannot be one of structural similarity, the task of elucidating this correspondence seems hopeless; and as a consequence, we may become suspicious of the concept of truth, and prefer not to use it. Tarski solved this apparently hopeless problem (with respect to formalized languages), by reducing the unmanageable idea of correspondence to a simpler idea (that of 'satisfaction' or 'fulfilment').

Owing to Tarski's teaching, I am no longer hesitant in speaking of 'truth' and 'falsity'. And like everybody else's views (unless he is a pragmatist), my views turned out, as a matter of course, to be consistent with Tarski's theory of absolute truth. Thus although my views on formal logic and its philosophy were revolutionized by Tarski's theory, my views on science and its philosophy were unaffected.

The current criticism of Tarski's theory seems to me completely off the mark. It is said that his definition is artificial and complex; but since he defines truth with respect to formalized languages, it has to be based on the definition of a well-formed formula in such a language; and it is of precisely the same degree of 'artificiality' or 'complexity' as this definition. It is also said that only propositions or statements can be true or false, but not sentences. Perhaps 'sentence' was not a good translation of Tarski's original terminology. (I personally prefer to speak of 'statement' rather than of 'sentence'; see for example my 'Note on Tarski's Definition of Truth', *Mind* **64**, 1955, p. 388, footnote 1.) But Tarski himself made it perfectly clear that an uninterpreted formula (or a string of symbols) cannot be said to be true or false, and that these terms only apply to interpreted formulae—to '*meaningful* sentences' (as the translation has it). Improvements in terminology are always welcome; but it is sheer obscurantism to criticize a theory on terminological grounds.

'*tautology*', '*contradiction*', '*conjunction*', '*implication*' and others of the kind. These are non-empirical concepts, logical concepts.[1] They describe or appraise a statement irrespective of any changes in the empirical world. Whilst we assume that the properties of physical objects (of 'genidentical' objects in Lewin's sense) change with the passage of time, we decide to use these logical predicates in such a way that the logical properties of statements become timeless: if a statement is a tautology, then it is a tautology once and for all. This same timelessness we also attach to the concepts 'true' and 'false', in agreement with common usage. It is not common usage to say of a statement that it was perfectly true yesterday but has become false today. If yesterday we appraised a statement as true which today we appraise as false, then we implicitly assert today that *we were mistaken yesterday*; that the statement was false even yesterday—timelessly false—but that we erroneously 'took it for true'.

Here one can see very clearly the difference between truth and corroboration. The appraisal of a statement as corroborated or as not coroborated is also a logical appraisal and therefore also timeless; for it asserts that a certain logical relation holds between a theoretical system and some system of accepted basic statements. But we can never simply say of a statement that it is as such, or in itself, 'corroborated' (in the way in which we may say that it is 'true'). We can only say that it is *corroborated with respect to some system of basic statements*—a system accepted up to a particular point in time. 'The corroboration which a theory has received up to yesterday' is *logically not identical* with 'the corroboration which a theory has received up to today'. Thus we must attach a subscript, as it were, to every appraisal of corroboration—a subscript characterizing the system of basic statements to which the corroboration relates (for example, by the date of its acceptance).*[2]

Corroboration is therefore not a 'truth value'; that is, it cannot be placed on a par with the concepts 'true' and 'false' (which are free from temporal subscripts); for to one and the same statement there may be any number of different corroboration values, of which indeed all can be 'correct' or 'true' at the same time. For they are values which are logically derivable from the theory and the various sets of basic statements accepted at various times.

(Added in 1934 in proof.)
[1] Carnap would probably say 'syntactical concepts' (*cf.* his *Logical Syntax of Language*).
*[2] *Cf.* note *1 to section 81.

The above remarks may also help to elucidate the contrast between my views and those of the pragmatists who propose to *define 'truth' in terms of the success of a theory—and thus of its usefulness, or of its confirmation or of its corroboration.* If their intention is merely to assert that a logical appraisal of the success of a theory can be no more than an appraisal of its corroboration, I can agree. But I think that it would be far from '*useful*' to identify the concept of corroboration with that of truth.*3 This is also avoided in ordinary usage. For one might well say of a theory that it has hardly been corroborated at all so far, or that it is still uncorroborated. But we should not normally say of a theory that it is hardly true at all so far, or that it is still false.

85. *The Path of Science.*

One may discern something like a general direction in the evolution of physics—a direction from theories of a lower level of universality to theories of a higher level. This is usually called the 'inductive' direction; and it might be thought that the fact that physics advances in this 'inductive' direction could be used as an argument in favour of the inductive method.

Yet an advance in the inductive direction does not necessarily consist of a sequence of inductive inferences. Indeed we have shown that it may be explained in quite different terms—in terms of degree of testability and corroborability. For a theory which has been well corroborated can only be superseded by one of a higher level of universality; that is, by a theory which is better testable and which, in addition, *contains* the old, well corroborated theory—or at least a good approximation to it. It may be better, therefore, to describe that trend—the advance towards theories of an ever higher level of universality—as 'quasi-inductive'.

The quasi-inductive process should be envisaged as follows. Theories of some level of universality are proposed, and deductively tested; after that, theories of a higher level of universality are proposed, and in their turn tested with the help of those of the previous levels of universality, and so on. The methods of testing are invariably

---

*3 Thus if we were to define 'true' as 'useful' or 'successful' or 'confirmed' or 'corroborated', we should only have to introduce a new 'absolute' and 'timeless' concept to play the role of 'truth'.

based on deductive inferences from the higher to the lower level;[1] on the other hand, the levels of universality are reached, in the order of time, by proceeding from lower to higher levels.

The question may be raised: 'Why not invent theories of the highest level of universality straight away? Why wait for this quasi-inductive evolution? Is it not perhaps because there is after all an inductive element contained in it?' I do not think so. Again and again suggestions are put forward—conjectures, or theories—of all possible levels of universality. Those theories which are on too high a level of universality, as it were (that is, too far removed from the level reached by the testable science of the day) give rise, perhaps, to a 'metaphysical system'. In this case, even if from this system statements should be deducible (or only semi-deducible, as for example in the case of Spinoza's system), which belong to the prevailing scientific system, there will be no *new* testable statement among them; which means that no crucial experiment can be designed to test the system in question.[2] If, on the other hand, a crucial experiment can be designed for it, then the system will contain, as a first approximation, some well corroborated theory, and at the same time also something new—and something that can be tested. Thus the system will not, of course, be 'metaphysical'. In this case, the system in question may be looked upon as a new advance in the quasi-inductive evolution of science. This explains why a link with the science of the day is as a rule established only by those theories which are proposed in an attempt to meet the current problem situation; that is, the current difficulties, contradictions, and falsifications. In proposing a solution to these difficulties, these theories may point the way to a crucial experiment.

To obtain a picture or model of this quasi-inductive evolution of science, the various ideas and hypotheses might be visualized as particles suspended in a fluid. Testable science is the precipitation of these particles at the bottom of the vessel: they settle down in layers (of universality). The thickness of the deposit grows with the

[1] The 'deductive inferences from the higher to the lower level' are, of course, *explanations* (in the sense of section 12); thus the hypotheses on the higher level are *explanatory* with respect to those on the lower level.

[2] It should be noted that I mean by a crucial experiment one that is designed to refute a theory (if possible) and more especially one which is designed to bring about a decision between two competing theories by refuting (at least) one of them—without, of course, proving the other. (See also note 1 to section 22, and appendix *ix.)

number of these layers, every new layer corresponding to a theory more universal than those beneath it. As the result of this process ideas previously floating in higher metaphysical regions may sometimes be reached by the growth of science, and thus make contact with it, and settle. Examples of such ideas are atomism; the idea of a single physical 'principle' or ultimate element (from which the others derive); the theory of terrestrial motion (opposed by Bacon as fictitious); the age-old corpuscular theory of light; the fluid-theory of electricity (revived as the electron-gas hypothesis of metallic conduction). All these metaphysical concepts and ideas may have helped, even in their early forms, to bring order into man's picture of the world, and in some cases they may even have led to successful predictions. Yet an idea of this kind acquires scientific status only when it is presented in falsifiable form; that is to say, only when it has become possible to decide empirically between it and some rival theory.

My investigation has traced the various consequences of the decisions and conventions—in particular of the criterion of demarcation—adopted at the beginning of this book. Looking back, we may now try to get a last comprehensive glimpse of the picture of science and of scientific discovery which has emerged. (What I have here in mind is not a picture of science as a biological phenomenon, as an instrument of adaptation, or as a roundabout method of production: I have in mind its epistemological aspects.)

Science is not a system of certain, or well-established, statements; nor is it a system which steadily advances towards a state of finality. Our science is not knowledge (*epistēmē*): it can never claim to have attained truth, or even a substitute for it, such as probability.

Yet science has more than mere biological survival value. It is not only a useful instrument. Although it can attain neither truth nor probability, the striving for knowledge and the search for truth are still the strongest motives of scientific discovery.

We do not know: we can only guess. And our guesses are guided by the unscientific, the metaphysical (though biologically explicable) faith in laws, in regularities which we can uncover—discover. Like Bacon, we might describe our own contemporary science—'the method of reasoning which men now ordinarily apply to nature'—as consisting of 'anticipations, rash and premature' and as 'prejudices'.[1]

---

[1] Bacon, *Novum Organum*, I, 26.

But these marvellously imaginative and bold conjectures or 'anticipations' of ours are carefully and soberly controlled by systematic tests. Once put forward, none of our 'anticipations' are dogmatically upheld. Our method of research is not to defend them, in order to prove how right we were. On the contrary, we try to overthrow them. Using all the weapons of our logical, mathematical, and technical armoury we try to prove that our anticipations were false—in order to put forward, in their stead, new unjustified and unjustifiable anticipations, new 'rash and premature prejudices', as Bacon derisively called them.[*1]

It is possible to interpret the ways of science more prosaically. One might say that progress can '. . . come about only in two ways: by gathering new perceptual experiences, and by better organizing those which are available already'.[2] But this description of scientific progress, although not actually wrong, seems to miss the point. It is too reminiscent of Bacon's induction: too suggestive of his industrious gathering of the 'countless grapes, ripe and in season',[3] from which he expected the wine of science to flow: of his myth of a scientific method that starts from observation and experiment and then proceeds to theories. (This legendary method, by the way, still inspires some of the newer sciences which try to practice it because of the prevalent belief that it is the method of experimental physics.)

The advance of science is not due to the fact that more and more

---

[*1] Bacon's term 'anticipation' ('*anticipatio*') means almost the same as 'hypothesis' (in my way of using this term). Bacon's view was that, to prepare the mind for the intuition of the true *essence* or *nature* of a thing, it has to be meticulously cleansed of all anticipations, prejudices, and idols. For the source of all error is the impurity of our own minds: Nature itself does not lie. The main function of eliminative induction is (as with Aristotle) to assist the purification of the mind. (*See also* my *Open Society*, chapter 24; note 59 to chapter 10; note 33 to chapter 11, where Aristotle's theory of induction is briefly described.) Purging the mind of prejudices is conceived as a kind of ritual prescribed for the scientist, analogous to the mystic's purification of his soul to prepare it for the vision of God. (*Cf.* section *4 of my *Postscript*.)

[2] P. Frank, *Das Kausalgesetz und seine Grenzen*, (1932). *The view that the progress of science is due to the accumulation of perceptual experiences is still widely held (*cf.* my second Preface, 1958). My denial of this view is closely connected with the rejection of the doctrine that science or knowledge is *bound* to advance, since our experiences are *bound* to accumulate. As against this, I believe that the advance of science depends upon the free competition of thought, and thus upon freedom, and that it must come to an end if freedom is destroyed (though it may well continue for some time in some fields, especially in technology). This view is more fully expounded in my *Poverty of Historicism* (section 32). I also argue there (in the Preface) that the growth of our knowledge is unpredictable by scientific means, and that, as a consequence, the future course of our history is also unpredictable.

[3] Bacon, *Novum Organum* I, 123.

perceptual experiences accumulate in the course of time. Nor is it due to the fact that we are making ever better use of our senses. Out of uninterpreted sense-experiences science cannot be distilled, no matter how industriously we gather and sort them. Bold ideas, unjustified anticipations, and speculative thought, are our only means for interpreting nature: our only organon, our only instrument, for grasping her. And we must hazard them to win our prize. Those among us who are unwilling to expose their ideas to the hazard of refutation do not take part in the scientific game.

Even the careful and sober testing of our ideas by experience is in its turn inspired by ideas: experiment is planned action in which every step is governed by theory. We do not stumble upon our experiences, nor do we let them flow over us like a stream. Rather, we have to be active: we have to *make* our experiences. It is we who always formulate the questions to be put to nature; it is we who try again and again to put these question so as to elicit a clear-cut 'yes' or 'no' (for nature does not give an answer unless pressed for it). And in the end, it is again we who give the answer; it is we ourselves who, after severe scrutiny, decide upon the answer to the question which we put to nature—after protracted and earnest attempts to elicit from her an unequivocal 'no'. 'Once and for all', says Weyl,[4] with whom I fully agree, 'I wish to record my unbounded admiration for the work of the experimenter in his struggle to wrest *interpretable facts* from an unyielding Nature who knows so well how to meet our theories with a decisive *No*—or with an inaudible *Yes*.'

The old scientific ideal of *epistēmē*—of absolutely certain, demonstrable knowledge—has proved to be an idol. The demand for scientific objectivity makes it inevitable that every scientific statement must remain *tentative for ever*. It may indeed be corroborated, but every corroboration is relative to other statements which, again, are tentative. Only in our subjective experiences of conviction, in our subjective faith, can we be 'absolutely certain'.[5]

With the idol of certainty (including that of degrees of imperfect certainty or probability) there falls one of the defences of obscurantism which bars the way of scientific advance, checking the boldness of our

<hr>

[4] Weyl, *Gruppentheorie und Quantenmechanik* (1931), p. 2. English translation by H. P. Robertson: *The Theory of Groups and Quantum Mechanics* (1931), p. xx.
[5] *Cf.* for example note 3 to section 30. This last remark is of course a psychological remark rather than an epistemological one; *cf.* sections 7 and 8.

questions, and endangering the rigour and the integrity of our tests. The wrong view of science betrays itself in the craving to be right; for it is not his *possession* of knowledge, of irrefutable truth, that makes the man of science, but his persistent and recklessly critical *quest* for truth.

Has our attitude, then, to be one of resignation? Have we to say that science can fulfil only its biological task; that it can, at best, merely prove its mettle in practical applications which may corroborate it? Are its intellectual problems insoluble? I do not think so. Science never pursues the illusory aim of making its answers final, or even probable. Its advance is, rather, towards the infinite yet attainable aim of ever discovering new, deeper, and more general problems, and of subjecting its ever tentative answers to ever renewed and ever more rigorous tests.

APPENDICES

Appendix i. *Definition of the Dimension of a Theory.* (*Cf.* sections 38 and 39.)

The definition which follows here should be regarded as only provisional.*[1] It is an attempt to define the dimension of a theory so as to make it agree with the dimension of the set of curves which results if the field of application of the theory is represented by a graph paper. A difficulty arises from the fact that we should not assume that either a metric or even a topology is defined for the field, to begin with; in particular, we should not assume that any neighbour-hood relations are defined. And I admit that this difficulty is circum-vented rather than overcome by the definition proposed. The possibility of circumventing it is connected with the fact that a theory always prohibits some '*homotypic*' *events*, as we have called them (*i.e.* a class of occurrences which differ only in their spatio-temporal co-ordinates; *cf.* sections 23 and 31). For this reason, spatio-temporal co-ordinates will, in general, appear in the schema which generates the field of application, and consequently the field of the relatively atomic statements will, in general, show a topological and even a metrical order.

The proposed definition says: A theory t is called 'd-dimensional with respect to the field of application F' if and only if the following relation holds between t and F: there is a number d such that (a) the theory does not clash with any d-tuple of the field and (b) any given d-tuple in conjunction with the theory divides all the remaining relatively atomic statements uniquely into two infinite sub-classes A

---

*[1] A simplified and slightly more general definition is this. Let A and X be two sets of statements. (Intuitively, A is a set of universal laws, X a set—usually infinite—of singular test statements.) Then we say that X is a (homogeneous) field of application with respect to A (in symbols: $X = F_A$) if and only if for every statement a in A, there exists a natural number $d(a) = n$ which satisfies the following two conditions: (i) any conjunction c_n of n different statements of X is compatible with a; (ii) for any such conjunction c_n there exist two statements x and y in X such that $x.c_n$ is incompatible with a and $y.c_n$ is derivable from $a.c_n$, but neither from a nor from c_n.

$d(a)$ is called the dimension of a, or the degree of composition of a, with respect to $X = F_A$; and $1/d(a)$ may be taken as a measure of the simplicity of a.

The problem is further developed in appendix *viii.

285

and B, such that the following conditions are satisfied: (α) every statement of the class A forms, when conjoined with the given d-tuple, a 'falsifying $d + 1$-tuple' *i.e.* a potential falsifier of the theory; (β) the class B on the other hand is the sum of one or more, but always a finite number, of infinite sub-classes $[B_i]$ such that the conjunction of any number of statements belonging to any one of these sub-classes $[B_i]$ is compatible with the conjunction of the given d-tuple and the theory.

This definition is intended to exclude the possibility of a theory's having two fields of application such that the relatively atomic statements of one field result from the conjunction of the relatively atomic statements of the other (this must be prevented if the field of application is to be identifiable with that of its graphic representation; *cf.* section 39). I may add that by means of this definition the problem of atomic statements (*cf.* note 2 to section 38) is solved in a manner which might be described as 'deductivist', since the theory itself determines which singular statements are *relatively atomic* (with respect to the theory). For it is the theory itself through which the field of application is defined—and with it the statements which because of their logical form have equal status with respect to the theory. Thus the problem of the atomic statements is not solved by the discovery of statements of some elementary form out of which the other, more composite, statements are built up inductively, or composed by the method of truth-functions. On the contrary, the relatively atomic statements—and along with them the singular statements—appear as a sort of precipitation, as it were, or as a (relatively) solid deposit, laid down by the universal statements of the theory.

Appendix ii. *The General Calculus of Frequency in Finite Classes. (Cf.* sections 52 and 53.)[*1]

The General Multiplication Theorem: We denote the finite reference class by 'α', and the two property classes by 'β' and 'γ'. Our first problem is to determine the frequency of those elements which belong both to β and to γ.

The solution is given by the formula

$$_{\alpha}F''(\beta.\gamma) = {_{\alpha}F''(\beta)} \cdot {_{\alpha.\beta}F''(\gamma)} \qquad (1)$$

or, since β and γ may be commuted:

$$_{\alpha}F''(\beta.\gamma) = {_{\alpha.\gamma}F''(\beta)} \cdot {_{\alpha}F''(\gamma)} \qquad (1')$$

The proof results immediately from the definition given in section 52. From (1) we obtain, by substitution in accordance with this definition:

$$\frac{N(\alpha.\beta.\gamma)}{N(\alpha)} = \frac{N(\alpha.\beta)}{N(\alpha)} \cdot \frac{N(\alpha.\beta.\gamma)}{N(\alpha.\beta)} \qquad (1,1)$$

which proves to be an identity after cancellation of '$N(\alpha.\beta)$'. (Contrast with this proof, and with the proof of (2_s), Reichenbach, 'Axiomatik der Wahrscheinlichkeitsrechnung', *Mathematische Zeitschrift* **34**, p. 593.)

If we assume *independence* (*cf.* section 53), *i.e.*

$$_{\alpha.\beta}F''(\gamma) = {_{\alpha}F''(\gamma)} \qquad (1^s)$$

we obtain from (1) the *special multiplication theorem*

$$_{\alpha}F''(\beta.\gamma) = {_{\alpha}F''(\beta)} \cdot {_{\alpha}F''(\gamma)} \qquad (1_s)$$

With the help of the equivalence of (1) and (1'), the symmetry of the relation of independence can now be proved (*cf.* also note 4 to section 53).

[*1] This appendix has since been developed by me into an axiomatic treatment of probability. See appendices *iii to *v.

The addition theorems deal with the frequency of those elements which belong either to β or to γ. If we denote the disjunctive combination of these classes by the symbol 'β + γ', where the sign '+', *if placed between class designations*, signifies not arithmetical addition but the *non-exclusive* 'or', then the general addition theorem is:

$$_\alpha F''(\beta+\gamma) = {}_\alpha F''(\beta) + {}_\alpha F''(\gamma) - {}_\alpha F''(\beta.\gamma) \qquad (2)$$

The proof of this statement is obtained from the definition in section 52 by the use of the universally valid formula of the calculus of classes

$$\alpha.(\beta+\gamma) = (\alpha.\beta)+(\alpha.\gamma), \qquad (2,2)$$

and the formula (which is also universally valid)

$$N(\beta+\gamma) = N(\beta) + N(\gamma) - N(\beta.\gamma) \qquad (2,1)$$

Under the assumption that, within α, β and γ have no member in common—a condition which can be symbolized by the formula

$$N(\alpha.\beta.\gamma) = 0 \qquad (2^s)$$

—we obtain from (2) the *special addition theorem*

$$_\alpha F''(\beta+\gamma) = {}_\alpha F''(\beta) + {}_\alpha F''(\gamma). \qquad (2_s)$$

The special addition theorem holds for *all* properties which are *primary properties* within a class α, since primary properties are mutually exclusive. The sum of the relative frequencies of these primary properties is of course always equal to 1.

The division theorems state the frequency of a property γ within a class *selected* from α with respect to the property β. The general formula is obtained immediately by inversion of (1).

$$_{\alpha.\beta} F''(\gamma) = \frac{{}_\alpha F''(\beta.\gamma)}{{}_\alpha F''(\beta)} \qquad (3)$$

If we transform the *general division theorem* (3) with the help of the special multiplication theorem we obtain

$$_{\alpha.\beta} F''(\gamma) = {}_\alpha F''(\gamma) \qquad (3^s)$$

In this formula we recognize again the condition (1^s); thus we see that *independence may be described as a special case of selection*.

The various theorems which may be connected with the name of

Bayes are all special cases of the division theorem. Under the assumption that $(\alpha.\gamma)$ is a sub-class of β, or in symbols

$$\alpha.\gamma \subset \beta \qquad\qquad (3^{bs})$$

we obtain from (3) the *first* (special) form of Bayes's rule

$$_{\alpha.\beta}F''(\gamma) = \frac{_\alpha F''(\gamma)}{_\alpha F''(\beta)}, \qquad\qquad (3_{bs})$$

We can avoid the assumption (3^{bs}) by introducing, in place of 'β', the sum of the classes $\beta_1, \beta_2, \ldots \beta_1$. We shall, in analogy to our use of the sign '$+$' *between* class designations, use the sign 'Σ' *in front of class designations*; we can then write a *second* (universally valid) form of Bayes's theorem as follows:

$$_{\alpha.\Sigma\beta_i}F''(\beta_i) = \frac{_\alpha F''(\beta_i)}{_\alpha F''(\Sigma\beta_i)}, \qquad\qquad (3_b)$$

To the numerator of this formula we can apply the special addition theorem (2') if we assume that the β_i have no members in common in α. This assumption can be written:

$$N(\alpha.\beta_i.\beta_j) = 0 \qquad (i \neq j) \qquad\qquad (3/2')$$

Under this assumption we obtain the *third* (special) form of Bayes's theorem, which in the case of primary properties β_i is always applicable.

$$_{\alpha.\Sigma\beta_i}F''(\beta_i) = \frac{_\alpha F''(\beta_i)}{\Sigma_\alpha F''(\beta_i)}. \qquad\qquad (3/2_s)$$

Appendix iii. Derivation of the First Form of the Binomial Formula (for finite sequences of overlapping segments, *cf.* section 56).

The first binomial formula[1]

$$_{\alpha_{(n)}}F''(m) = {}^nC_m\, p^m q^{n-m} \qquad (1)$$

where $p = {}_\alpha F''(1)$, $q = {}_\alpha F''(0)$, $m \leqslant n$, may be said to be proved under the assumption that α is (at least) $n-1$-free (neglecting errors arising at the last term; *cf.* section 56), if we can show that

$$_{\alpha_{(n)}}F''(\sigma_m) = p^m q^{n-m} \qquad (2)$$

where 'σ_m' denotes one particular n-tuple (although an arbitrarily chosen one) which contains m ones. (The symbol is intended to indicate that what is given is the complete arrangement of this n-tuple, i.e. not only the number of ones but also their positions in the n-tuple.) For assume that (2) holds for all n, m and σ (*i.e.* the various arrangements of the ones). Then there will be, according to a well-known combinatorial theorem, ${}^nC_m$ distinct ways of distributing m ones in n places; and in view of the special addition theorem, we could then assert (1).

Now suppose (2) to be proved for any one n, *i.e.* for *one* particular n and for every m and every σ which are compatible with this n. We now show that given this assumption it must also hold for $n+1$, i.e. we shall prove

$$_{\alpha_{(n+1)}}F''(\sigma_{m+0}) = p^m q^{n+1-m} \qquad (3,0)$$

and

$$_{\alpha_{(n+1)}}F''(\sigma_{m+1}) = p^{m+1} q^{(n+1)-(m+1)} \qquad (3,1)$$

where 'σ_{m+0}' or 'σ_{m+1}' respectively signify those sequences of the $n+1$ length which result from σ_m by adding to its end a zero or a one.

---

[1] Note that $\binom{n}{m}$ is an alternative way of writing the binomial coefficient ${}^nC_m$, *i.e.* the number of ways in which m things may be arranged in n places, provided $m \leqslant n$.

Let it be assumed, for every length n of the n-tuples (or segments) considered, that α is (at least) $n-1$-free (from after-effect); thus for a segment of the length $n+1$, α has to be regarded as being at least n-free. Let '$\acute{\sigma}_m$' denote the property of being a successor of an n-tuple σ_m. Then we can assert

$$_\alpha F''(\acute{\sigma}_m.0) = {}_\alpha F''(\acute{\sigma}_m)._\alpha F''(0) = {}_\alpha F''(\acute{\sigma}_m).q \qquad (4,0)$$

$$_\alpha F''(\acute{\sigma}_m.1) = {}_\alpha F''(\acute{\sigma}_m)._\alpha F''(1) = {}_\alpha F''(\acute{\sigma}_m).p \qquad (4,1)$$

We now consider that there must obviously be just as many σ_m, i.e. successors of the sequence 'σ_m' in α, as there are sequences σ_m in $\alpha_{(n)}$, and hence that

$$_\alpha F''(\acute{\sigma}_m) = {}_{\alpha_{(n)}} F''(\sigma_m) \qquad (5)$$

With this we can transform the right hand side of (4). For the same reason we have

$$_\alpha F''(\acute{\sigma}_m.0) = {}_{\alpha_{(n+1)}} F''(\sigma_{m+0}) \qquad (6,0)$$

$$_\alpha F''(\acute{\sigma}_m.1) = {}_{\alpha_{(n+1)}} F''(\sigma_{m+1}). \qquad (6,1)$$

With these we can transform the left hand side of (4). That is to say, we obtain by substituting (5) and (6) in (4)

$$_{\alpha_{(n+1)}} F''(\sigma_{m+0}) = {}_{\alpha_{(n)}} F''(\sigma_m).q \qquad (7,0)$$

$$_{\alpha_{(n+1)}} F''(\sigma_{m+1}) = {}_{\alpha_{(n)}} F''(\sigma_m).p \qquad (7,1)$$

Thus we see that, assuming that (2) holds for some one n (and all the arrangements σ_m belonging to it), we can derive (3) by mathematical induction. That (2) is in fact valid for $n=2$ and for all σ_m (where $m \leqslant 2$) is seen by assuming first $m=1$ and then $m=0$. Thus we can assert (3) and consequently (2) and (1).

Appendix iv. *A Method of Constructing Models of Random Sequences.*
(*Cf.* sections 58, 64, and 66.)

We assume (as in section 55) that for every given finite number n a generating period can be constructed which is n-free (from after effect) and which shows equal distribution. In every such period, every combinatorially possible x-tuple (for $x \leqslant n + 1$) of ones and zeros will appear at least once.[*1]

(a) We construct a model sequence which is 'absolutely free' (from after effect) in the following way. We write down an n-free period for an arbitrarily chosen n. This period will have a finite number of terms—say n_1. We now write down a new period which is at least $n_1 - 1$-free. Let the new period have the length n_2. In this new period, at least one sequence must occur which is identical with the previously given period of length n_1; and we rearrange the new period in such a way that it begins with this sequence (this is always possible, in accordance with the analysis of section 55). This we call the second period. We now write down another new period which is at least $n_2 - 1$-free and seek in this third period that sequence which is identical with the *second period* (after rearrangement), and then so rearrange the third period that it begins with the second, and so on. In this way

[*1] There are various construction methods applicable to the task of constructing a generating period for an n-free sequence with equidistribution. A simple method is the following. Putting $x = n+1$, we first construct the *table* of all the 2^x possible x-tuples of ones and zeros (ordered by some lexicographic rule—say, according to magnitude). Then we commence our period by writing down the last of these x-tuples, consisting of x ones, checking it off our table. We then proceed according to the following rule: always add a zero to the commencing segment *if permissible*; if not, add a one instead; and always check off the table whatever is the last created x-tuple of the commencing period. (Here '*if permissible*' means 'if the thereby created last x-tuple of the commencing period has not yet occurred, and thus not yet been checked off the table'.) Proceed in this manner until all the x-tuples of the list have been checked off: the result is a sequence of the length $2^x + x - 1$, consisting of (a) a generating period, of the length $2^x = 2^{n+1}$, of an n-free alternative to which (b) the first n elements of the next period have been added. A sequence constructed in this way may be called a '*shortest*' n-free sequence, since it is easily seen that there can be no shorter generating period of a periodic n-free sequence than one of the length 2^{n+1}.

Proofs of the validity of the rule of construction here given were found by Dr. L. R. B. Elton and myself. We intend to publish jointly a paper on the subject.

we obtain a sequence whose length increases very quickly and whose commencing period is the period which was written down first. This period, in turn, becomes the commencing sequence of the second period, and so on. By prescribing a particular commencing sequence together with some further conditions, *e.g.* that the periods to be written down must never be longer than necessary (so that they must be *exactly* $n_i - 1$-free, and not merely *at least* $n_i - 1$-free), this method of construction may be so improved as to become *unambiguous* and to define a definite sequence, so that we can calculate for every term of the sequence whether it is a one or a zero.*² We thus have a

*² To take a concrete example of this construction—the construction of a *shortest random-like sequence*, as I now propose to call it—we may start with the period

$$0\ 1 \tag{0}$$

of the length $n_0 = 2$. (We could say that this period generates a 0-free alternative). Next we have to construct a period which is n_0-1-free, that is to say, 1-free. The method of note *1, above, yields '1100' as generating period of a 1-free alternative; and this has now to be so re-arranged as to begin with the sequence '01' which I have here called (0). The result of the arrangement is

$$0\ 1\ 1\ 0 \tag{1}$$

with $n_1 = 4$. We next construct the n_1-1-free (*i.e.* 3-free) period determined by the method of note *1. It is

$$1\ 1\ 1\ 1\ 0\ 0\ 0\ 0\ 1\ 0\ 0\ 1\ 1\ 0\ 1\ 0$$

We re-arrange this so that it begins with our commencing sequence (1), which yields

$$0\ 1\ 1\ 0\ 1\ 0\ 1\ 1\ 1\ 1\ 0\ 0\ 0\ 0\ 1\ 0 \tag{2}$$

Since $n_2 = 16$, we have next to construct, by the method of note *1, a 15-free period (3) of the length $2^{16} = 65,536$. Once we have constructed this 15-free period (3), we must be able to discover where, in this long period, our sequence (2) occurs. We then re-arrange (3) so that it commences with (2), and proceed to construct (4), of the length $2^{65,536}$.

A sequence constructed in this way may be called a '*shortest* random-like sequence' (i) because every step of its construction consists in the construction, for some *n*, of a shortest *n*-free period (*cf.* note *1 above), and (ii) because the sequence is so constructed that, whatever the stage of its construction, it always *begins* with a shortest *n*-free period. As a consequence, this method of construction ensures that every beginning piece of the length

$$m = 2^{2 \cdot \cdot^{\cdot^2}}$$

is a shortest *n*-free period for the largest possible *n* (*i.e.* for $n = (\log_2 m) - 1$).

This property of 'shortness' is very important; for we can always obtain *n*-free, or absolutely free, sequences with equidistribution which commence with a finite segment of *any* chosen length *m* such that this finite segment has no random character but consists of say, only zeros, or only ones, or of any other intuitively 'regular' arrangement; which shows that for applications, the demand for *n*-freedom, or even absolute freedom, is not enough, and must be replaced by something like a demand for *n*-freedom *becoming manifest from the beginning*; which is, precisely, what a 'shortest' random-like sequence achieves, in the most radical fashion possible. Thus they alone can set an ideal standard for randomness. For these 'shortest' sequences, *convergence can be proved immediately*, as opposed to the examples under (b) and (c) below. See also appendix *vi.

(definite) sequence, constructed according to a mathematical rule, with frequencies whose limits are,

$$_\alpha F'(1) = {}_\alpha F'(0) = \tfrac{1}{2}.$$

By using the procedure employed in the proof of the third form of the binomial formula (section 60) or of the theorem of Bernoulli (section 61), it can be shown (with any degree of approximation) *for whatever frequency value we may chose* that there exist sequences which are 'absolutely free'—provided only that we make the assumption (which we have just proved) that at least one sequence exists which is absolutely free.

(b) An analogous method of construction can now be used to show that sequences exist which possess an 'absolutely free' middle frequency (*cf.* section 64) even though they have no frequency-limit. We need only change procedure (a) in such a way that after some given number of increases in length, we always add to the sequence a finite 'block' (or 'iteration')—of ones, for example. This block is made so long that some given frequency p which is different from $\tfrac{1}{2}$ is reached. After attaining this frequency the whole sequence now written down (it may now have the length m_i) is regarded as the commencing sequence of a period which is $m_i - 1$-free (with equal distribution), and so on.

(c) Finally, it is possible to construct in an analogous way a model of a sequence which has *more than one* 'absolutely free' middle frequency. According to (a), there are sequences which do not have equal distribution and are 'absolutely free'. Thus all we have to do is to combine two such sequences, (A) and (B) (with the frequencies p and q), in the following way. We write down some commencing sequence of (A), then search (B) until we find in it this sequence, and rearrange the period of (B) preceding this point in such a way that it begins with the sequence written down; we then use this whole rearranged period of (B) as commencing sequence. Next we search (A) until we find this new written-down sequence, rearrange (A), and so on. In this way we obtain a sequence in which again and again terms occur up to which it is n_i-free for the relative frequency p of the sequence (A), but in which also again and again terms occur up to which the sequence is n_i-free for the frequency q of (B). Since in this case the numbers n_i increase without limit, we obtain a mode of construction for a sequence which has two distinct 'middle frequencies' both of which are 'absolutely free'. (For we did determine (A) and (B) in such a way that their frequency limits are distinct.)

Note. The applicability of the special multiplication theorem to the classical problem of throwing two dice X and Y at a time (and related problems) is assured if, for example, we make the hypothetical estimate that the 'combination sequence' (as we may call it)—*i.e.* the sequence α that has the throws with X as its odd terms and the throws with Y as its even terms—is random.

Appendix v. *Examination of an Objection. The Two–Slit Experiment.*
(*Cf.* section 76.)*¹

The imaginary experiment described below under (a) is intended to refute my assertion that arbitrarily exact simultaneous (non–predictive) measurements of the position and momentum of a particle are compatible with the quantum theory.

(a) Let *A* be a radiating atom, and let light from it fall on a screen *S* after passing through two slits, *Sl₁* and *Sl₂*. According to Heisenberg we can in this case measure exactly either the position of *A* or the momentum of the radiation (but not both). If we measure the position exactly (an operation that 'blurs' or 'smears' the momentum) then we can assume that light is emitted from *A* in spherical waves. But if we measure the momentum exactly, for example by measuring the recoils due to the emission of photons (thereby 'blurring' or 'smearing' the position), then we are able to calculate the exact direction and the momentum of the emitted photons. In this case we shall have to regard the radiation as corpuscular ('needle–radiation'). Thus to the two measuring operations there correspond two different kinds of radiation, so that we obtain two different experimental results. For if we measure the position exactly we obtain an interference-pattern on the screen: a point-like source of light—and one whose position can be exactly measured *is* point-like—emits coherent light. If on the other hand we measure the momentum exactly, we get no interference pattern. (Flashes of light, or scintillations, without interference pattern, appear on the screen after the photons have passed through the slits, consonantly with the fact that the position is 'blurred' or 'smeared' and that a non–point-like source of light does not emit coherent light.) If we were to suppose that we could measure both the position and

*¹ See also appendix *xi and my *Postscript*, chapter *v, section *110. My present view is that the two–slit experiment is to be treated differently, but that the interpretation proposed in this appendix is still of some interest. The remarks under (e) seem to me still to contain a valid criticism of the attempt to explain the dualism of particle and wave in terms of 'complementarity'—an attempt which, it seems, has been abandoned by some physicists more recently.

the momentum exactly, then the atom would have to emit, on the one hand, according to the wave theory, continuous spherical waves that would produce interference patterns; and it would have to emit, on the other hand, an incoherent corpuscular beam of photons. (If we were able to calculate the path of each photon we should never get anything like 'interference', in view of the fact that photons neither destroy one another nor otherwise interact.) The assumption of exact measurements of position *and* momentum taken simultaneously leads thus to two mutually contradictory predictions. For on the one hand it leads to the prediction that interference patterns will appear, and on the other hand to the prediction that no interference patterns will appear.

(b) I shall now re-interpret this imaginary experiment statistically. I shall deal first with the attempt to measure position exactly. I replace the *single* radiating atom by a cluster of atoms in such a way that coherent light is emitted by them, and propagated in the form of spherical waves. This result is obtained by the use of a second screen pierced by a very small aperture A so placed between the atom-cluster and the first screen that the aperture A is in exactly the place previously occupied by the single radiating atom A. The atom-cluster emits light which, after undergoing selection according to a given position by passing through the aperture A, spreads in the form of continuous spherical waves. Thus we replace the single atom, whose position is exactly determined, by a statistical case of pure positional selection.

(c) In a similar way, the atom with exactly measured momentum but blurred or smeared position will be replaced by a pure selection according to a given momentum; or, in other words, by a mono-chromatic beam of photons travelling along parallel lines from some (non-point-like) source of light.

In each of the two cases we obtain the correct experimental result: interference patterns in case (b), and no interference patterns in case (c).

(d) How are we to interpret the third case, which is supposed to lead to two mutually contradictory predictions? To discover this we imagine that we have exactly observed the path of the atom A, which means both its position and its momentum. We should then find that the atom emits single photons, and recoils with each emission. Each recoil shifts it to another position, and each time the shift is in a new direction. Assuming that the atom radiates in this way for a period of time (we do not raise the question whether it also absorbs energy during this period), it will take up a number of different positions during this period, ranging over a considerable volume of space. For this reason we are not allowed to replace the atom by a point-like cluster of atoms: we can only replace it by a cluster of atoms distributed over a considerable volume of space. Furthermore, since the atom radiates in all directions, we have to replace it by a cluster of atoms radiating in all directions. Thus we do not obtain a pure case; nor do we get coherent radiation. And we do not get interference patterns.

Objections similar to the one here examined may be re-interpreted statistically along the lines of this example.

(e) In connection with our analysis of this imaginary experiment I should like to say that argument (a), contrary to what might be supposed at first, would be in any case quite insufficient to elucidate the so-called problem of complementarity (or the dualism of wave and particle). It tries to do so by showing that the atom is able to emit only *either* coherent waves *or* incoherent photons, and that *therefore* no contradiction arises, because the two experiments are mutually exclusive. But it is simply not true that the two experiments are mutually exclusive, for we can of course combine a not too exact measurement of position with a not too exact measurement of momentum; and in this case the atom neither emits completely coherent waves nor completely incoherent photons. My own statistical interpretation has, clearly, no difficulty whatever in treating such intermediate cases, even though it was never intended to solve the problem of the dualism of waves and particles. I suppose that a really satisfactory solution of this problem will hardly be possible within the framework of statistical quantum physics (Heisenberg's and Schrödinger's particle theory as interpreted by Born in 1925–1926), but I think it might perhaps be solved within the framework of the quantum physics of wave-fields or the 'second quantization' (Dirac's emission and absorbtion theory and the wave-field theory of matter by Dirac, Jordan, Pauli, Klein, Mie, Wiegner, 1927–1928. *Cf.* note 2 to my introduction to section 73).

Appendix vi. *Concerning a Non-Predictive Procedure of Measuring.*
(*Cf.* section 77.)*1

We suppose that a non-monochromatic beam of particles—for instance a beam of light—moving along parallel paths in the x-direction is subjected to selection of their momenta by the interposition of a filter. (If the beam consists of electrons we shall have to use instead of a filter an electric field perpendicular to the direction of the beam in order to analyse its spectrum.) We assume with Heisenberg that this procedure leaves unaltered the momenta (or, more precisely, their components in the x-direction) and consequently also the *velocities* (or their x-components) of the selected particles.

Behind the filter we put a Geiger-counter (or a moving strip of photographic film) in order to measure the time of arrival of the

---

*1 Heisenberg—who speaks of *measuring* or *observing* rather than of *selecting*—presents the situation, in form of a description of an imaginary experiment, as follows: if we wish to observe the *position* of the electron, we must use high frequency light which will strongly interact with it, and thus *disturb* its momentum. If we wish to observe its momentum, then we must use low frequency light which does leave its momentum (practically) unchanged, but cannot help us to determine its position. It is important that in this discussion *the uncertainty of the momentum is due to disturbance, while the uncertainty of the position is not due to anything of the sort.* Rather it is the result of *avoiding* any strong disturbance of the system. (See appendix *xi, point 9.)

My old argument (which was based upon this observation) proceeded now as follows. Since a determination of the momentum leaves the momentum unchanged because it interacts weakly with the system, it must also leave its position unchanged, although it fails to *disclose* this position. But the undisclosed position may later be disclosed by a second measurement; and since the first measurement left the state of the electron (practically) unchanged, we can calculate the past of the electron not only *between* the two measurements, but also before the first measurement.

I do not see that Heisenberg can avoid this conclusion without essentially modifying his argument. (In other words, I still believe that my argument and my experiment of section 77 can be used to point out an inconsistency in Heisenberg's discussion of the observation of an electron.) But I now believe that I was wrong in assuming that what holds for Heisenberg's imaginary 'observations' or 'measurements' would also hold for my 'selections'. As Einstein shows (in appendix *xii), it does not hold for a filter acting upon a photon. Nor does it hold for the electric field perpendicular to the direction of a beam of electrons, mentioned (like the filter) in the first paragraph of the present appendix. For the width of the beam must be considerable if the electrons are to move parallel to the x-axis, and as a consequence, their position before their entry into the field cannot be calculated with precision after they have been deflected by the field. This invalidates the argument of this appendix and the next, and of section 77.

particles; and this allows us—since the velocities of the particles are known—to calculate their x-co-ordinates for any instant preceding their time of arrival. Now we may consider two possible assumptions. If, on the one hand, it is assumed that the x-co-ordinates of the positions of the particles were not interfered with by the measuring of their momenta, then the measurement of position and momentum can be validly extended to the time before the momentum was selected (by the filter). If, on the other hand, it is assumed that a selection according to the momentum does interfere with the x-co-ordinates of the positions of the particles, then we can calculate their paths exactly only for the time-interval *between* the two measurements.

Now the assumption that the position of the particles in the direction of their flight might be disturbed in some unpredictable way by a selection according to a given momentum means the same as that the position co-ordinate of a particle would be altered in some incalculable way by this selection. But since the velocity of the particle has remained unchanged, this assumption must be equivalent to the assumption that, owing to that selection, the particle must have jumped *discontinuously* (with super-luminal velocity) to a different point of its path.

This assumption, however, is incompatible with the quantum theory as at present accepted. For although the theory permits discontinuous jumps, it permits them only in the case of particles within an atom (within the range of discontinuous *Eigen*-values, but not in the case of free particles within the range of continuous *Eigen*-values).

It is possible, presumably, to design a theory (in order to escape the conclusions reached above, or to preserve the principle of indeterminacy) which alters the quantum theory in such a way that the assumption of a disturbance of the position by selecting the momentum is compatible with it; but even this theory—which I might call a 'theory of indeterminacy'—could derive only statistical consequences from the principle of indeterminacy, and could therefore be corroborated only statistically. Within this theory, the principle of indeterminacy would only be a formally singular probability statement, although its content would go beyond what I have called the 'statistical scatter relations'. For, as will be shown below with the help of an example, these are compatible with the assumption that selecting the momentum does not disturb the position. *Thus this latter assumption does not allow us to infer the existence of a 'super-pure case' such as is forbidden by the scatter relations.* This statement shows that the method of measuring I have examined does not affect the statistically interpreted formulae of Heisenberg. It may thus be said to occupy, within my statistical interpretation, the same 'logical place', as it were, as (within his interpretation) Heisenberg's statement denying the 'physical reality' of exact measurements; in fact one might regard my statement as the translation of Heisenberg's statement into the statistical language.

That the statement in question is correct may be seen from the following considerations. We might try to obtain a 'super-pure case' by reversing the order of the steps in the experiment; by first selecting, say, a position in the x-direction (the flight direction) with the help of a fast shutter, and only afterwards selecting the momentum with the help of a filter. This might be thought feasible; for as a result of the position-measurement, all sorts of momenta would appear, out of which the filter—without disturbing the position—will select only those which happened to fall within some small range. But these considerations would be mistaken. For if a group of particles is selected by an 'instantaneous shutter', in the way indicated, then Schrödinger's wave-packets (obtained by superimposition of waves of various frequencies) give us only *probabilities*, to be interpreted statistically, of the occurrence of particles in this group which have the given momentum. For any given finite range of momenta Δp_x, this probability tends towards o provided we make the length of the wave-train infinitely short, *i.e.* measure the position with arbitrary precision (by opening the instantaneous shutter for an arbitrarily short time). In the same way, the probability tends towards o for any finite period during which the instantaneous shutter is open, *i.e.* for any value of the position range Δx, provided Δp_x tends towards o. The more exactly we select the position and the momentum, the more improbable it will be that we shall find any particles at all behind the filter. But this means that only among a very great number of experiments will there be some in which any particles are found behind the filter—and this without our being able to predict in advance in which of the experiments particles will be found there. Thus we cannot by any means prevent these particles from appearing only at intervals scattered at random; and consequently we shall not be able to produce in this way an aggregate of particles which is more homogeneous than a pure case.

There appears to be a comparatively simple crucial experiment for deciding between the 'theory of indeterminacy' (described above) and the quantum theory. According to the former theory, photons would arrive on a screen behind a highly selective filter (or spectrograph) even after the extinction of the source of light, for a period of time; and further, this 'after-glow' produced by the filter would last the longer the more highly selective the filter was.[*2]

[*2] This is precisely what will happen, according to Einstein's remarks printed here in appendix *xii.

Appendix vii. *Remarks Concerning an Imaginary Experiment.* (*C*ᵢ. section 77.)*¹

We may start from the assumption that $\mathbf{a_1}$ and $|\mathbf{b_1}|$ are measured, or selected, with an arbitrary degree of precision. In view of the result obtained in appendix vi, we may assume that the absolute momentum $|\mathbf{a_2}|$ of the particle arriving at X from the direction PX can be measured with an arbitrary degree of precision. Accordingly, $|\mathbf{b_2}|$ may also be determined as precisely as we choose (by using the principle of conservation of energy). Moreover, the position of Sl and X, and the instants of the arrival of the $[A]$-particles at X, can be measured with arbitrary precision. Thus we need only investigate the situation with respect to the indeterminacies $\Delta\mathbf{a_2}$ and $\Delta\mathbf{b_2}$, which arise in consequence of indeterminacies of the corresponding *directions*, and the vector $\Delta\mathbf{P}$ connected with the indeterminacy of the position of P which also arises in consequence of the indeterminacy of a *direction*, *viz.* the direction PX.

If the beam PX passes through a slit at X, then a directional indeterminacy φ will occur, in consequence of the diffraction at the slit. This angle φ can be made as small as we like by making $|\mathbf{a_2}|$ sufficiently large; for we have

$$\varphi \cong \frac{h}{r.|\mathbf{a_2}|} \tag{1}$$

where r is the width of the slit. But it is impossible by this method to decrease $|\Delta\mathbf{a_2}|$; it would decrease only by increasing r which would lead to an increase of $|\Delta\mathbf{P}|$; for we have

$$|\Delta\mathbf{a_2}| \cong \varphi|\mathbf{a_2}| \tag{2}$$

which in view of (1) leads to

$$|\Delta\mathbf{a_2}| \cong \frac{h}{r} \tag{3}$$

showing that $|\Delta\mathbf{a_2}|$ does not depend upon $|\mathbf{a_2}|$.

*¹ For a criticism of some of the assumptions underlying section 77 and this appendix, see note *1 to appendix vi.

Owing to the fact that for any chosen r we can make φ as small as we like by increasing $|\mathbf{a_2}|$, we can also make the component $\Delta\mathbf{a_2}$ in the PX-direction, which we denote by '$(\Delta\mathbf{a_2})_x$', as small as we like; and we can do this without interfering with the precision of the measurement of the position of P, since this position too becomes more precise with increasing $|\mathbf{a_2}|$, and decreasing r. Now we wish to show that a corresponding argument holds for $(\Delta\mathbf{b_2})_y$, *i.e.* for the SY-component of $\Delta\mathbf{b_2}$.

Since we may put $\Delta\mathbf{a_1} = 0$ (according to our assumption), we obtain from the conservation of momenta

$$\Delta\mathbf{b_2} = \Delta\mathbf{b_1} - \Delta\mathbf{a_2} \tag{4}$$

For any given $\mathbf{a_1}$, $|\mathbf{b_1}|$ and $|\mathbf{a_2}|$, $\Delta\mathbf{b_1}$ depends directly upon φ, which means that we can have an arrangement such that

$$|\Delta\mathbf{b_1}| \cong |\Delta\mathbf{a_2}| \cong \frac{h}{r} \tag{5}$$

holds, and therefore also

$$|\Delta\mathbf{b_1}| - |\Delta\mathbf{a_2}| \cong \frac{h}{r} \tag{6}$$

Moreover, we obtain, in analogy to (2),

$$|\Delta\mathbf{b_2}| \cong \psi.|\mathbf{b_2}|, \tag{7}$$

where 'ψ' denotes the indeterminacy of the direction of $\mathbf{b_2}$. Accordingly we obtain in view of (4) and (5)

$$\psi \cong \frac{|\Delta\mathbf{b_1} - \Delta\mathbf{a_2}|}{\mathbf{b_2}} \cong \frac{h}{r.|\mathbf{b_2}|} \; ; \tag{8}$$

But this means: however small we make r, we always can make ψ and with it $(\Delta\mathbf{b_2})_y$ as small as we like by using sufficiently high values for the momentum $|\mathbf{b_2}|$; and this, again, without interfering with the precision of the measurement of the position P.

This shows that it is possible to make each of the two factors of the product $(\Delta\mathbf{P})_y.(\Delta\mathbf{b_2})_y$ as small as we like, independently of the other. But for the refutation of Heisenberg's assertion as to the limits of attainable precision, it would have been sufficient to show that one of these two factors can be made as small as we like without making the other grow beyond all bounds.

In addition it may be noted that by an appropriate choice of the PX-direction it is possible to determine the *distance PX* in such a way that ΔP and $\Delta \mathbf{b_2}$ are parallel, and thus (for sufficiently small φ) normal to PY.[1] In consequence, the precision of the momentum in this direction, and moreover the precision of the position (in the same direction), both become *independent of the precision of the measurement of the position of P.* (The latter, if we use high values of $|\mathbf{a_2}|$, depends mainly upon the smallness of r.) *They are both dependent only upon the precision of the measurements of position and momentum of the particle arriving at X from the direction PX*, and upon the smallness of ψ. (This corresponds to the fact that the precision $(\Delta \mathbf{a_2})_x$ of the particle arriving at X depends upon the smallness of φ.)

One sees that with respect to the precision of the measurements, the situation of the apparently non-predictive measurement of the [A]-particle arriving at X and of the prediction of the path of the [B]-particle leaving P are completely *symmetrical*.

[1] The fact that an examination of the degree of the exactness of measurement taken in a direction perpendicular to Δs can be relevant, was pointed out to me by Schiff during a discussion of my imaginary experiment.

I wish to offer here my warmest thanks to Dr. K. Schiff for fruitfully collaborating throughout the better part of a year.

NEW APPENDICES

Although I found, to my surprise, that I could still agree with almost all the philosophical views expressed in the book, and even with most of those on probability—a field in which my ideas have changed more than in any other—I felt that I had to append to it some of the new material accumulated through the years. There was a considerable amount of this, for I never stopped working on the problems raised in the book; and it was therefore not possible to include in these new appendices all my relevant results. I must mention especially one result that is missing here. It is *the propensity interpretation of probability* (as I call it). The exposition and discussion of this interpretation grew, quite against my intentions, until it became the main part of a new book.

The title of this new book is *Postscript: After Twenty Years*. It is a sequel to the present book, and it contains much that is closely related to it, quite apart from probability theory. In this connection I may also refer to two papers of mine which I might have included among these appendices had I not been reluctant to add to them any further. They are 'Three Views Concerning Human Knowledge', and 'Philosophy of Science: A Personal Report'.[1]

The first two of my new appendices contain three short notes, published between 1933 and 1938, and closely connected with the book. They do not read well, I am afraid: they are unduly compressed, and I was unable to make them more readable without changes that would have diminished their value as documents.

Appendices *ii to *v are somewhat technical—too much so for my taste, at least. But these technicalities are necessary, it seems to me, in order to solve the following philosophical problem. *Is the degree of corroboration or acceptability of a theory a probability*, as so many philosophers have thought? Or in other words, *Does it obey the rules of the probability calculus?*

I had answered this question in my book and my answer was, '*No*'. To this some philosophers replied, 'But I mean by probability

---

[1] Published respectively in *Contemporary British Philosophy* 3, ed. by H. D. Lewis, 1956, pp. 355–388, and in *British Philosophy in the Mid-Century*, ed. by C. A. Mace, 1957, pp. 153–191. Both are included in my *Conjectures and Refutations*.

(or by corroboration, or by confirmation) something different from what you mean'. To justify my rejection of this evasive reply (which threatens to reduce the theory of knowledge to mere verbalism), it was necessary to go into technicalities: the rules ('axioms') of the probability calculus had to be formulated, and the part played by each of them had to be found. For in order not to prejudge the issue whether or not degree of corroboration is one of the possible interpretations of the calculus of probability, this calculus had to be taken in its widest sense, and only such rules admitted as were essential to it. I began these investigations in 1935, and a brief report of some of my earlier investigations is contained in appendix *ii. An outline of my more recent results is given in appendices *iv and *v. In all these appendices it is asserted that, apart from the classical, the logical, and the frequency interpretations of probability, which were all dealt with in the book, *there are many different interpretations of the idea of probability, and of the mathematical calculus of probability*. They thus prepare the way for what I have later called the *propensity interpretation* of probability.[2]

Yet I had not only to examine the rules of the probability calculus: I had also to formulate *rules for the evaluation of tests*—that is, for degree of corroboration. This was done in a series of three papers, here reprinted in appendix *ix. Appendices *vii and *viii form a kind of link between my treatment of probability and of corroboration.

The remaining appendices will be, I hope, of interest to both philosophers and scientists; especially those on objective disorder, and on imaginary experiments. Appendix *xii consists of a letter from Albert Einstein, here published for the first time, with the kind permission of his literary executors.

[2] *Cf.* my paper, 'The Propensity Interpretation of Probability and the Quantum Theory' in *Observation and Interpretation*, ed. by S. Körner, 1957, pp. 65–70, and 88 *f.* See also the two papers mentioned in the foregoing footnote, especially pp. 388 and 188, respectively.

*Appendix *i. Two Notes on Induction and Demarcation, 1933–1934.*

The first of the two notes here republished is a Letter to the Editor of *Erkenntnis*. The second is a contribution to a discussion at a philosophical conference in Prague, 1934. It was published in *Erkenntnis* in 1935, as part of the report on the conference.

I

The letter to the Editor was first published in 1933, in *Erkenntnis*, 3 (*i.e. Annalen der Philosophie*, II) no. 4–6, p. 426 *f.* I have broken up some of the paragraphs, for easier reading.

The letter was evoked by the fact that my views, at the time, were being widely discussed by members of the Vienna Circle, even in print (*cf.* note 3), although none of my own manuscripts (which had been read by some members of the Circle) had been published, partly because of their length: my book, *Logik der Forschung*, had to be cut to a fraction of its original length, to be acceptable for publication. The emphasis, in my letter, upon the difference between the problem of a criterion of *demarcation* and the pseudo-problem of a criterion of *meaning* (and upon the contrast between my views and those of Schlick and Wittgenstein) was provoked by the fact that even in those days my views were discussed, by the Circle, under the misapprehension that I was advocating the replacement of the verifiability criterion of meaning by a falsifiability criterion of meaning, whereas in fact I was not concerned with the problem of *meaning*, but with the problem of *demarcation*. As my letter shows, I tried to correct this misinterpretation of my views as early as 1933. I have tried to do the same in my *Logik der Forschung*, and I have been trying ever since. But it appears that my positivist friends still cannot quite see the difference. These misunderstandings led me, in my letter, to point out, and to dwell upon, the contrast between my views and those of the Vienna Circle; and as a consequence, some people were led to assume, wrongly, that I had developed my views originally as a criticism of Wittgenstein. In fact, I had formulated the problem of demarcation and the falsifiability or testability criterion in the autumn of 1919, years before Wittgenstein's views became a topic of discussion in Vienna. (*Cf.* my paper 'Philosophy of Science: A Personal Report', now in my *Conjectures and Refutations*.) This explains why, as soon as I heard of the Circle's new verifiability criterion of *meaning*, I contrasted this with my falsifiability criterion—a criterion of *demarcation*, designed to demarcate systems of scientific statements from perfectly meaningful systems of metaphysical statements. (As to meaningless nonsense, I do not pretend that my criterion is applicable to it.)

Here is the letter of 1933:

A Criterion of the Empirical Character of Theoretical Systems.

(1) *Preliminary Question. Hume's Problem of Induction*—the question of the validity of natural laws—arises out of an apparent contradiction between the principle of empiricism (the principle that only 'experience' can decide about the truth or falsity of a factual statement), and Hume's realization that inductive (or generalizing) arguments are invalid.

Schlick,[1] influenced by Wittgenstein, believes that this contradiction could be resolved by adopting the assumption that natural laws 'are not genuine statements' but, rather, 'rules for the transformation of statements';*[1] that is to say, that they are a particular kind of 'pseudo-statement'.

This attempt to solve the problem (the solution seems to me to be verbal anyway) shares with all the older attempts, such as *apriorism*, conventionalism, *etc.* a certain unfounded assumption, *viz.*, the assumption that all genuine statements must be, in principle, completely decidable, *i.e.* verifiable and falsifiable; more precisely, that for all genuine statements, an (ultimate) empirical verification, and an (ultimate) empirical falsification must both be logically possible.

If this assumption is dropped, then it becomes possible to resolve in a simple way the contradiction which constitutes the problem of

---

[1] Schlick, *Die Naturwissenschaften* **19** (1931), No. 7, p. 156.

*[1] In order to get Schlick's intended meaning, it might be better to say 'rules for the formation or transformation of statements'. The German reads: '*Anweisungen zur Bildung von Aussagen*'. Here '*Anweisungen*' may be translated, clearly, by 'rules'; but 'Bildung' had, at that time, hardly yet any of the technical connotations which have since led to the clear differentiation between the 'formation' and the 'transformation' of statements.

induction. We can, quite consistently, interpret natural laws or theories as genuine statements which are *partially decidable*, *i.e.* which are, for logical reasons, not verifiable but, *in an asymmetrical way, falsifiable only*: they are statements which are tested by being submitted to systematic attempts to falsify them.

The solution suggested here has the advantage of preparing the way also for a solution of the second and more fundamental of two problems of the theory of knowledge (or of the theory of the empirical method); I have in mind the following:

(2) *Main Problem*. This, the *problem of demarcation* (Kant's problem of the limits of scientific knowledge) may be defined as the problem of finding a criterion by which we can distinguish between assertions (statements, systems of statements) which belong to the empirical sciences, and assertions which may be described as 'metaphysical'.

According to a solution proposed by Wittgenstein,[2] this demarcation is to be achieved with the help of the idea of 'meaning' or 'sense': every meaningful or senseful proposition must be a truth function of 'atomic' propositions, *i.e.*, it must be logically completely reducible to (or deducible from) singular observation statements. If some alleged statement turns out not to be so reducible, then it is 'meaningless' or 'nonsensical' or 'metaphysical' or a 'pseudo-proposition'. Thus *metaphysics is meaningless nonsense*.

It may appear as if the positivists, by drawing this line of demarcation, had succeeded in annihilating metaphysics more completely than the older anti-metaphysicists. However, it is not only metaphysics which is annihilated by these methods, but natural science as well. For the laws of nature are no more reducible to observation statements than metaphysical utterances. (Remember the problem of induction!) They would seem, if Wittgenstein's criterion of meaning is applied consistently, to be 'meaningless pseudo-propositions', and consequently to be 'metaphysical'. Thus this attempt to draw a line of demarcation collapses.

The dogma of meaning or sense, and the pseudo-problems to which it has given rise, can be eliminated if we adopt, as our criterion of demarcation, the *criterion of falsifiability*, *i.e.* of an (at least) unilateral or asymmetrical or *one-sided* decidability. According to this criterion, statements, or systems of statements, convey information about the

---

[2] Wittgenstein, *Tractatus Logico Philosophicus* (1922).

empirical world only if they are capable of clashing with experience; or more precisely, only if they can be *systematically tested*, that is to say, if they can be subjected (in accordance with a 'methodological decision') to tests which *might* result in their refutation.[3]

In this way, the recognition of unilaterally decidable statements allows us to solve not only the problem of induction (note that there is only one type of argument which proceeds in an inductive direction: the deductive *modus tollens*), but also the more fundamental problem of demarcation, a problem which has given rise to almost all the other problems of epistemology. For our criterion of falsifiability distinguished with sufficient precision the theoretical systems of the empirical sciences from those of metaphysics (and from conventionalist and tautological systems), without asserting the meaninglessness of metaphysics (which from a historical point of view can be seen to be the source from which the theories of the empirical sciences spring).

Varying and generalizing a well-known remark of Einstein's,[4] one might therefore characterize the empirical sciences as follows: *In so far as a scientific statement speaks about reality, it must be falsifiable: and in so far as it is not falsifiable, it does not speak about reality.*

A logical analysis would show that the rôle of (one-sided) *falsifiability* as a criterion for *empirical science* is formally analogous to that of *non-contradictoriness* for *science in general*. A contradictory system fails to single out, from the set of all possible statements, a proper sub-set; similarly, a non-falsifiable system fails to single out, from the set of all possible 'empirical' statements (of all singular synthetic statements), a proper sub-set.[5]

---

[3] This testing procedure is reported by Carnap in *Erkenntnis* **3**, p. 223 *ff.*, 'procedure B'.—See also Dubislav, *Die Definition*, 3rd edition, p. 100 *ff.* ∗ Added 1957: This reference will be found not to be one to Carnap's but to some of my own work which Carnap reported and accepted in the article referred to. Carnap made full acknowledgment of the fact that I was the author of what he there described as 'procedure B' ('*Verfahren* B').

[4] Einstein, *Geometrie und Erfahrung*, p. 3 *f.* ∗Added 1957: Einstein said: 'In so far as the statements of geometry speak about reality, they are not certain, and in so far as they are certain, they do not speak about reality.'

[5] A fuller exposition will be published soon in book form (in: *Schriften zur wissenschaftlichen Weltauffassung*, ed. by Frank and Schlick, and published by Springer in Vienna). ∗Added 1957: The reference was to my book, *Logik der Forschung*, then in process of being printed. (It was published in 1934, but—in accordance with a continental custom—with the imprint '1935'; and I myself have, therefore, often quoted it with this imprint.)

2

The second note consists of some remarks which I made in a discussion of a paper read by Reichenbach at a philosophical conference in Prague, in the summer of 1934 (when my book was in page proofs). A report on the conference was later published in *Erkenntnis*, and my contribution, here published in translation, was printed in *Erkenntnis* 5, 1935, p. 170 *ff.*

On the so-called 'Logic of Induction' and the 'Probability of Hypotheses'.

I do not think that it is possible to produce a satisfactory theory of what is traditionally—and also by Reichenbach, for example— called 'induction'. On the contrary, I believe that any such theory— whether it uses classical logic or a probability logic—must for purely logical reasons either lead to an infinite regress, or operate with an *aprioristic* principle of induction, a synthetic principle which cannot be empirically tested.

If we distinguish, with Reichenbach, between a 'procedure of finding' and a 'procedure of justifying' a hypothesis, then we have to say that the former—the procedure of finding a hypothesis—cannot be rationally reconstructed. Yet the analysis of the procedure of justifying hypotheses does not, in my opinion, lead us to anything which may be said to belong to an inductive logic. For a theory of induction is superfluous. It has no function in a logic of science.

Scientific theories can never be 'justified', or verified. But in spite of this, a hypothesis *A* can under certain circumstances achieve more than a hypothesis *B*—perhaps because *B* is contradicted by certain results of observations, and therefore 'falsified' by them, whereas *A* is not falsified; or perhaps because a greater number of predictions can be derived with the help of *A* than with the help of *B*. The best we can say of a hypothesis is that up to now it has been able to show its worth, and that it has been more successful than other hypotheses although, in principle, it can never be justified, verified, or even shown to be probable. This appraisal of the hypothesis relies solely upon *deductive* consequences (predictions) which may be drawn from the hypothesis: *There is no need even to mention 'induction'.*

The mistake usually made in this field can be explained historically: science was considered to be a system of knowledge—of knowledge as certain as it could be made. 'Induction' was supposed to guarantee the truth of this knowledge. Later it became clear that absolutely certain truth was not attainable. Thus one tried to get in its stead at least some kind of watered-down certainty or truth; that is to say, 'probability'.

But speaking of 'probability' instead of 'truth' does not help us to escape either from the infinite regress or from *apriorism*.[1]

From this point of view, one sees that it is useless and misleading to employ the concept of probability in connection with scientific hypotheses.

The concept of probability is used in physics and in the theory of games of chance in a definite way which may be satisfactorily defined with the help of the concept of relative frequency (following von Mises).[2] Reichenbach's attempts to extend this concept so as to include the so-called 'inductive probability' or the 'probability of hypotheses' are doomed to failure, in my opinion, although I have no objection whatever against the idea of a 'truth-frequency' within a sequence of statements[3] which he tries to invoke. For hypotheses cannot be satisfactorily interpreted as sequences of statements;[4] and even if one accepts this interpretation, nothing is gained: one is only led to various utterly unsatisfactory definitions of the probability of a hypothesis. For example, one is led to a definition which attributes the probability 1/2—instead of 0—to a hypothesis which has been falsified a thousand times; for this attribution would have to be made if the hypothesis is falsified by every second result of its tests. One might perhaps consider the possibility of interpreting the hypothesis not as a sequence of statements but rather as an *element* of a sequence of hypotheses,[5] and of attributing to it a certain probability value *qua* element of such a sequence (though not on the basis of a 'truth-frequency', but rather on the basis of a 'falsity-frequency' within that sequence). But this

---

[1] *Cf.* Popper, *Logik der Forschung*, for example pp. 188 and 195 f. *(of the original edition); that is, sections 80 and 81.

[2] *Op. cit.*, p. 94 ff. *(that is, sections 47 to 51).

[3] This concept is due to Whitehead.

[4] Reichenbach interprets 'the assertions of the natural sciences' as sequences of statements in his *Wahrscheinlichkeirslogik*, p. 15. (*Ber. d. Preuss. Akad., phys.-math. Klasse*, 29, 1932, p. 488.)

[5] This would correspond to the view upheld by Grelling in our present discussion; *cf. Erkenntnis* 5, p. 168 f.

attempt is also quite unsatisfactory. Simple considerations lead to the result that it is impossible in this way to arrive at a probability concept which would satisfy even the very modest demand that a falsifying observation should produce a marked decrease in the probability of the hypothesis.

I think that we shall have to get accustomed to the idea that we must not look upon science as a 'body of knowledge', but rather as a system of hypotheses; that is to say, as a system of guesses or anticipations which in principle cannot be justified, but with which we work as long as they stand up to tests, and of which we are never justified in saying that we know that they are 'true' or 'more or less certain' or even 'probable'.

Appendix *ii.   *A Note on Probability*, 1938.

The following note, 'A Set of Independent Axioms for Probability ,
was first published in *Mind*, N.S., **47**, 1938, p. 275 *ff*. It is brief, but,
unfortunately, it is badly written. It was my first publication in the
English language; moreover, the proofs never reached me. (I was then
in New Zealand.)

The introductory text of the note, which alone is here reprinted,
clearly states—and I believe for the first time—that the mathematical
theory of probability should be constructed as a '*formal*' *system*; that
is to say, a system which should be susceptible of many different interpretations, among them, for example, (1) the classical interpretation,
(2) the frequency interpretation, and (3) the logical interpretation (now
sometimes called the 'semantic' interpretation).

One of the reasons why I wanted to develop a formal theory
which would not depend upon any particular choice of an interpretation was that I hoped later to show that what I had called in my book
'degree of corroboration' (or of 'confirmation' or of 'acceptibility')
was not a 'probability': that its properties were incompatible with the
formal calculus of probability. (*Cf.* appendix *ix, and my *Postscript*,
sections *27 to *32).

Another of my motives for writing this note was that I wanted to
show that what I had called in my book 'logical probability' was the
logical interpretation of an 'absolute probability'; that is to say, of
a probability $p(x, y)$, with tautological y. Since a tautology may be
written not-(x and not-x), or $\overline{x\bar{x}}$, in the symbols used in my note,
we can define the absolute probability of x (for which we may write
'$p(x)$' or '$pa(x)$') in terms of relative probability as follows:

$$p(x) = p(x, \overline{x\bar{x}}), \text{ or } pa(x) = p(x, \overline{x\bar{x}}) = p(x, \overline{y\bar{y}})$$

A similar definition is given in my note.

When I wrote this note I did not know Kolmogorov's book
Foundations of Probability, although it had been first published in German
in 1933. Kolmogorov had very similar aims; but his system is less

'formal' than mine, and therefore susceptible to fewer interpretations. The main point of difference is this. He interprets the arguments of the probability functor as *sets*; accordingly, he assumes that they have members (or 'elements'). No corresponding assumption was made in my system: *in my theory, nothing whatever is assumed concerning these arguments (which I call 'elements') except that their probabilities behave in the manner required by the axioms.* Kolmogorov's system can be taken, however, as one of the interpretations of mine. (See also my remarks on this topic in appendix *iv.)

The actual axiom system at the end of my note is somewhat clumsy, and very shortly after its publication I replaced it by a simpler and more elegant one. Both systems, the old and the new, were formulated in terms of *product* (or conjunction) and *complement* (negation), as were also my later systems.*¹ At that time, I had not succeeded in deriving the distributive law from simpler ones (such as the associative law), and I had therefore to state it as an axiom. But, written in terms of product and complement, the distributive law is very clumsy. I have therefore here omitted the end of the note, with the old axiom system; instead I will restate here my simpler system (*cf. Brit. Journal Phil. Sc., loc. cit.*), based, like the old system, on absolute probability. It is, of course, derivable from the system based on relative probability given in appendix *iv. I am stating the system here in an order corresponding to that of my old note.

A1	$p(xy) \geqslant p(yx)$	(Commutation)
A2	$p((xy)z) \geqslant p(x(yz))$	(Association)
A3	$p(xx) \geqslant p(x)$	(Tautology)
A4	There are at least one x and one y such that	
	$p(x) \neq p(y)$	(Existence)
B1	$p(x) \geqslant p(xy)$	(Monotony)
B2	$p(x) = p(xy) + p(x\bar{y})$	(Complement)
B3	For every x there is a y such that	
	$p(y) \geqslant p(x)$, and $p(xy) = p(x)p(y)$	(Multiplication)

Here follows my old *note* of 1938, with a few slight stylistic corrections.

---

*¹ Two of these I published in the *British Journal for the Philosophy of Science* **6**, 1955, pp. 51–57, 176, and 351; and a further improvement in the Appendix to 'Philosophy of Science: A Personal Report', in *British Philosophy in Mid-Century*, ed. by A. C. Mace, 1956. My final system (which, I think, can hardly be further simplified) is to be found in appendix *iv.

A Set of Independent Axioms for Probability.

From the formal point of view of 'axiomatics', probability can be described as a two-termed functor[1] (*i.e.*, a numerical function of two arguments which themselves need not have numerical values), whose arguments are variable or constant *names* (which can be interpreted, *e.g.*, as names of predicates or as names of statements[1] according to the interpretation chosen). If we desire to accept for both of the arguments the same rules of substitution and the same interpretation, then this functor can be denoted by

$$'p(x_1, x_2)'$$

which can be read as 'the probability of x_1 with regard to x'_2.

It is desirable to construct a system of axioms, s_1, in which '$p(x_1, x_2)$' appears as (undefined) primitive variable and which is constructed in such a way that it can be equally interpreted by any of the proposed interpretations. The three interpretations which have been most widely discussed are: (1) the classical definition[2] of probability as the ratio of the favourable to the equally possible cases, (2) the frequency theory[3] which defines probability as the relative frequency of a certain class of occurrences within a certain other class, and (3) the logical theory[4] which defines probability as the degree of a logical relation between statements (which equals 1 if x_1 is a logical consequence of x_2, and which equals 0 if the negation of x_1 is a logical consequence of x_2).

In constructing such a system s_1, capable of being interpreted by any of the interpretations mentioned (and by some others too), it is advisable to introduce, with the help of a special group of axioms (see below, Group A), certain undefined functions of the arguments, *e.g.*, the conjunction ('x_1 and x_2', symbolized here by 'x_1x_2') and the negation ('non-x_1', symbolized by '$\bar{x}_1$'). Thus we can express symbolically an idea like 'x_1 and not x_1' with the help of '$x_1\bar{x}_1$', and its negation by '$\overline{x_1\bar{x}_1}$'. (If (3), *i.e.*, the logical interpretation, is adopted, '$x_1\bar{x}_1$' is to be interpreted as the name of the statement which is the conjunction of the statement named 'x_1' and its negation.)

[1] For the terminology see Carnap, *Logical Syntax of Language* (1937); and Tarski, *Erkenntnis* **5**, 175 (1935).

[2] See *e.g.*, Levy-Roth, *Elements of Probability*, p. 17 (1936).

[3] See Popper, *Logik der Forschung*, 94–153 (1935).

[4] See Keynes, *A Treatise on Probability* (1921); a more satisfactory system has been given recently by Mazurkiewicz, *C.R. Soc. d. Sc. et de L.*, Varsovie, 25, Cl. III (1932); see Tarski, *l.c.*

Supposing the rules of substitution are suitably formulated it can be proved for any x_1, x_2, and x_3:

$$p(x_1, \overline{x_2 \bar{x}_2}) = p(x_1, \overline{x_3 \bar{x}_3}).$$

Thus the value of $p(x_1, x_2 x_2)$ depends on the one real variable x_1 only. This justifies[5] the following explicit definition of a new one-termed functor '$pa(x_1)$', which I may call '*absolute* probability':

$$pa(x_1) = p(x_1, \overline{x_2 \bar{x}_2}) \qquad\qquad \text{Df}_1$$

(An example of an interpretation of '$pa(x_1)$' in the sense of (3), *i.e.* of the logical interpretation, is the concept 'logical probability' as used by me in a previous publication.[6])

Now it is possible to proceed with the whole construction from the other end: instead of introducing '$p(x_1, x_2)$' as primitive concept (primitive functor) of an axiom system s_1 and defining '$pa(x_1)$' explicitly, we can construct another axiom system s_2 in which '$pa(x_1)$' appears as (undefined) primitive variable, and we can then proceed to define '$p(x_1, x_2)$' explicitly, with the help of '$pa(x_1)$'; as follows.

$$p(x_1, x_2) = \frac{pa(x_1 x_2)}{pa(x_2)} \qquad\qquad \text{Df}_2$$

The formulae adopted in s_1 as axioms (and Df$_1$) now become theorems within s_2, *i.e.* they can be deduced with the help of the new system of axioms s_2.

It can be shown that the two methods described—the choice of s_1 and Df$_1$, or s_2 and Df$_2$ respectively—are not equally convenient from the viewpoint of formal axiomatics. The second method is superior to the first in certain respects, the most important of which is that it is possible to formulate in s_2 an axiom of uniqueness which is much stronger than the corresponding axiom of s_1 (if the generality of s_1 is not restricted). This is due to the fact that if $pa(x_1) = 0$, the value of $p(x_1, x_2)$ becomes indeterminate.[*1]

---

[5] See Carnap, *l.c.*, 24. *It would have been simpler to write Df$_1$ (without 'justification') as follows: $pa(x_1) = p(x_1, \overline{x_1 \bar{x}_1})$.

[6] See Popper, *l.c.*, 71, 151.

[*1] The absolute system (s_1) has an advantage over the relative system (s_2) only as long as the relative probability $p(x, y)$ is considered as indeterminate if $pa(y) = 0$. I have since developed a system (see appendix *iv) in which relative probabilities are determinate even in case $pa(y) = 0$. This is why I now consider the relative system superior to the absolute system. (I may also say that I now consider the term 'axiom of uniqueness' as badly chosen. What I intended to allude to was, I suppose, something like postulate 2 or axiom A2 of the system of appendix *iv.)

A system of independent axioms, s_2, as described above, is here subjoined. (It is easy to construct a system s_1 with the help of it.) Combined with definition Df_2 it is sufficient for the deduction of the mathematical theory of probability. The axioms can be divided into two groups. Group A is formed by the axioms for the junctional operations—conjunction and negation—of the argument, and is practically an adaptation of the system of postulates for the so-called 'Algebra of Logic'.[7] Group B gives the axioms peculiar to the measurement of probability. The axioms are:

(Here followed—with several misprints—the complicated axiom system which I have since replaced by the simpler one given above.) Christchurch, N.Z., *November 20th*, 1937.

[7] See Huntington, *Trans. Amer. Mathem. Soc.*, 5, 292 (1904), and Whitehead-Russell, *Principia Mathematica*, I, where the five propositions 22·51, 22·52, 22·68, 24·26, 24·1 correspond to the five axioms of Group A, as given here.

Appendix *iii. On the Heuristic Use of the Classical Definition of Probability, Especially for Deriving the General Multiplication Theorem.*

The classical definition of probability as the number of favourable cases divided by the number of equally possible cases has considerable heuristic value. Its main drawback is that it is applicable to homogeneous or symmetrical dice, say, but not to biased dice; or in other words, that it does not make room for *unequal weights of the possible cases*. But in some special cases there are ways and means of getting over this difficulty; and it is in these cases that the old definition has its heuristic value: every satisfactory definition will have to agree with the old definition where the difficulty of assigning weights can be overcome, and therefore, *a fortiori*, in those cases in which the old definition turns out to be applicable.

(1) The classical definition will be applicable in all cases in which we conjecture that we are faced with equal weights, or equal possibilities, and therefore with equal probabilities.

(2) It will be applicable in all cases in which we can transform our problem so as to obtain equal weights or possibilities or probabilities.

(3) It will be applicable, with slight modifications, whenever we can assign a weight function to the various possibilities.

(4) It will be applicable, or it will be of heuristic value, in most cases where an over-simplified estimate that works with equal possibilities leads to a solution approaching to the probabilities zero or one.

(5) It will be of great heuristic value in cases in which weights can be introduced in the form of probabilities. Take, for example, the following simple problem: we are to calculate the probability of throwing with a die an even number when the throws of the number six are *not counted, but considered as 'no throw'*. The classical definition leads, of course, to 2/5. We may now assume that the die is biased, and that the (unequal) probabilities $p(1)$, $p(2)$, ... ,$p(6)$ of its sides are given. We can then still calculate the required probability as equal to

$$\frac{p(2) + p(4)}{p(1) + p(2) + p(3) + p(4) + p(5)} = \frac{p(2) + p(4)}{1 - p(6)}$$

That is to say, we can modify the classical definition so as to yield the following simple rule:

Given the probabilities of all the (mutually exclusive) possible cases, the required probability is the sum of the probabilities of all the (mutually exclusive) favourable cases, divided by the sum of the probabilities of all the (mutually exclusive) possible cases.

It is clear that we can also express this rule, for exclusive or non-exclusive cases, as follows.

The required probability is always equal to the probability of the disjunction of all the (exclusive or non-exclusive) favourable cases, divided by the probability of the disjunction of all the (exclusive or non-exclusive) possible cases.

(6) These rules can be used for a heuristic derivation of the definition of relative probability, and of the general multiplication theorem.

For let us symbolize, in the last example, 'even' by 'a' and 'other than a six' by 'b'. Then our problem of determining the probability of an even throw if we disregard throws of a six is clearly the same as the problem of determining $p(a, b)$, that is to say, the probability of a, given b, or the probability of finding an a among the b's.

The calculation can then proceed as follows. Instead of writing '$p(2) + p(4)$' we can write, more generally, '$p(ab)$', that is to say, the probability of an even throw other than a six. And instead of writing '$p(1) + p(2) + \ldots + p(5)$' or, what amounts to the same, '$1 - p(6)$', we can write '$p(b)$', that is to say, the probability of throwing a number other than six. It is clear that these calculations are quite general, and assuming $p(b) \neq 0$, we are led to the formula,

(1) $$p(a, b) = p(ab) \,/\, p(b)$$

or to the formula (more general because it remains meaningful even if $p(b) = 0$),

(2) $$p(ab) = p(a, b)\, p(b).$$

This is the general multiplication theorem for the absolute probability of a product ab.

By substituting 'bc' for 'b', we obtain from (2):[1]

$$p(abc) = p(a, bc)\, p(bc)$$

[1] I omit brackets round 'bc' because my interest is here heuristic rather than formal, and because the problem of the law of association is dealt with at length in the next two appendices.

and therefore, by applying (2) to $p(bc)$:

$$p(abc) = p(a, bc)\, p(b, c)\, p(c)$$

or, assuming $p(c) \neq 0$,

$$p(abc) / p(c) = p(a, bc)\, p(c).$$

This, in view of (1), is the same as

(3) $$p(ab, c) = p(a, bc)\, p(b, c).$$

This is the general multiplication theorem for the *relative* probability of a product ab.

(7) The derivation here sketched can be easily formalized. The formalized proof will have to proceed from an axiom system rather than from a definition. This is a consequence of the fact that our heuristic use of the classical definition consisted in introducing weighted possibilities—which is practically the same as probabilities—into what was the classical *definiens*. The result of this modification cannot any longer be regarded as a proper definition; rather it must establish relations between various probabilities, and it therefore amounts to the construction of an axiom system. If we wish to formalize our derivation—which makes implicit use of the laws of association and of the addition of probabilities—then we must introduce rules for these operations into our axiom system. An example is our axiom system for absolute probabilities, as described in appendix *ii.

If we thus formalize our derivation of (3), we can get (3) at best only with the condition 'provided $p(bc) \neq 0$', as will be clear from our heuristic derivation.

But (3) may be meaningful even without this proviso, if we can construct an axiom system in which $p(a, b)$ is generally meaningful, even if $p(b) = 0$. It is clear that we cannot, in a theory of this kind, derive (3) in the way here sketched; but we may instead adopt (3) itself as an axiom, and take the present derivation (see also formula (1) of my old appendix ii) as a heuristic justification for introducing this axiom. This has been done in the system described in the next appendix (appendix *iv).

Appendix *iv.   *The Formal Theory of Probability.*

In view of the fact that a probability statement such as '$p(a, b) = r$' can be interpreted in many ways, it appeared to me desirable to construct a purely 'formal' or 'abstract' or 'autonomous' system, in the sense that its 'elements' (represented by 'a', 'b', . . .) can be interpreted in many ways, so that we are not bound to any particular one of these interpretations. I proposed first a formal axiom system of this kind in a Note in *Mind* in 1938 (here re-printed in appendix *ii). Since then, I have constructed many simplified systems.[1]

There are three main characteristics which distinguish a theory of this kind from others. (i) It is formal; that is to say, it does not assume any particular interpretation, although allowing for at least all known interpretations. (ii) It is autonomous; that is to say, it adheres to the principle that probability conclusions can be derived only from probability premises; in other words, to the principle that the calculus of probabilities is a method of transforming probabilities into other probabilities. (iii) It is symmetrical; that is to say, it is so constructed that whenever there is a probability $p(a, b)$—i.e. a probability of a given b—then there is always a probability $p(b, a)$ also—even when the

---

[1] In *Brit. Journ. Phil. of Science* **6**, 1955, pp. 53 and 57 f., and in the first footnote to the Appendix to my paper 'Philosophy of Science: A Personal Report', in *British Philosophy in Mid-Century*, ed. by C. A. Mace, 1956.

It should be noted that the systems here discussed are 'formal' or 'abstract' or 'autonomous' in the sense explained, but that for a complete 'formalization', we should have to embed our system in some mathematical formalism. (Tarski's 'elementary algebra' would suffice.)

The question may be asked whether a decision procedure might exist for a system consisting, say, of Tarski's elementary algebra and our system of formulae A1, B, and C+. The answer is, no. For formulae may be added to our system which express how many elements a, b, . . . there are in S. Thus we have in our system a theorem.

There exists an element a in S such that $p(a, \bar{a}) \neq p(\bar{a}, a)$.

We may now add the formula:

(o) For every element a in S, $p(a, \bar{a}) \neq p(\bar{a}, a)$.

But if this formula is added to our system, then it can be proved that there are *exactly two* elements in S. The examples by which we prove, below, the consistency of our axioms show however that there may be any number of elements in S. This shows that (o), and all similar formulae determining the number of elements in S, cannot be derived. Thus our system is incomplete.

326

absolute probability of b, $p(b)$, equals zero; that is, even when $p(b) = p(b, \overline{aa}) = 0$.

Apart from my own attempts in this field, a theory of this kind, strange to say, does not seem to have existed hitherto. Some other authors have intended to construct an 'abstract' or 'formal' theory—for example Kolmogorov—but in their constructions they have always assumed a more or less specific *interpretation*. For example, they assumed that in an equation like

$$p(a, b) = r$$

the 'elements' a and b are *statements*, or systems of statements; or they assumed that a and b are *sets*, or systems of sets; or perhaps properties; or perhaps finite classes (ensembles) of things.

Kolmogorov writes [2] 'The theory of probability, as a mathematical discipline, can and should be developed from axioms in exactly the same way as geometry and algebra'; and he refers to 'the introduction of basic geometric concepts in the *Foundations of Geometry* by Hilbert', and to similar abstract systems.

And yet, he assumes that, in '$p(a, b)$'—I am using my own symbols, not his—a and b are *sets*; thereby excluding, among others, the logical interpretation according to which a and b are statements (or 'propositions', if you like). He says, rightly, 'what the members of this set represent is of no importance'; but this remark is not sufficient to establish the formal character of the theory at which he aims; for in some interpretations, a and b have *no members*, nor anything that might correspond to members.

All this has grave consequences in connection with the actual construction of the axiom system itself.

Those who interpret the elements a and b as statements or propositions very naturally assume that the calculus of statement-composition (the propositional calculus) holds for these elements. Similarly, Kolmogorov assumes that the operations of addition, multiplication, and complementation of *sets* hold for his elements, since they are interpreted as sets.

More concretely, it is always presupposed (often only tacitly), that such algebraic laws as the law of association

(a) $$(ab)c = a(bc)$$

---

[2] The quotations here are all from p. 1 of A. Kolmogorov, *Foundation of the Theory of Probability*, 1950. (First German edition 1933.)

or the law of commutation

(b) $$ab = ba$$

or the law of indempotence

(c) $$a = aa$$

hold for the elements of the system; that is to say, for the arguments of the function $p(\,.\,.\,,\,.\,.\,)$.

Having made this assumption either tacitly or explicitly, a number of axioms or postulates are laid down for relative probability,

$$p(a, b)$$

that is to say for the probability of a, given the information b; or else for absolute probability,

$$p(a)$$

that is to say, for the probability of a (given no information, or only tautological information).

But this procedure is apt to veil the surprising and highly important fact that some of the adopted axioms or postulates for relative probability, $p(a, b)$, alone *guarantee that all the laws of Boolean algebra hold for the elements*. For example, a form of the law of association is entailed by the following two formulae (*cf.* the preceding appendix *iii),

(d) $$p(ab) = p(a, b)p(b)$$

(e) $$p(ab, c) = p(a, bc)p(b, c)$$

of which the first, (d), also gives rise to a kind of definition of relative probability in terms of absolute probability,

(d′) $\qquad$ If $p(b) \neq 0$ then $p(a, b) = p(ab) / p(b)$,

while the second, the corresponding formula for relative probabilities, is well known as the 'general law of multiplication'.

These two formulae, (a) and (b), entail, without any further assumption (except substitutivity of equal *probabilities*) the following form of the law of association:

(f) $$p((ab)c) = p(a(bc)).$$

But this interesting fact[3] remains unnoticed if (f) is introduced by

---

[3] The derivation is as follows:

(1)	$p((ab)c) = p(ab, c)p(c)$	d
(2)	$p((ab)c) = p(a, bc)p(b,c)p(c)$	1, e
(3)	$p(a(bc)) = p(a, bc)p(bc)$	d
(4)	$p(a(bc)) = p(a, bc)p(b, c)p(c)$	3, d
(5)	$p((ab)c) = p(a(bc))$	2, 4

way of *assuming* the algebraic identity (a)—the law of association—before even starting to develop the calculus of probability; for from

(a) $$(ab)c = a(bc)$$

we may obtain (f) merely by substitution into the identity

$$p(x) = p(x).$$

Thus the derivability of (f) from (d) and (e) remains unnoticed. Or in other words, it is not seen that the assumption (a) is completely redundant if we operate with an axiom system which contains, or implies, (d) and (e); and that by assuming (a), in addition to (d) and (e), we prevent ourselves from finding out *what kind of relations are implied by our axioms or postulates*. But to find this out is one of the main points of the axiomatic method.

In consequence, it has also not been noticed that (d) and (e), although implying (f), *i.e.* an equation in terms of *absolute* probability, do not alone imply (g) and (h), which are the corresponding formulae in terms of *relative* probability:

(g) $$p((ab)c, d) = p(a(bc), d)$$

(h) $$p(a, (bc)d) = p(a, b(cd)).$$

In order to derive these formulae (see appendix *v, (41) to (62)), much more is needed than (d) and (e); a fact which is of considerable interest from an axiomatic point of view.

I have given this example in order to show that Kolmogorov fails to carry out his programme. The same holds for all other systems known to me. In my own systems of postulates for probability, all theorems of Boolean algebra can be deduced; and Boolean algebra, in its turn, can of course be interpreted in many ways: as an algebra of sets, or of predicates, or of statements (or propositions), *etc.*

Another point of considerable importance is the problem of a 'symmetrical' system. As mentioned above, it is possible to define relative probability in terms of absolute probability by (d'), as follows:

(d') $$\text{If } p(b) \neq 0 \text{ then } p(a, b) = p(ab) / p(b).$$

Now the antecedent 'If $p(b) \neq 0$' is unavoidable here since division by 0 is *not a defined operation*. As a consequence, most formulae of relative probability can be asserted, in the customary systems, only in conditional form, analogous to (d'). For example, in most systems,

(g) is invalid, and should be replaced by the much weaker conditional formula (g⁻):

(g⁻) If $p(d) \neq 0$ then $p((ab)c, d) = p(a(bc), d)$

and an analogous condition should be prefixed to (h).

This point has been overlooked by some authors (for example, Jeffreys, and von Wright; the latter uses conditions amounting to $b \neq 0$, but this does not ensure $p(b) \neq 0$, especially since his system contains an 'axiom of continuity'). Their systems, accordingly, are inconsistent as they stand, although they can sometimes be mended. Other authors have seen the point. But as a consequence, their. systems are (at least as compared with my system) very weak: it can occur in their systems that

$$p(a, b) = r$$

is a meaningful formula, while at the same time, and for the same elements

$$p(b, a) = r$$

is not meaningful, *i.e.* not properly defined, and not even definable, because $p(a) = 0$.

But a system of this kind is not only weak; it is also for many interesting purposes *inadequate*: it cannot, for example, be properly applied to statements whose absolute probability is zero, although this application is very important: universal laws, for example, have, we may here assume (*cf.* appendices *vii and *viii), zero probability. If we take two universal theories, s and t, say, such that s is deducible from t, then we should like to assert that

$$p(s, t) = 1$$

But if $p(t) = 0$, we are prevented from doing so, in the customary systems of probability. For similar reasons, the expression

$$p(e, t)$$

where e is evidence in favour of the theory t, may be undefined; but this expression is very important. (It is Fisher's 'likelihood' of t on the given evidence e; see also appendix *ix.)

Thus there is a need for a probability calculus in which we may operate with second arguments of zero absolute probability. It is, for example, indispensable for any serious discussion of the theory of corroboration or confirmation.

This is why I have tried for some years to construct a calculus of relative probability in which, whenever

$$p(a, b) = r$$

is a well-formed formula *i.e.* true or false,

$$p(b, a) = r$$

is also a well formed formula, even if $p(a) = 0$. A system of this kind may be labelled 'symmetrical'. I published the first system of this kind only in 1955.[4] This symmetrical system turned out to be much simpler than I expected. But at that time, I was still pre-occupied with the peculiarities which every system of this kind must exhibit. I am alluding to such facts as these: in every satisfactory symmetrical system, rules such as the following are valid:

$$p(a, b\bar{b}) = 1$$
$$\text{If } p(\bar{b}, b) \neq 0 \quad \text{then } p(a, b) = 1$$
$$\text{If } p(a, \bar{a}b) \neq 0 \quad \text{then } p(a, b) = 1$$

These formulae are either invalid in the customary systems, or else (the second and third) *vacuously* satisfied, since they involve second arguments with zero absolute probabilities. I therefore believed, at that time, that some of them would have to appear in my axioms. But I found later that my axiom system could be simplified; and in simplifying it, I found that all these unusual formulae could be derived from formulae having a completely 'normal' look. I published the resulting simplified system first in my paper 'Philosophy of Science: A Personal Report'.[5] It is the same system of six axioms which is more fully presented in the present appendix.

The system is surprisingly simple and intuitive, and its power, which far surpasses that of any of the customary systems, is merely due to the fact that I omit from all the formulae except one (axiom C), any condition like 'If $p(b) \neq 0$ then ...'. (In the customary systems, these conditions either are present, or they ought to be present, in order to avoid inconsistencies.)

I intend to explain, in the present appendix, first the axiom system, with proofs of consistency and independence, and afterwards a few definitions based upon the system, among them that of a Borel field of probabilities.

[4] In the *British Journal for the Philosophy of Science*, 6, 1955, pp. 56 f.
[5] In *British Philosophy in the Mid-Century*, ed. by C. A. Mace, 1956, p. 191. The six axioms given there are B1, C, B2, A3, A2, and A1 of the present appendix; they are there numbered B1, B2, B3, C1, D1, and E1, respectively.

First the axiom system itself.

Four undefined concepts appear in our postulates: (i) S, the universe of discourse, or the system of admissible elements; the *elements* of S are denoted by lower case italics, 'a', 'b', 'c', ... etc.; (ii) a binary numerical function of these elements, denoted by '$p(a, b)$', etc.; that is to say, the probability a given b; (iii) a binary operation on the elements, denoted by 'ab', and called the *product* (or meet or conjunction) of a and b; (iv) the complement of the element a, denoted by '$\bar{a}$'.

To these four undefined concepts we may add a fifth—one that can be treated, according to choice, as an undefined or as a defined concept. It is the 'absolute probability of a', denoted by '$p(a)$'.

Each of the undefined concepts is introduced by a *Postulate*. For an intuitive understanding of these postulates, it is advisable to keep in mind that $p(a, a) = 1 = p(b, b)$ for all elements a and b of S, as can of course be formally proved with the help of the postulates.

Postulate 1. The number of elements in S is at most denumerably infinite.

Postulate 2. If a and b are in S, then $p(a, b)$ is a real number, and the following axioms hold:

A1 There are elements c and d in S such that $p(a, b) \neq p(c, d)$

(Existence)

A2 If $p(a, c) = p(b, c)$ for every c in S, then $p(d, a) = p(d, b)$ for every d in S.

(Substitutivity)

A3 $p(a, a) = p(b, b)$

(Reflexivity)

Postulate 3. If a and b are in S, then ab is in S; and if, moreover, c is in S (and therefore also bc) then the following axioms hold:

B1 $p(ab, c) \leqslant p(a, c)$

(Monotony)

B2 $p(ab, c) = p(a, bc)p(b, c)$

(Multiplication)

Postulate 4. If a is in S, then $\bar{a}$ is in S; and if, moreover, b is in S, then the following axiom holds:

C $p(a, b) + p(\bar{a}, b) = p(b, b)$, unless $p(b, b) = p(c, b)$ for every c in S.

(Complementation).

This concludes the 'elementary' system ('elementary' as opposed to its extension to Borel fields). We may, as indicated, add here the *definition of absolute probability* as a fifth postulate, called 'Postulate AP'; alternatively, we may regard this as an explicit definition rather than as a postulate.

Postulate AP. If a and b are in S, and if $p(a, a) = p(b, c)$ for every c in S, then $p(a) = p(a, b)$ (Definition of Absolute Probability)

The system of postulates and axioms given here will be shown below to be *consistent* and *independent*.[6]

The following comments may be made upon this system of postulates:

The six axioms—A1, A2, A3, B1, B2, and B3—are explicitly used in the actual operations of deriving the theorems. The remaining (existential) parts of the postulates can be taken for granted, as in the paper in which I first published the system here presented.[7]

At the price of introducing a fourth variable, 'd', in Postulates 3 and 4, these six axioms may be replaced by a system of only four axioms, consisting of A1, A2, and the following two:

B$^+$ If $p(a, bc)p(b, c) \neq p(d, c)$ provided that $p(a, c) \geqslant p(d, c)$, then $p(ab, c) \neq p(d, c)$

C$^+$ If $p(a, b) + p(\bar{a}, b) \neq p(c, c)$ then $p(c, c) = p(d, b)$

In this system, B$^+$ is equivalent to the conjunction of B1 and B2,

---

[6] An alternative system would be the following: the postulates are the same as in the text, and so are axioms A1 and A2, but the axioms A3 and B1 are replaced by the following two:

A3′ $p(a, a) = 1$
A4′ $p(a, b) \geqslant 0$

Axiom B2 remains as in the text, and axiom C is replaced by
C′   If $p(a, b) \neq 1$, then $p(c, b) + p(\bar{c}, b) = 1$

This system looks very much like some of the customary systems (except for the omission of antecedents in the axioms other than C′, and the form of the antecedent of C′); and it is remarkable that it yields for the elements $a, b, \ldots$, as does the system in the text, the theorems of Boolean algebra which ordinarily are separately assumed. Nevertheless it is unnecessarily strong; not only because it introduces the numbers 1 and 0 (thus hiding the fact that these need not be mentioned in the axioms) but also because A3, B1, and C follow immediately from A3′, A4′, and C′, while for the opposite derivations, all the axioms of the system given in the text except A2 are indispensable. (For these derivations, see appendix *v.)

Within the system of axioms here described, and also within the system given in the text, the conjunction of the axioms A4′ and B1′ is replaceable by B1, and *vice versa*. My independence proofs (given below) are applicable to the system here described.

The derivation of B1 from A4′ and B1′, in the presence of the axioms A3 or A3′, C or C′, and B2, is as follows:

(1) $0 \leqslant p(a, b) \leqslant p(a, a)$ A4′; C or C′; A3 or A3′
(2) $p(a, a) \geqslant p((aa)a, a) = p(aa, aa)p(a, a) = p(a, a)^2$ 2, B2; A3 or A3′
(3) $0 \leqslant p(a, b) \leqslant p(a, a) \leqslant 1$ 1, 2
(4) $p(ba, c) \leqslant p(a, c)$ B2, 3
Now we apply B1′
(5) $p(ab, c) \leqslant p(a, c)$ 4, B1′
For the derivation of A4′ and B1′ from B1, see appendix *v.

[7] *Cf.* note 1, above.

and C$^+$, similarly, to that of A3 and C.[8] The resulting system of four axioms is very brief, and shares many of the advantages of the longer system: product and complement occur separately, so that all the axioms except those lettered 'B' are free of the product, and the complement occurs once only. But personally I prefer the longer system of six axioms.[9]

The following comments may be made upon the various postulates and axioms of the system.

Postulate 1 (which only belongs to the *elementary* theory) may be dispensed with. This is shown by the fact that, in order to prove its independence, we may construct a system S which is non-denumerable. (All other postulates are satisfied if we interpret S as the set of all finite sums of half-open sub-intervals $[x, y)$ of the unit interval $[0, 1)$, where x and y are real numbers rather than rational numbers; we may then interpret $p(a)$ as the length of these intervals, and $p(a, b)$ as equal to $p(ab)/p(a)$ provided $p(b) \neq 0$, and as equal to 1 provided $b=0$; otherwise as $\lim p(ab)/p(b)$, provided this limit exists). The function of Postulate 1 is merely to characterize the *elementary* systems: a postulate of this kind is often assumed in an axiomatic treatment of Boolean algebra or the logic of statements or propositions; and we wish to be able to show that, in the *elementary* theory, S is a (denumerable) Boolean algebra.

In Postulate 2, A1 is needed to establish that *not all probabilities are equal* (say, equal to 0 or equal to 1). The function of A2 is to allow us to prove '$p(x, a)=p(x, b)$' for all elements a and b whose probabilities,

---

[8] C$^+$ follows immediately from A3 and C. The converse can be shown by deriving A3 from C$^+$ as follows:

(1) $p(c, b) + p(c, b) \neq p(b, b) \rightarrow p(b, b) = p(d, b) = p(c, b) = p(\bar{c}, b)$ C$^+$
(2) $p(a, a) \neq p(b, b) \rightarrow p(a, a) = p(c, b) + p(\bar{c}, b) \neq p(b, b) = p(c, b) = p(\bar{c}, b)$ C$^+$, 1
(3) $p(a, a) \neq p(b, b) \rightarrow p(a, a) = 2p(b, b)$ 2
(4) $p(b, b) \neq (p(a, a) \rightarrow p(b, b) = 2p(a, a) = 4p(b, b) = 0 = p(a, a)$ 3
(5) $p(a, a) = p(b, b)$.   4

C$^+$ may also be replaced, for example, by the slightly stronger formula
C^s $p(a, a) \neq p(b, c) \rightarrow p(a, c) + p(\bar{a}, c) = p(d, d)$.

B$^+$ is merely an 'organic' way of writing the simpler but 'inorganic' formula
B^s $p(ab, c) = p(a, bc)p(b, c) \leqslant p(a, c)$.

[9] Three of the reasons why I prefer the system of six axioms to the system of four are these: (i) the axioms of the longer system are a little less unusual and thus a little more intuitive, especially in the form mentioned in footnote 6, above; (ii) introducing an additional variable is too high a price to pay for a reduction in the number of axioms; (iii) The 'organicity' of B$^+$ is achieved by a kind of mechanical trick and is therefore of little value.

given *any* condition c, are equal. This can be done *without* A2, but only under the assumption $p(a) \neq 0 \neq p(b)$. Thus A2 is to enable us to extend the probabilistic equivalence of a and b to the second argument even in those cases in which a and b have *zero absolute probability*.

A2 may be replaced by the following stronger formula:

A2$^+$ If $p(a, a) = p(b, c) = p(c, b)$, then $p(a, b) = p(a, c)$, for every c in S;

or by

B3 If $p(ab, c) = p(ba, c)$, then $p(c, ab) = p(c, ba)$.

Obviously, it can therefore also be replaced by the formula (which is simpler but much stronger):

B3$^+$ $p(a, bc) = p(a, cb)$.

But since B3$^+$ is stronger than necessary—in fact, $p(a,(bc)(cb)) = p(a,(cb)(bc))$, though weaker, would suffice—it is a little misleading: its adoption would veil the fact that with the help of the other axioms alone, the law of commutation can be proved for the first argument. A2$^+$ is preferable to the other formulae here mentioned in so far as it avoids (like the much weaker A2) using the product of a and b.

However, we can make use of the facts here stated in order to reduce the number of our axioms to three, *viz.* A1, C$^+$, and the following axiom B which combines B3$^+$ with B$^+$:

B If $p(ab, c) \neq p(a, d)p(b, c)$ provided that $p(a, c) \geqslant p(a, d)p(b, c)$ and $p(a, d) = p(a, bc)$, then $p(a, cb) \neq p(a, d)$.

Apart from being stronger than one might wish it to be, this system of only three axioms has all the advantages of the system of four axioms A1, A2, B$^+$, and C$^+$.

A3 is needed to prove that $p(a, a) = 1$, for every element a of S, as has been indicated. But it may be omitted if we strengthen C: as may be seen from axiom C$^+$, A3 becomes redundant if we replace in C the two occurrences of '$p(b, b)$' by '$p(d, d)$' (or only the second occurrence).

Postulate 3 demands the existence of a product (or meet, or intersection) of any elements a and b in S. It characterizes exhaustively all the properties of the product (such as idempotence, commutation, and association) by two simple axioms of which the first is intuitively obvious; the second has been discussed in appendix *iii.

Axiom B1 is, in my opinion, the intuitively most obvious of all our axioms. It is preferable to both A4′ and B1′ (*cf.* note 6 above) which together may replace it. For A4′ may be mistaken for a convention, as opposed to B1; and B1′ does not, as does B1, characterize an intuitive metrical aspect of *probability*, but rather the product or conjunction *ab*.

As shown by formula B above, axiom B2 can be combined with B1 and A2$^+$; there are other possible combinations, among them some in which the product appears only once. They are very complicated, but have the advantage that they may be given a form analogous to that of a definition. One such definitional form may be obtained from the following axiom BD (which, like B, may replace A2, B1, and B2) by inserting the symbol '(*a*)' twice, once at the beginning and a second time before '(E*b*)', and by replacing the first arrow (conditional) by a double arrow (for the bi-conditional). I am using here the abbreviations explained in the beginning of appendix *v.

BD $p(xy, a) = p(z, a) \rightarrow (\mathrm{E}b)\,(c)\,(d)\,(\mathrm{E}e)\,(\mathrm{E}f)\,(\mathrm{E}g)\,(p(x, a) \geqslant p(z, a) =$
$= p(x, b)p(y, a) \;\&\; p(a, c) \geqslant p(b, c) \leqslant p(y, c) \;\&\; (p(a, e) \geqslant p(c, e)$
$\leqslant p(y, e) \rightarrow p(c, d) \leqslant p(b, d)) \;\&\; (p(a, f) = p(y, f) \rightarrow p(x, a) = p(x, b) =$
$= p(x, y)) \;\&\; (p(x, g) \geqslant p(c, g) \leqslant p(y, g) \rightarrow p(c, a) \leqslant p(z, a))).$

Postulate 4 demands the existence of a complement, $\bar{a}$, for every *a* in *S*, and it characterizes this complement by (a weakened conditional form of) what appears to be an obvious formula, '$p(a, c) + p(\bar{a}, c) = 1$', considering that $1 = p(a, a)$. The condition which precedes this formula is needed, because in case *c* is, say $a\bar{a}$, (the 'empty element': $a\bar{a} = 0$) we obtain $p(a, c) = 1 = p(\bar{a}, c)$ so that, in this limiting case, the apparently 'obvious' formula breaks down.

This postulate, or the axiom C, has the character of a definition of $p(\bar{a}, b)$, in terms of $p(a, b)$ and $p(a, a)$, as can be easily seen if we write C as follows. (Note that (ii) follows from (i).)

(i) $p(\bar{a}, b) = p(a, a) - p(a, b)$, provided there is a *c* such that $p(c, b) \neq p(a, a)$

(ii) $p(\bar{a}, b) = p(a, a)$, provided there is no such *c*.

The following formula CD is analogous to BD and may be transformed into a bi-conditional exactly like BD:

CD $p(\bar{x}, a) = p(y, a) \rightarrow (b)(c)(p(x, a) + p(y, a) \neq p(b, b) \rightarrow p(c, a) = p(b, b))$

The system consisting of AI, BD, and CD is, I think, slightly preferable to AI, B, and C^+, in spite of the complexity of BD.

Postulate AP, ultimately, can be replaced by the simple definition

(.) $$p(a) = p(a, \bar{\bar{a}}a)$$

which, however, uses complementation and the product, and accordingly presupposes both Postulates 3 and 4. Formula (.) will be derived below in appendix *v as formula 75.

Our axiom system can be proved to be *consistent*: we may construct systems of elements S (with an infinite number of different elements: for a finite S, the proof is trivial) and a function $p(a, b)$ such that all the axioms are demonstrably satisfied. Our system of axioms may also be proved to be *independent*. Owing to the weakness of the axioms, these proofs are quite easy.

A trivial consistency proof for a finite S is obtained by assuming that $S = \{ 1, 0 \}$; that is to say, that S consists of the two elements, 1 and 0; product or meet and complement are taken to be equal to arithmetical product and complement (with respect to 1). We define $p(0, 1) = 0$, and in all other cases put $p(a, b) = 1$. Then all the axioms are satisfied.

Two further finite interpretations of S will be given before proceeding to a denumerably infinite interpretation. Both of these satisfy not only our axiom system but also, for example, the following existential assertion (E).

(E) There are elements a, b, and c in S such that
$$p(a, b) = 1 \text{ and } p(a, bc) = 0.$$

A similar assertion would be

(E') There is an element a in S such that
$$p(a) = p(a, \bar{a}) = p(\bar{a}, a) = 0 \neq p(a, a) = 1.$$

This assertion (E) is not satisfied by our first example, nor can it be satisfied in any system of probability known to me (except, of course, some of my own systems).

The first example satisfying our system and (E) consists of four elements. $S = \{ 0, 1, 2, 3 \}$. Here ab is defined as the smaller of the two numbers a and b, except that $1.2 = 2.1 = 0$. We define: $\bar{a} = 3 - a$, and $p(a) = p(a, 3) = 0$ whenever $a = 0$ or 1, and $p(a) = p(a, 3) = 1$ whenever $a = 2$ or 3; $p(a, 0) = 1$; $p(a, 1) = 0$ unless $a = 1$ or $a = 3$,

337

in which case $p(a, 1) = 1$. In the remaining cases $p(a, b) = p(ab)/p(b)$. Intuitively, the element 1 may be identified with a universal law of zero absolute probability, and 2 with its existential negation. In order to satisfy (E), we take $a = 2$, $b = 3$, and $c = 1$.

The example just described may be represented by way of the following two 'matrices'. (This method, I believe, was first introduced by Huntington in 1904.)

ab	0	1	2	3	$\bar{a}$
0	0	0	0	0	3
1	0	1	0	1	2
2	0	0	2	2	1
3	0	1	2	3	0

$p(a, b)$	0	1	2	3
0	1	0	0	0
1	1	1	0	0
2	1	0	1	1
3	1	1	1	1

The second example is a generalization of the first, showing that the idea underlying the first example can be extended to a number of elements exceeding any chosen number, provided these elements form a Boolean Algebra, which means that the number of elements has to be equal to 2^n. Here n may be taken to be the number of the smallest exclusive areas or classes into which some universe of discourse is divided. We can freely correlate with each of these classes some positive fraction, $0 \leqslant r \leqslant 1$, as its absolute probability, taking care that their sum equals 1. With any of the Boolean sums, we correlate the arithmetical sum of their probabilities, and with any Boolean complement, the arithmetical complement with respect to 1. We may assign to one or several of the smallest (non-zero) exclusive areas or classes the probability 0. If b is such an area or class, we put $p(a, b)=0$ in case $ab=0$; otherwise $p(a, b) = 1$. We also put $p(a, 0) = 1$; and in all other cases, we put $p(a, b) = p(ab)/p(b)$.

In order to show that our system is consistent even under the assumption that S is denumerably infinite, the following interpretation may be chosen. (It is of interest because of its connection with the frequency interpretation.) Let S be the class of rational fractions in diadic representation, so that, if a is an element of S, we may write a as a sequence, $a = a_1, a_2 \ldots$, where a_i is either 0 or 1. We interpret ab as the sequence $ab = a_1b_1, a_2b_2, \ldots$ so that $(ab)_i = a_ib_i$, and $\bar{a}$ as the sequence $\bar{a} = 1 - a_1, 1 - a_2, \ldots$, so that $\bar{a}_i = 1 - a_i$. In order to define $p(a, b)$, we introduce an auxiliary expression, A_n, defined as follows:

$$A_n = \sum_n a_i$$

so that we have

$$(AB)_n = \sum_n a_ib_i;$$

moreover, we define an auxiliary function, q:

$$q(a_n, b_n) = 1 \text{ whenever } B_n = 0$$
$$q(a_n, b_n) = (AB)_n / B_n, \text{ whenever } B_n \neq 0.$$

Now we can define,

$$p(a, b) = \lim q(a_n, b_n).$$

This limit exists for all elements a and b of S, and it can be easily shown to satisfy all our axioms.

So much about the *consistency* of our axiom systems.

In order to show the *independence* of A1 we may take $p(a, b) = 0$ for every a and b in S. Then all the axioms except A1 are satisfied.

In order to show the independence of A2 we [10] take S to consist of three elements, $S = \{ 0, 1, 2 \}$. We can easily show that the product ab must be non-commutative; it may be defined as follows: $1.2 = 2$; and in all other cases, including 2.1, ab is equal to $\min(a, b)$, *i.e.* to the smallest of its two components a and b. We also define: $\bar{a} = 1$ if and only if $a = 0$; otherwise $\bar{a} = 0$; and we define $p(0, 2) = 0$; in all other cases, $p(a, b) = 1$. It is now easy to show that for every b, $p(1, b) = p(2, b)$ while $p(0, 1) = 1$ and $p(0, 2) = 0$. Thus A2 is not satisfied. But all the other axioms are.

[10] In view of what has been said above about A2 it is clear that the problem of proving its independence amounts to that of constructing an example (a matrix) which is non-commutative, combined with a numerical rule about the p-values which ensures that the law of commutation is violated only for the second argument. The independence proof for A2 here described, designed to satisfy these conditions, was found at the same time by Dr. J. Agassi and by myself. This matrix satisfies definition (.) on p. 337, though *not* postulate AP; but it does so if, in AP, we put a bar over the '*b*'.

We can illustrate this interpretation by writing the non-commutative matrix as follows:

ab	0	1	2	$\bar{a}$
0	0	0	0	1
1	0	1	2	0
2	0	1	2	0

$p(0, 2) = 0;$
in all other cases
$p(a, b) = 1$

In order to show that A3 is independent, we take, as in our first consistency proof, $S = \{\,0, 1\,\}$, with logical products and complements equal to the arithmetical ones. We define $p(1, 1) = 1$, and in all other cases $p(a, b) = 0$. Then $p(1, 1) \neq p(0, 0)$, so that A3 fails. The other axioms are satisfied.

In order to show that B1 is independent, we take $S = \{\,-1, 0, +1\,\}$; we take ab to be the arithmetical product of a and b; $\bar{a} = -a$; and $p(a, b) = a.(1 - |b|)$. Then all axioms are satisfied except B1 which fails for $a = -1$, $b \neq +1$, and $c = 0$. The matrices may be written:

ab	−1	0	+1	$\bar{a}$
−1	+1	0	−1	+1
0	0	0	0	0
+1	−1	0	+1	−1

$p(a, b)$	−1	0	+1
−1	0	−1	0
0	0	0	0
+1	0	+1	0

This example also proves the independence of A4' (cf. note 6, above). A second example, proving the independence of B1 and also of B1', is based upon the following non-commutative matrix:

ab	0	1	2	$\bar{a}$
0	0	1	0	2
1	0	1	1	0
2	0	1	2	0

$p(0, 2) = 0;$
in all other cases
$p(a, b) = 1$

B1 fails for $a = 0$, $b = 1$, and $c = 2$.

In order to show that B2 is independent, we take the same S as for A3, and define $p(0, 1) = 0$; in all other cases, $p(a, b) = 2$. B2 fails because $2 = p(1.1, 1) \neq p(1, 1.1)\, p(1, 1) = 4$, but all other axioms are satisfied.

(Another example showing the independence of B2 can be obtained if we consider that B2 is needed to prove '$p(ba, c) \leqslant p(a, c)$', that is to say, that dual of B1. This suggests that we may use the second example for B1, changing only the value of 1.0 from 0 to 1, and that of 0.1 from 1 to 0. Everything else may be left unchanged. B2 fails for $a = 1$, $b = 0$, and $c = 2$.)

Ultimately, for showing that C is independent, we take again the same S, but assume that $\bar{a} = a$. If we now take $p\,(0, 1) = 0$ and in all other cases $p(a, b) = 1$, then C fails, because $p(\bar{0}, 1) \neq p(1, 1)$. The other axioms are satisfied.

This concludes the proofs of the independence of the operational *axioms*.

As to the non-operational parts of the *postulates*, a proof of the independence of postulate 1 has been given above (when I commented upon this postulate).

Postulate 2 requires (in its non-operational part) that whenever a and b are in S, $p(a, b)$ is a real number. In order to show the independence of this requirement—which we may briefly refer to as 'postulate 2'—we first consider a *non-numerical Boolean interpretation* of S. To this end, we interpret S as an at most denumerable and non-numerical Boolean algebra (such as a set of statements, so that 'a', 'b', etc. are variable *names of statements*). And we stipulate that '$\bar{x}$' is to denote, if x is a number, the same as '$-x$'; and if x is a Boolean element (say, a statement) then '$\bar{x}$' is to denote the Boolean complement (negation) of x. Similarly, we stipulate that 'xy'; '$x + y$'; '$x = y$'; '$x \neq y$'; and '$x \leqslant y$', have their usual arithmetical meaning if x and y are numbers, and their well-known Boolean meanings whenever x and y are Boolean elements. (If x and y are statements, '$x \leqslant y$' should be interpreted as 'x entails y'.) In order to prove the independence of postulate 2, we now merely add one more stipulation: we interpret '$p(a, b)$' as synonymous with '$a + \bar{b}$'. Then postulate 2 breaks down while A1, A2, A3 and all the other axioms and postulates turn into well-known theorems of Boolean algebra.[11]

---

[11] A slight variant of this interpretation transforms all the axioms into tautologies of the propositional calculus, satisfying all the postulates except postulate 2.

The proofs of the independence of the existential parts of postulates 3 and 4 are almost trivial. We first introduce an auxiliary system $S' = \{0, 1, 2, 3\}$ and define product, complement, and absolute probability by the matrix:

ab	0	1	2	3	$\bar{a}$	$p(a)$
0	0	0	0	0	3	0
1	0	1	0	1	2	1/2
2	0	0	2	2	1	1/2
3	0	1	2	3	0	1

Relative probability is defined by

$$p(a, b) = 1 \text{ whenever } p(b) = 0$$
$$p(a, b) = p(ab)/p(b) \text{ whenever } p(b) \neq 0.$$

This system S' satisfies all our axioms and postulates. In order to show the independence of the existential part of postulate 3, we now take S to be confined to the elements 1 and 2 of S', leaving everything else unchanged. Obviously, postulate 3 fails, because the product of the elements 1 and 2 is not in S; everything else remains valid. Similarly, we can show the independence of postulate 4 by confining S to the elements 0 and 1 of S'. (We may also choose 2 and 3, or any combination consisting of three of the four elements of S' except the combination consisting of 1, 2, and 3.)

The proof of the independence of postulate AP is even more trivial: we only need to interpret S and $p(a, b)$ in the sense of our first consistency proof and take $p(a) = constant$ (a constant such as 0, or 1/2, or 1, or 2) in order to obtain an interpretation in which postulate AP fails.

Thus we have shown that every single assertion made in our axiom system is independent. (To my knowledge, no proofs of independence for axiom systems of probability have been published before. The reason, I suppose, is that the known systems—provided they are otherwise satisfactory—are not independent.)

The redundancy of the usual systems is due to the fact that they all postulate, implicitly or explicitly, the validity of some or all of the rules of Boolean algebra for the elements of S; but as we shall prove at the end of appendix *v, these rules are all derivable from our system if we define Boolean equivalence, '$a = b$', by the formula

(∗) $a = b$ if, and only if, $p(a, c) = p(b, c)$ for every c in S.

The question may be asked whether any of our axioms would become redundant if we *postulate* that ab is a Boolean product and $\bar{a}$ a Boolean complement; that they both obey all the laws of Boolean algebra; and that (∗) is valid. The answer is that none of our axioms would become redundant. (Only if we were to postulate, in addition, that any two elements for which Boolean equivalence can be proved may be substituted for each other *in the second argument* of the p-function, then *one* of our axioms would become redundant, *i.e.* our axiom of substitutivity, A2, which serves precisely the same function as this additional postulate.) That our axioms remain non-redundant can be seen from the fact that their independence (except that of A2, of course), can be proved with the help of examples that satisfy Boolean algebra. I have given such examples for all of them except B1, for which a simpler example has been given. An example of a Boolean algebra that shows the independence of B1 (and of A4′) is the following. (0 and 1 are the Boolean zero and universal elements, and $\bar{a} = 1 - a$; the example is, essentially, the same as the last one, but with the probabilities — 1 and 2 attached to the two elements which are neither empty nor universal.)

ab	-1	0	1	2	$\bar{a}$
-1	-1	0	-1	0	2
0	0	0	0	0	1
1	-1	0	1	2	0
2	0	0	2	2	-1

$p(a) = a;$
$p(a, 0) = 1;$
in all other cases,
$p(a, b) = p(ab)/p(b,) = ab/b$

B1 is violated because $2 = p(1.2, 1) > p(1, 1) = 1$

The fact that our system remains independent even if we postulate Boolean algebra *and* (∗) may be expressed by saying that our system

is 'autonomously independent'. If we replace our axiom B1 by A4'
and B1' (see note 6 above), then our system ceases, of course, to be
autonomously independent. Autonomous independence seems to
me an interesting (and desirable) property of axiom systems for the
calculus of probability.

In conclusion I wish to give a definition, in the 'autonomous'
i.e. probabilistic terms of our theory, of an 'admissible system' S, and
of a *'Borel field of probabilities'* S. The latter term is Kolmogorov's;
but I am using it in a sense slightly wider than his. I will discuss the
difference between Kolmogorov's treatment of the subject and mine
in some detail because it seems to me illuminating.

I first define, in probabilistic terms, what I mean by saying that
a is a super-element of b (and wider than, or equal to, b) or that b
is a sub-element of a (and logically stronger than, or equal to, a).
The definition is as follows. (See also the end of appendix *v.)

a is a super-element of b, or b is a sub-element of a—in symbols,
$a \geqslant b$—if, and only if, $p(a, x) \geqslant p(b, x)$ for every x in S.

Next I define what I mean by the product-element a of an infinite
sequence, $A = a_1, a_2, \ldots$, all of whose members a_n are elements of S.

Let some or perhaps all elements of S be ordered in an *infinite
sequence* $A = a_1, a_2, \ldots$, such that any element of S is permitted to
recur in the sequence. For example, let S consist only of the two
elements, 0 and 1; then $A = 0, 1, 0, 1, \ldots$, and $B = 0, 0, 0, \ldots$, will
both be infinite sequences of elements of S, in the sense here intended.
But the more important case is of course that of an infinite sequence
A such that all, or almost all, of its members are *different* elements of
S which, accordingly, will contain infinitely many elements.

A case of special interest is a *decreasing* (or rather non-increasing)
infinite sequence, that is to say, a sequence $A = a_1, a_2, \ldots$, such that
$a_n \geqslant a_{n+1}$ for every consecutive pair of members of A.

We can now define the (Boolean, as opposed to set-theoretical)
product element a of the infinite sequence $A = a_1, a_2, \ldots$, as the widest
among those elements of S which are sub-elements of every element
a_n belonging to the sequence A; or in symbols:

$a = \pi a_n$ if, and only if, a satisfies the following two conditions
(i) $p(a_n, x) \geqslant p(a, x)$ for all elements a_n of A, and for every element x
of S.
(ii) $p(a, x) \geqslant p(b, x)$ for all elements x of S and for every element b of

S that satisfies the condition $p(a_n, y) \geqslant p(b, y)$ for all elements a_n and for every element y of S.

In order to show the difference between our (Boolean) product element a of A and the set-theoretical (inner) product or meet of A, we will now confine our discussion to examples S, satisfying our postulates 2 to 5, whose elements x, y, z, . . . are sets, such that xy is their set-theoretic product.

Our main example S_1 to which I shall refer as 'the example of the missing half-interval' is the following.

S_1 is a system of certain half-open sub-intervals of the universal interval $u = (0, 1]$. S_1 contains, precisely, (a) the decreasing sequence A such that $a_n = (0, \frac{1}{2} + 2^{-n}]$, and in addition (b) the set-theoretic products of any two of its elements and the set-theoretic complements of any one of its elements.

Thus S_1 does not contain the 'half-interval' $h = (0, \frac{1}{2}]$, nor any non-empty sub-interval of h.

Since the missing half-interval $h = (0, \frac{1}{2}]$ is the set-theoretic product of the sequence A, it is clear that S_1 does not contain the set-theoretic product of A. But S_1 does contain the (Boolean) 'product-element' of A, as here defined. For the empty interval trivially satisfies condition (i); and since it is the widest interval satisfying (i), it also satisfies (ii).

It is clear, moreover, that if we add to S_1, say any of the intervals $b_1 = (0, \frac{1}{8}]$, or $b_2 = (0, \frac{3}{16}]$, etc., then the largest of these will be the product element of A in the (Boolean) sense of our definition, although none of them will be the set-theoretic product of A.

One might think, for a moment, that owing to the presence of an empty element in every S, every S will contain, like S_1, a product element (in the sense of our definition) of any A in S; for if it does not contain any wider element satisfying (i), the empty element will always fill the bill. That this is not so is shown by an example S_2 containing, in addition to the elements of S_1, the elements (plus the set-theoretic products of any two elements and the set-theoretic complement of any one element) of the sequence $B = b_1$, b_2, where $b_n = (0, (2^n - 1) / 2^{n+2}]$. It will be easily seen that, although each b_n satisfies condition (i) for the product-element of A, none of them satisfies condition (ii); so that in fact, there is *no widest element* in S_2 that satisfies the condition (i) for the product-element of A.

Thus S_2 contains neither the set-theoretic product of A, nor a product-element in our (Boolean) sense. But S_1, and all the systems obtained by adding to S_1 a finite number of new intervals (plus products and complements) will contain a product-element of A in our sense but not in the set-theoretic sense, unless, indeed, we add to S_1 the missing half-interval $h = (0, \frac{1}{2}]$.

Remembering that the emptiness of an element a may be characterized in our system by $p(\bar{a}, a) \neq 0$, we can now define an 'admissible system S' and a 'Borel field of probabilities S', as follows.

(i) A system S that satisfies our postulates 2 to 4 is called an *admissible system* if, and only if, S satisfies our set of postulates and in addition the following *defining condition*.

Let $bA = a_1 b, a_2 b, \ldots$ be any decreasing sequence of elements of S. (We say in this case that $A = a_1, a_2, \ldots$ is 'decreasing relative to b'.) Then if the product element ab of this sequence is in S,[12] then

$$\lim p(a_n, b) = p(a, b)$$

(ii) An admissible system S is called a *Borel field of probabilities* if, and only if, there is in S a product-element of any (absolutely or relatively) decreasing sequence of elements of S.

Of these two definitions, (i) corresponds precisely to Kolmogorov's so-called 'axiom of continuity', while (ii) plays a part in our system analogous to Kolmogorov's definition of Borel fields of probability.

It can now be shown that *whenever S is a Borel field of probabilities in Kolmogorov's sense, it is also one in the sense here defined, with probability as a countably additive measure function of the sets which are the elements of S.*

The definitions of admissible systems and Borel fields of probabilities are framed in such a way that all systems S satisfying our postulates and containing only a finite number of different elements are admissible systems and Borel fields; accordingly, our definitions are interesting only in connection with systems S *containing an infinite number of different elements*. Such infinite systems may, or may not,

---

[12] I might have added here 'and if $p(\overline{ab}, ab) \neq 0$, so that ab is empty': this would have approximated my formulation still more closely to Kolmogorov's. But this condition is not necessary. I wish to point out here that I have received considerable encouragement from reading A. Rényi's most interesting paper 'On a New Axiomatic Theory of Probability', *Acta Mathematica Acad. Scient. Hungaricae* **6**, 1955, pp. 286-335. Although I had realized for years that Kolmogorov's system ought to be relativized, and although I had on several occasions pointed out some of the mathematical advantages of a relativized system, I only learned from Rényi's paper how fertile this relativization could be. The relative systems published by me since 1955 are more general still than Rényi's system which, like Kolmogorov's, is set-theoretical, and non-symmetrical; and it can be easily seen that these further generalizations may lead to considerable simplifications in the mathematical treatment.

satisfy the one or the other or both of our defining conditions; in other words, for infinite systems our defining conditions are non-redundant or independent.

This non-redundancy can be proved for (i) most easily in that form of it which is mentioned in footnote 12, with the help of the example of the missing half-interval, S_1, given above. All we have to do is to define probability $p(x)$ as equal to $l(x)$, that is to say, the length of the interval x. Our first definition, (i), is then violated since $\lim p(a_n) = \frac{1}{2}$ while for the product-element (in S) of A, $p(a) = 0$. Definition (ii) is violated by our example S_2 (which vacuously satisfies the first definition).

While the first of these examples establishes the independence or more precisely the non-redundancy of our first definition—by violating it—it does not, as it stands, establish the independence of Kolmogorov's 'axiom of continuity' which is clearly satisfied by our example. For the missing half-interval, $h = (0, \frac{1}{2}]$, whether in S or not, is the only set-theoretic product of A, so that $a = h$ is true for the set-theorist, whether or not a is in S. And with $a = h$, we have $\lim p(a_n) = p(a)$. Thus Kolmogorov's axiom is satisfied (even if we omit the condition $p(\bar{a}, a) \neq 0$; cf. footnote 12).

It should be mentioned, in this connection, that Kolmogorov fails, in his book, to offer an independence proof for his 'axiom of continuity' although he claims independence for it. But it is possible to re-frame our proof of independence so that it becomes applicable to Kolmogorov's axiom and his set-theoretic approach. This may be done by choosing, instead of our S_1 a system of intervals S_3, exactly like S_1 but based upon a sequence $C = c_1, c_2, \ldots$, defined by $c_n = (0, 2^{-n}]$ rather than upon the sequence $A = a_1, a_2, \ldots$, with $a_n = (0, \frac{1}{2} + 2^{-n}]$. We can now show the independence of Kolmogorov's axiom by defining the probabilities of the elements of the sequence A as follows:

$$p(c_n) = l(c_n) + \frac{1}{2} = p(a_n)$$

Here $l(c_n)$ is the length of the interval c_n. This definition is highly counter-intuitive, since, for example, it assigns to both the intervals $(0, \frac{1}{2}]$ and $(0, 1]$ the probability one, and therefore to the interval $(\frac{1}{2}, 1]$ the probability zero; and the fact that it violates Kolmogorov's axiom (thereby establishing its independence) is closely connected with its counter-intuitive character. For it violates the axiom because

$\lim p(c_n) = \frac{1}{2}$, even though $p(c) = 0$. Because of its counter-intuitive character, the *consistency* of this example is far from self-evident; and so the need arises to prove its consistency in order to establish the validity of this independence proof of Kolmogorov's axiom.

But this consistency proof is easy in view of our previous independence proof—the proof of the independence of our own first definition with the help of the example S_1. For the probabilities $p(a_n)$ and $p(c_n)$ of the two examples S_1 and S_3 coincide. And since by correlating the two sequences, A and C, we may establish a one-one correspondence between the elements of S_1 and S_3, the consistency of S_1 proves that of S_3.

It is clear that *any* example proving the independence of Kolmogorov's axiom must be equally counter-intuitive, so that its consistency will be in need of proof by some method similar to ours. In other words, the proof of the independence of Kolmogorov's axiom will have to utilize an example which is, essentially, based upon a (Boolean) definition of product such as ours, rather than upon the set-theoretic definition.

Although every Borel field of probabilities in Kolmogorov's sense is also one in our sense, the opposite is not the case. For we can construct a system S_4 which is exactly like S_1, with $h = (a, \frac{1}{2}]$ still missing and containing in its stead the *open* interval $g = (a, \frac{1}{2})$, with $p(g) = \frac{1}{2}$. We define, somewhat arbitrarily, $\bar{g} = u - g = (\frac{1}{2}, 1]$, and $u - (g + \bar{g}) = u\bar{u}$ (rather than the point $\frac{1}{2}$). It is easily seen that S_4 is a Borel field in our sense, with g as the product-element of A. But S_4 is not a Borel field in Kolmogorov's sense since it does not contain the set-theoretic product of A: our definition allows an *interpretation by a system of sets* which is not a Borel system of sets, and in which product and complement are not exactly the set-theoretic product and complement. Thus our definition is wider than Kolmogorov's.

Our independence proofs of (i) and (ii) seem to me to shed some light upon the functions performed by (i) and (ii). The function of (i) is to exclude systems such as S_1, in order to ensure measure-theoretical adequacy of the product (or limit) of a decreasing sequence: the limit of the measures must be equal to the measure of the limit. The function of (ii) is to exclude systems such as S_2, with increasing sequences without limits. It is to ensure that every decreasing sequence has a produce in S and every increasing sequence a sum.

Appendix *v.   Derivations in the Formal Theory of Probability.*

In this appendix I propose to give the most important derivations from the system of postulates which has been explained in appendix *iv. I am going to show how the laws of the upper and lower bounds, of idempotence, commutation, association, and distribution are obtained, as well as a simpler definition of absolute probability. I will also indicate how Boolean algebra is derivable in the system. A fuller treatment will be given elsewhere.

As an abbreviation for 'if ... then ... ', I am going to use an arrow '... → ...'; a double arrow '... ←→ ...', for '... if and only if ...', '(E*a*) ... ' for 'there is an *a* in *S* such that ... '; and '(*a*) ... ' for 'for all *a* in *S*, ... '.

I first re-state postulate 2 and the six operational axioms which will all be cited in the proofs. (The other postulates will be used implicitly; even postulate 2 will be cited only once, in the proof of 5.) In reading the axioms A3 and C, it should be kept in mind that I shall soon prove—see formula 25—that $p(a, a) = 1$.

Postulate 2.   If *a* and *b* are in *S*, then $p(a, b)$ is a real number.

A1	$(Ec)(Ed)\ p(a, b) \neq p(c, d),$
A2	$((c)(p(a, c) = p(b, c))\ \rightarrow\ p(d, a) = p(d, b),$
A3	$p(a, a) = p(b, b)\ .$
B1	$p(ab, c) \leqslant p(a, c),$
B2	$p(ab, c) = p(a, bc)p(b, c)\ .$
C	$p(a, a) \neq p(b, a) \rightarrow p(a, a) = p(c, a) + p(\bar{c}, a)\ .$

I now proceed to the derivations.

(1)	$p(a, a) = p(b, b) = k$	Abbreviation based upon A3
(2)	$p((aa)a, a) \leqslant p(aa, a) \leqslant p(a, a) = k$	B1, 1
(3)	$p((aa)a, a) = p(aa, aa)p(a, a) = k^2$	B2, 1
(4)	$k^2 \leqslant k$	2, 3
(5)	$0 \leqslant k \leqslant 1$	4 (and Postulate 2)

349

(6)	$k \neq p(a, b) \;\rightarrow\; k = k + p(\bar{b}, b)$	C, 1
(7)	$k \neq p(a, b) \;\rightarrow\; p(\bar{b}, b) = 0$	6

(8)	$p(a\bar{b}, b) = p(a, \bar{b}b)p(\bar{b}, b)$	B2
(9)	$k \neq p(a, b) \;\rightarrow\; 0 = p(a\bar{b}, b) \leqslant p(a, b)$	7, 8, B1
(10)	$k \neq p(a, b) \;\rightarrow\; 0 \leqslant p(a, b)$	9
(11)	$0 > p(a, b) \;\rightarrow\; k = p(a, b)$	10
(12)	$k = p(a, b) \;\rightarrow\; 0 \leqslant p(a, b)$	5
(13)	$0 > p(a, b) \;\rightarrow\; 0 \leqslant p(a, b)$	11, 12
(14)	$0 \leqslant p(a, b)$	13 (or 10, 12)

(15)	$0 \leqslant p(\bar{a}, b)$	14
(16)	$k \neq p(a, b) \;\rightarrow\; k \geqslant p(a, b)$	C, 1, 15
(17)	$p(a, b) \leqslant k \leqslant 1$	16, 5

(18)	$0 \leqslant p(a, b) \leqslant k \leqslant 1$	14, 17

(19)	$k = p(aa, aa) \leqslant p(a, aa) \leqslant k$	1, B1, 17
(20)	$k = p(a(aa), a(aa)) \leqslant p(a, a(aa)) \leqslant k$	1, B1, 17
(21)	$k = p(aa, aa) = p(a, a(aa))p(a, aa) = k^2$	1, B2, 19, 20
(22)	$k = k^2$	21
(23)	$(Ea)\,(Eb)\; p(a, b) \neq 0 \;\rightarrow\; k = 1$	18, 22

(24)	$(Ea)\,(Eb)\; p(a, b) \neq 0$	A1
(25)	$p(a, a) = k = 1$	1, 23, 24

(26)	$(Eb)\,(Ea)\; p(b, a) \neq k$	A1, 1
(27)	$(Ea)\; p(\bar{a}, a) = 0$	7, 26

We have now established all the laws of the upper and lower bounds: (14) and (17), summed up in (18), show that probabilties are bounded by 0 and 1. (25) and (27) show that these bounds are actually reached. We now turn to the derivation of the various laws usually taken either from Boolean algebra or from the propositional calculus. First we derive the law of idempotence.

(28)	$1 = p(ab, ab) \leqslant p(a, ab) = 1$	25, B1, 17
(29)	$p(aa, b) = p(a, ab)p(a, b)$	B2
(30)	$p(aa, b) = p(a, b)$	28, 29

This is the law of idempotence, sometimes also called the 'law of tautology'. We now turn to the derivation of the law of commutation.

$$(31) \quad p(a, bc) \leqslant 1 \qquad\qquad\qquad\qquad\qquad 17$$
$$(32) \quad p(ab, c) \leqslant p(b, c) \qquad\qquad\qquad\qquad \text{B2, 31, 14}$$

This is the second law of monotony, analogous to B1.

$$(33) \quad p(a(bc), a(bc)) = 1 \qquad\qquad\qquad\qquad\qquad 25$$
$$(34) \quad p(bc, a(bc)) = 1 \qquad\qquad\qquad\qquad\qquad 33, 32, 17$$
$$(35) \quad p(b, a(bc)) = 1 \qquad\qquad\qquad\qquad\qquad 34, \text{B1}, 17$$
$$(36) \quad p(ba, bc) = p(a, bc) \qquad\qquad\qquad\qquad\qquad 35, \text{B2}$$
$$(37) \quad p((ba)b, c) = p(ab, c) \qquad\qquad\qquad\qquad\qquad 36, \text{B2}$$
$$(38) \quad p(ba, c) \geqslant p(ab, c) \qquad\qquad\qquad\qquad\qquad 37, \text{B1}$$
$$(39) \quad p(ab, c) \geqslant p(ba, c) \qquad\qquad\qquad\qquad\qquad 38 \text{ (subst.)}$$
$$(40) \quad p(ab, c) = p(ba, c) \qquad\qquad\qquad\qquad\qquad 38, 39$$

This is the law of commutation for the first argument. (In order to extend it to the second argument, we should have to use A2.) It has been derived from (25), merely by using the two laws of monotony (B1 and 32) and B2. We now turn to the derivation of the law of association.

$$(41) \quad p(ab, d((ab)c)) = 1 \qquad\qquad\qquad\qquad\qquad 35 \text{ (subst.)}$$
$$(42) \quad p(a, d((ab)c)) = 1 = p(b, d((ab)c)) \qquad\qquad 41, \text{B1}, 17, 32$$
$$(43) \quad p(a, (bc)((ab)c)) = 1 \qquad\qquad\qquad\qquad\qquad 42 \text{ (subst.)}$$
$$(44) \quad p(a(bc), (ab)c) = p(bc, (ab)c) \qquad\qquad\qquad\qquad 43, \text{B2}$$
$$(45) \quad p(bc, (ab)c) = p(b, c((ab)c))p(c, (ab)c) \qquad\qquad \text{B2}$$
$$(46) \quad p(b, c((ab)c)) = 1 \qquad\qquad\qquad\qquad\qquad 42 \text{ (subst.)}$$
$$(47) \quad p(c, (ab)c) = 1 \qquad\qquad\qquad\qquad\qquad 25, 32, 17$$
$$(48) \quad p(a(bc), (ab)c) = 1 \qquad\qquad\qquad\qquad\qquad 44 \text{ to } 47$$

This is a preliminary form of the law of association. (62) follows from it by A2^{+} (and B2), but I avoid where possible using A2 or A2^{+}.

$$(49) \quad p(a(b(cd)), d) = p(cd, b(ad))p(b, ad)p(a, d) \qquad 40, \text{B2}$$
$$(50) \quad p(a(bc), d) = p(c, b(ad))p(b, ad)p(a, d) \qquad\qquad 40, \text{B2}$$
$$(51) \quad p(a(bc), d) \geqslant p(a(b(cd)), d) \qquad\qquad\qquad\qquad 49, 50 \text{ B1}$$

This is a kind of weak generalization of the first monotony law, B1.

$$(52) \quad p(a(b(cd)), (ab)(cd)) = 1 \qquad\qquad\qquad\qquad 48 \text{ (subst.)}$$
$$(53) \quad p((a(b(cd)))(ab), cd) = p(ab, cd) \qquad\qquad\qquad 52, \text{B2}$$

(54) $p(a(b(cd)), cd) \geqslant p(ab, cd)$ 53, B1
(55) $p((a(b(cd)))c, d) \geqslant p((ab)c, d)$ 54, B2
(56) $p(a(b(cd)), d) \geqslant p((ab)c, d)$ 55, B1
(57) $p(a(bc), d) \geqslant p((ab)c, d)$ 51, 56

This is one half of the law of association.

(58) $p((bc)a, d) \geqslant p((ab)c, d)$ 57, 40
(59) $p((ab)c, d) \geqslant p(b(ca), d)$ 58 (subst.), 40
(60) $p((bc)a, d) \geqslant p(b(ca), d)$ 58, 59
(61) $p((ab)c, d) \geqslant p(a(bc), d)$ 60, (subst.)

This the second half of the law of association.

(62) $p((ab)c, d) = p(a(bc), d)$ 57, 61

This is the complete form of the law of association, for the first argument (see also formula (g) at the beginning of appendix *iv). The law for the second argument can be obtained by applying A2. (Applying B2 twice to each side of (62) only leads to a conditional form with '$p(bc, d) \neq 0 \rightarrow$' as antecedent.)

I now turn to a generalization of the axiom of complementation, C. I shall be a little more concise in my derivations from now on.

(63) $p(\bar{b}, b) \neq 0 \rightarrow p(c, b) = 1$ 7, 25
(64) $p(a, b) + p(\bar{a}, b) = 1 + p(\bar{b}, b)$ C, 25, 63

This is an unconditional form of the principle of complementation, C, which I am now going to generalize.

In view of the fact that (64) is unconditional, and that 'a' does not occur on the right-hand side, we can substitute 'c' for 'a' and assert

(65) $p(a, b) + p(\bar{a}, b) = p(c, b) + p(\bar{c}, b)$ 64
(66) $p(a, bd) + p(\bar{a}, bd) = p(c, bd) + p(\bar{c}, bd)$ 65

By multiplying with $p(b, d)$ we get:

(67) $p(ab, d) + p(\bar{a}b, d) = p(cb, d) + p(\bar{c}b, d)$ 66

This is a generalization of (65). By substitution, we get:

(68) $p(ab, c) + p(\bar{a}b, c) = p(cb, c) + p(\bar{c}b, c)$.   67

In view of

(69) $p(\bar{c}b, c) = p(\bar{c}, c)$,   7, B1, 25, 63

we may also write (68) more briefly, and in analogy to (64),

(70) $p(ab, c) + p(\bar{a}b, c) = p(b, c) + p(\bar{c}, c)$.   68, 69[1]

This is the generalization of the unconditional form of C and of formula (64).

(71) $p(aa, b) + p(\bar{a}a, b) = p(a, b) + p(\bar{b}, b)$ 70

(72) $p(\bar{a}a, b) = p(a\bar{a}, b) = p(\bar{b}, b)$ 40, 71, 30

(73) $p(\bar{a}a, b) + p(\overline{\bar{a}a}, b) = p(a\bar{a}, b) + p(\overline{a\bar{a}}, b) = 1 + p(\bar{b}, b)$ 64

(74) $p(\bar{a}a, b) = 1 = p(\overline{a\bar{a}}, b)$ 72, 73

This establishes the fact that the elements $\overline{a\bar{a}}$ satisfy the condition of Postulate AP. We obtain, accordingly,

(75) $p(a) = p(a, \overline{a\bar{a}}) = p(a, \overline{\bar{a}a}) = p(a, \overline{b\bar{b}}) = p(a, \overline{\bar{b}b})$;   25, 74, AP

that is, a definition of absolute probability in a more workable form.

We next derive the general law of addition.

(76) $p(a\bar{b}, c) = p(a, c) - p(ab, c) + p(\bar{c}, c)$ 70, 40

(77) $p(\overline{a}\bar{b}, c) = p(\bar{a}, c) - p(\bar{a}b, c) + p(\bar{c}, c)$ 76

(78) $p(\overline{a\bar{b}}, c) = 1 - p(a,c) - p(b,c) + p(ab,c) + p(\bar{c},c)$ 77, 76, 64, 40

(79) $p(\overline{\overline{a\bar{b}}}, c) = p(a, c) + p(b, c) - p(ab, c)$ 78, 64

This is a form of the general law of addition, as will be easily seen

---

[1] In the derivation of (70) we also need the following formula
$$p(cb, c) = p(b, c),$$
which may be called '(29′)'. Its derivation, in the presence of (40) and (32) is analogous to the steps (28) and (29):

(28′) $p(ab, ab) = 1 = p(b, ab)$ 25, 32, 17

(29′) $p(ba, b) = p(b, ab)p(a, b) = p(a, b)$.   B2, 28′

To this we may add the law of idempotence for the second argument

(30′) $p(ab, b) = p(a, bb) = p(a, b)$.   B2, 25, 29′, 40

Moreover, from (28) we obtain by substitution

(31′) $p(a, a\bar{a}) = 1$ 28

and likewise from (28′)

(32′) $p(\bar{a}, a\bar{a}) = 1$ 28′

This yields, by C,

(33′) $p(a, b\bar{b}) = 1$ 31′, 32′, C

We therefore have

(34′) $(Eb)(a) \, p(a, b) = 1$ 33′

(35′) $(Ea) \, p(\bar{a}, a) = 1$ 34′

See also (27). Formulae (31′) to (35′) *do not belong to the theorems of the usual systems.*

if it is remembered that '$\overline{\overline{ab}}$' means the same in our system as '$a + b$' in the Boolean sense. It is worth mentioning that (79) has the usual form: it is unconditional *and* free of the unusual '$+ p(\bar{c}, c)$'. (79) can be further generalized:

$$(80) \quad p(\overline{\overline{bc}}, ad) = p(b, ad) + p(c, ad) - p(bc, ad) \qquad\qquad 79$$

$$(81) \quad p(a\,\overline{\overline{bc}}, d) = p(ab, d) + p(ac, d) - p(a(bc), d) \qquad\quad \text{80, B2, 40}$$

This is a generalization of (79).

We now proceed to the derivation of the law of distribution. It may be obtained from (79), (81), and a simple lemma (84) which I propose to call the 'distribution lemma', and which is a generalization of (30):

$$(82) \quad p(a(bc), d) = p(a, (bc)d)p(bc, d) = p((aa)(bc), d) \qquad\qquad \text{B2, 30}$$
$$(83) \quad p(((aa)b)c, d) = p(a(ab), cd)p(c, d) = p(((ab)a)c, d) \qquad \text{B2, 62, 40}$$
$$(84) \quad p(a(bc), d) = p((ab)(ac), d) \qquad\qquad\qquad\qquad \text{82, 83, 62}$$

This is the 'distribution lemma'.

$$(85) \quad p(\overline{\overline{ab}\ \overline{ac}}, d) = p(ab, d) + p(ac, d) - p((ab)(ac), d) \qquad \text{79 (subst.)}$$

To this formula and (81) we can now apply the 'distribution lemma'; and we obtain:

$$(86) \quad p(a\,\overline{\overline{bc}}, d) = p(\overline{\overline{ab}\ \overline{ac}}, d) \qquad\qquad\qquad\qquad \text{81, 85, 84}$$

This is a form of the first law of distribution. It can be applied to the left side of the following formula

$$(87) \quad p(\overline{\overline{b}\,\overline{\overline{b}}a}, c) = p(\overline{b}\,\overline{\overline{b}}, ac)p(a, c) = p(a, c) \qquad\qquad \text{B2, 74}$$

We then obtain,

$$(88) \quad p(\overline{\overline{ab}\ \overline{ab}}, c) = p(a, c)\,. \qquad\qquad\qquad\qquad \text{86, 87, 40}$$

It may be noted that

$$(89) \quad p(\bar{\bar{a}}b, c) = p(ab, c)\,, \qquad\qquad\qquad\qquad\qquad \text{64, B2}$$
$$(90) \quad p(a, c) = p(b, c) \;\rightarrow\; p(\bar{a}, c) = p(\bar{b}, c) \qquad\qquad\qquad 64$$

Consequently, we have

$$(91) \quad p(\overline{\overline{a\,\overline{\overline{b}}\,\bar{c}}}, d) = p(\overline{\overline{\bar{a}\,\overline{\overline{b}}\,c}}, d) \qquad\qquad\qquad\qquad \text{62, 89, 40}$$
$$(92) \quad p(\overline{\overline{a\,\overline{\overline{b}}\,\bar{c}}}, d) = p(\overline{\overline{\bar{a}\,\overline{\overline{b}}\,c}}, d) \qquad\qquad\qquad\qquad 90, 91$$

This is the law of association for the Boolean sum. By substituting in (40) the complements of a and b, we find

(93) $\quad p(\overline{\overline{a}\,\overline{b}}, c) = p(\overline{\overline{b}\,\overline{a}}, c)$ 40, 90

This is the law of commutation for the Boolean sum. In the same way we get

(94) $\quad p(\overline{\overline{a}\,\overline{a}}, b) = p(a, b)$ 30, 89, 90

This is the law of idempotence for the Boolean sum. From (87) we obtain

(95) $\quad p(a, b) = p(a, b\overline{c}\overline{c})$, 87, 40, A2

(96) $\quad p(a, b)p(b) = p(ab)$ 95, B2, 75

This may also be written

(97) $\quad p(b) \neq 0 \;\rightarrow\; p(a, b) = p(ab)/p(b)$ 96

This formula shows that our generalized concept of relative probability coincides, for $p(b) \neq 0$, with the usual concept, and that our calculus is a generalization of the usual calculus. That the generalization is a genuine one can be seen from the examples, given in the preceding appendix ✴iv, showing the consistency of our system with the following formula (E):

(E) $\qquad\qquad$ (Ea)(Eb)(Ec) $p(a, b) = 1$ and $p(a, bc) = 0$

—a formula which is invalid in many finite interpretations of our S but valid in its normal infinite interpretations.

In order to prove now that, in any consistent interpretation, S must be a Boolean algebra, we note that

(98) $\quad ((x)p(a, x) = p(b, x)) \;\rightarrow\; p(ay, z) = p(by, z)$ B2

(99) $\quad ((x)p(a, x) = p(b, x)) \;\rightarrow\; p(y, az) = p(y, bz)$ 98, A2

It is interesting that (99) needs A2: it does not follow from 98, 40, and B2, since it is possible that $p(a, z) = p(b, z) = 0$. (This will be the case, for example, if $\overline{a} = z \neq x\overline{x}$.)

(100) $\quad ((x)(p(a, x) = p(b, x) \ \& \ p(c, x) = p(d, x))) \;\rightarrow\; p(ac, y) =$
$\qquad\quad p(bd, y)$ 99, B2

With the help of (90), of (100), and of A2, it can now easily be shown at once that whenever the condition

(✴) $\qquad\qquad p(a, x) = p(b, x)$ for every x in S

is satisfied, any name of the element a may be substituted for some or all occurrences of names of the element b in any well-formed formula of the calculus without changing its truth value; or in other words, the condition (*) guarantees the *substitutional equivalence* of a and b.

In view of this result, we now define the Boolean equivalence of two elements, a and b, as follows.

(D1) $a = b \longleftrightarrow (x)p(a, x) = p(b, x)$

From this definition we obtain at once the formulae

(A) $a = a$
(B) $a = b \rightarrow b = a$
(C) $(a = b \ \& \ b = c) \rightarrow a = c$
(D) $a = b \rightarrow a$ may replace b in some or all places of any formula without affecting its truth value.   A2, 90, 100

We may also introduce a second definition

(D2) $a = b + c \longleftrightarrow a = \overline{\overline{b}\,\overline{c}}$

Then we obtain:

(i)   If a and b are in S, then $a + b$ is in S (Postulate 3, D2, D1, 90, 100)
(ii)   If a is in S then $\bar{a}$ is in S (Postulate 4)
(iii) $a + b = b + a$ 93, D2
(iv) $(a + b) + c = a + (b + c)$ 92, D2
(v) $a + a = a$ 94, D2
(vi) $ab + a\bar{b} = a$ 88, D2
(vii) $(Ea)(Eb)\ a \neq b$ 27, 74, 90, D1

But the system (A) to (D2) and (i) to (vi) is a well-known axiom system for Boolean algebra, due to Huntington;[2] and it is known that all valid formulae of Boolean algebra are derivable from it.

Thus S is a Boolean algebra. And since Boolean algebra may be interpreted as a logic of derivation. we may assert that *in its logical interpretation, the probability calculus is a genuine generalization of the logic of derivation.*

More particularly, we may interpret

$$a \geqslant b$$

which is definable by '$ab = b$', to mean, in logical interpretation, 'a follows from b' (or 'b entails a'). It can be easily proved that

(+) $a \geqslant b \rightarrow p(a, b) = 1$

[2] *Cf.* E. V. Huntington, *Transactions Am. Math. Soc.* **35**, 1933, p. 274–304. The system (i) to (vi) is Huntington's 'fourth set', and is described on p. 280. On the same page may be found (A) to (D), and (D2). Formula (v) is redundant, as Huntington showed on pp. 557 *f* of the same volume. (vii) is also assumed by him.

This is an important formula[3] which is asserted by many authors, but which is nevertheless invalid in the usual systems—provided they are consistent. For in order to make it valid, we must make allowance for[4]

$$p(a, a\bar{a}) + p(\bar{a}, a\bar{a}) = 2,$$

even though we have

$$p(a + \bar{a}, a\bar{a}) = 1.$$

That is to say, such formulae as $p(a + \bar{a}, b) = p(a, b) + p(\bar{a}, b)$ must not be unconditionally asserted in the system. (*Cf.* our axiom *C*; see also footnote 1, above.)

The converse of $(^{+})$, that is to say,

$$p(a, b) = 1 \rightarrow a \geqslant b$$

must *not* be demonstrable, of course, as our second and third examples proving consistency show. (*Cf.* also the formula (E) in the present and in the preceding appendices.) But there are other valid equivalences in our system such as

$$a \geqslant b \longleftrightarrow p(a, \bar{a}b) \neq 0$$
$$a \geqslant b \longleftrightarrow p(a, \bar{a}b) = 1$$

None of these can hold in the usual systems in which $p(a, b)$ is undefined unless $p(b) \neq 0$. It seems to be quite clear, therefore, that the usual systems of probability theory are wrongly described as generalizations of logic: they are formally inadequate for this purpose, since they do not even entail Boolean algebra.

The formal character of our system makes it possible to interpret it, for example, as a many-valued propositional logic (with as many values as we choose—either discrete, or dense, or continuous), or as a system of modal logic. There are in fact many ways of doing this; for example, we may define 'a necessarily implies b' by '$p(b, a\bar{b}) \neq 0$', as just indicated, or 'a is logically necessary' by '$p(a, \bar{a}) = 1$'. Even the problem whether a necessary statement is necessarily necessary finds a

[3] It is asserted, for example, by H. Jeffreys, *Theory of Probability*, § 1.2 'Convention 3'. But if it is accepted, his Theorem 4 becomes at once contradictory, since it is asserted without a condition such as our '$p(b) \neq 0$'. Jeffreys improved, in this respect, the formulation of Theorem 2 in his second edition, 1948: but as shown by Theorem 4 (and many others) his system is still inconsistent (even though he recognized, in the second edition, p. 35, that two contradictory propositions entail any proposition; *cf.* note *2 to section 23, and my answer to Jeffreys in *Mind* **52**, 1943, p. 47 *ff.*).

[4] See formulae 31′ *ff.* in footnote 1, above.

natural place in probability theory: it is closely connected with the relation between primary and secondary probability statements which plays an important part in probability theory (as shown in appendix *ix, point *13 of the Third Note). Roughly, if we write '$\vdash x$' for 'x is necessary (in the sense of demonstrable)' and 'h' for '$p(a, \bar{a}) = 1$', we may show that

$$\vdash a \;\longleftrightarrow\; \vdash h,$$

and therefore we find that

$$\vdash a \;\rightarrow\; \vdash \text{'}p(h, \bar{h}) = 1\text{'},$$

which may be taken to mean that $\vdash a$ entails that a is necessarily necessary; and since this means something like

$$\vdash a \;\rightarrow\; \vdash \text{'}p(\text{'}p(a, \bar{a}) = 1\text{'}, \overline{\text{'}p(a, \bar{a}) = 1\text{'}}) = 1\text{'},$$

we obtain (secondary) probability statements about (primary) probability statements.

But there are of course other possible ways of interpreting the relation between a primary and a secondary probability statement. (Some interpretations would prevent us from treating them as belonging to the same linguistic level, or even as belonging to the same language.)

*Appendix *vi. On Objective Disorder or Randomness.*

It is essential for an objective theory of probability and its application to such concepts as entropy (or molecular disorder) to give an
objective characterization of *disorder or randomness, as a type of order.*

In this appendix, I intend to indicate briefly some of the general
problems this characterization may help to solve, and the way in which
they may be approached.

(1) The distribution of velocities of the molecules of a gas in
equilibrium is supposed to be (very nearly) *random*. Similarly, the
distribution of nebulae in the universe appears to be *random*, with a
constant over-all density of occurrence. The occurrence of rain on
Sundays is *random*: in the long run, each day of the week gets equal
amounts of rain, and the fact that there was rain on Wednesday (or
any other day) may not help us to predict whether or not there will be
rain on Sunday.

(2) We have certain statistical *tests* of randomness.

(3) We can describe randomness as 'absence of regularity'; but
this is not, as we shall see, a helpful description. For there are no tests for
presence or absence of regularity in general, only tests for presence or
absence of some given or proposed *specific* regularity. Thus our tests of
randomness are never tests which exclude the presence of all regularity:
we may test whether or not there is a significant correlation between
rain and Sundays, or whether a certain given formula for predicting
rain on Sundays works, such as 'at least once in three weeks'; but
though we may reject this formula in view of our tests, we cannot determine, by our tests, whether or not there exists some better formula.

(4) Under these circumstances, it seems tempting to say that
randomness or disorder is not a type of order which can be described
objectively and that it must be interpreted as *our lack of knowledge* as to
the order prevailing, if any order prevails. I think that this temptation
should be resisted, and that we can develop a theory which allows us
actually to construct ideal types of disorder (and of course also ideal
types of order, and of all degrees in between these extremes).

(5) The simplest problem in this field, and the one which, I believe, I have solved, is the construction of a *one-dimensional ideal type of disorder*—an ideally disordered sequence.

The problem of constructing a sequence of this kind arises immediately from any frequency theory of probability which operates with infinite sequences. This may be shown as follows.

(6) According to von Mises, a sequence of o's and 1's with equi-distribution is random if it admits of *no gambling system*, that is to say, of no system which would allow us to select in advance a subsequence in which the distribution is unequal. But of course, von Mises admits that any gambling system may, 'accidentally', work for some time; it is only postulated that it will break down *in the long run*—or more precisely, in an infinite number of trials.

Accordingly, a Mises collective may be extremely regular *in its commencing segment*: provided they become irregular in the end, von Mises's rule is incapable of excluding collectives which start off very regularly, say with

$$00 \; 11 \; 00 \; 11 \; 00 \; 11 \ldots.$$

and so on, for the first five hundred million places.

(7) It is clear that we cannot empirically test *this* kind of deferred randomness; and it is clear that whenever we do test randomness in a sequence, we have a different type of randomness in mind: a sequence which *from the very beginning* behaves in a 'reasonably random-like' fashion.

But this phrase, 'from the very beginning', creates its own problem. Is the sequence 010110 random-like? Clearly, it is *too short* for us to say yes or no. But if we say that we need a *long sequence* for deciding a question of this kind then, it seems, we unsay what we have said before: it seems that we retract the phrase 'from the very beginning'.

(8) The solution of this difficulty is the construction of an *ideally random sequence*—one which for each beginning segment, whether short or long, is as random as the length of the segment permits; or in other words, a sequence whose degree n of randomness (that is, its n-freedom from after-effects) grows with the length of the sequence as quickly as is mathematically possible.

How to construct a sequence of this kind has been shown in appendix iv of the book. (See especially note *1 to appendix iv, with a reference to an as yet unpublished paper by Dr. L. R. B. Elton and myself.)

(9) The infinite set of all sequences answering this description may be called *the ideal type of random alternatives* with equal distribution.

(10) Although no more is postulated of these sequences than that they are 'strongly random'—in the sense that the finite commencing segments would pass all tests of randomness—*they can easily be shown to possess frequency limits*, in the sense usually demanded by frequency theories. This solves in a simple manner one of the central problems of my chapter on probability—elimination of the limit axiom, by way of a reduction of the limit-like behaviour of the sequences to their random-like behaviour in finite segments.

(11) The construction may quite easily be extended into both directions of the one-dimensional case, by correlating the first, second, ... of the odd numbered elements with the first, second, ... place of the positive direction, and the first, second, ... of the even numbered elements with the first, second, ... place of the negative direction; and by similar well-known methods, we can extend our construction to the cells of an *n*-dimensional space.

(12) While other frequency theorists—especially von Mises, Copeland, Wald, and Church—were mainly interested in defining random sequences in the most severe way by excluding 'all' gambling systems in the widest possible sense of the word 'all' (that is, in the widest sense compatible with a proof that random systems so defined exist), my aim has been quite different. I wished from the beginning to answer the objection that randomness is compatible with *any finite commencing segment*; and I wished to describe sequences that arise from *random-like finite sequences*, by a transition to infinity. I hoped by this method to achieve two things: to keep close to that type of sequence which would pass statistical tests of randomness, and to *prove* the limit theorem. Both have been done now, as here indicated under point (8), with the help of the construction given in my old appendix iv. But I have meanwhile found that the 'measure-theoretical approach' to probability is preferable to the frequency interpretation (see my *Postscript*, chapter *iii), both for mathematical and philosophical reasons. (The decisive point is connected with the propensity inter-pretation of probability, fully discussed in my *Postscript*.) I therefore do not think any longer that the elimination of the limit axiom from the frequency theory is very important. Still, it can be done: we can

build up the frequency theory with the help of the ideal type of the random sequences constructed in appendix iv; and we can say that an empirical sequence is random to the extent to which tests show its statistical similarity to an ideal sequence.

The sequences admitted by von Mises, Copeland, Wald, and Church are not necessarily of this kind, as mentioned above. But it is a fact that any sequence ever rejected on the basis of statistical tests for being not random may later turn into an admissible random sequence in the sense of these authors.

(13) Today, some years after having solved my old problems in a way which would have satisfied me in 1934, I no longer quite believe in the importance of the undoubted fact that a frequency theory can be constructed which is free of all the old difficulties. Yet I still believe that it is important to be able to show that randomness or disorder may be described as a type of order, and that objective models of randomness or disorder can be constructed.

*Appendix *vii. Zero Probability and the Fine-Structure of Probability
and of Content.*

In the book, a sharp distinction is made between the idea of the *probability* of a hypothesis, and its *degree of corroboration*. It is asserted that if we say of a hypothesis that it is well corroborated, we do not say more than that it has been severely tested (it must be thus a hypothesis with a high degree of testability) and that it has stood up well to the severest tests we were able to design so far. And it is further asserted that *degree of corroboration cannot be a probability*, because it cannot satisfy the laws of the probability calculus. For the laws of the probability calculus demand that, of two hypotheses, the one that is logically stronger, or more informative, or better testable, and thus the one which can be *better corroborated*, is always *less probable*—on any given evidence—than the other. (See especially sections 82 and 83.)

Thus a higher degree of corroboration will, in general, be combined with a lower degree of probability; which shows not only that we must distinguish sharply between probability (in the sense of the probability calculus) and degree of corroboration or confirmation, but also that *the probabilistic theory of induction, or the idea of an inductive probability, is untenable.*

The impossibility of an inductive probability is illustrated in the book (sections 80, 81, and 83) by a discussion of certain ideas of Reichenbach's, Keynes's and Kaila's. One result of this discussion is that *in an infinite universe* (it may be infinite with respect to the number of distinguishable things, or of spatio-temporal regions), *the probability of any (non-tautological) universal law will be zero.*

(Another result was that we must not uncritically assume that scientists ever aim at a high degree of probability for their theories. They have to choose between high probability and high informative content, since *for logical reasons they cannot have both*; and faced with this choice, they have so far always chosen high informative conte nt in preference to high probability—provided that the theory has stoo d up well to its tests.)

By 'probability', I mean here either the *absolute* logical probability of the universal law, or its probability *relative to some evidence*; that is to say, relative to a singular statement, or to a finite conjunction of singular statements. Thus if *a* is our law, and *b* any empirical evidence, I assert that

(1) $$p(a) = 0$$

and also that

(2) $$p(a, b) = 0$$

These formulae will be discussed in the present appendix.

The two formulae, (1) and (2), are equivalent. For as Jeffreys and Keynes observed, if the 'prior' probability (the absolute logical probability) of a statement *a* is zero, then so must be its probability relative to any finite evidence *b*, since we may assume that for any finite evidence *b*, we have $p(b) \neq 0$. For $p(a) = 0$ entails $p(a, b) = 0$, and since $p(a, b) = p(ab)/p(b)$, we obtain (2) from (1). On the other hand, we may obtain (1) from (2); for if (2) holds for any evidential *b*, however weak or 'almost tautological', we may assume that it also holds for the zero-evidence, that is to say, for the tautology $t = \overline{b}\overline{b}$; and $p(a)$ may be defined as equal to $p(a, t)$.

There are many arguments in support of (1) and (2). First, we may consider the classical definition of probability as the number of the *favourable* possibilities divided by that of *all* (equal) possibilities. We can then derive (2), for example, if we identify the favourable possibilities with the favourable evidence. It is clear that, in this case, $p(a, b) = 0$; for the favourable evidence can only be finite, while the possibilities in an infinite universe must be clearly infinite. (Nothing depends here on 'infinity', for any sufficiently large universe will yield, with any desired degree of approximation, the same result; and we know that our universe is extremely large, compared with the amount of evidence available to us.)

This simple consideration is perhaps a little vague, but it can be considerably strengthened if we try to derive (1), rather than (2), from the classical definition. We may to this end interpret the universal statement *a* as entailing an infinite product of singular statements, each endowed with a probability which of course must be less than unity. In the simplest case, *a* itself may be interpreted as such an infinite product; that is to say, we may put *a* = 'everything has the property

A'; or in symbols, '$(x)Ax$', which may be read 'for whatever value of x we may choose, x has the property A'.[1] In this case, a may be interpreted as the infinite product $a = a_1 a_2 a_3 \ldots$ where $a_1 = Ak_1$, and where k_1 is the name of the ith individual of our infinite universe of discourse.

We may now introduce the name 'a^n' for the product of the first n singular statements, $a_1 a_2 \ldots a_n$, so that a may be written

$$a = \lim_{n \to \infty} a^n$$

and

(3)
$$p(a) = \lim_{n \to \infty} p(a^n)$$

It is clear that we may interpret a^n as the assertion that, within the finite sequence of elements k_1, k_2, $\ldots k_n$, all elements possess the property A. This makes it easy to apply the classical definition to the evaluation of $p(a^n)$. There is only *one possibility that is favourable* to the assertion a^n: it is the possibility that all the n individuals, k_i without exception, possess the property A rather than the property non-A. But there are in all 2^n possibilities, since we must assume that it is possible for any individual k_i either to possess the property A or the property non-A. Accordingly, the classical theory gives

(4^c)
$$p(a^n) = 1/2^n$$

But from (3) and (4^c), we obtain immediately (1).

The 'classical' argument leading to (4^c) is not entirely adequate, although it is, I believe, essentially correct.

The inadequacy lies merely in the assumption that A and non-A are equally probable. For it may be argued—correctly, I believe—

---

[1] 'x' is here an individual variable ranging over the (infinite) universe of discourse. We may choose, for example, $a = $ 'All swans are white' = 'for whatever value of x we may choose, x has the property A' where 'A' is defined as 'being white or not being a swan'. We may also express this slightly differently, by assuming that x ranges over the spatio-temporal regions of the universe, and that 'A' is defined by 'not inhabited by a non-white swan'. Even laws of more complex form—say of a form like '$(x)(y)(xRy \to xSy)$' may be written '$(x)Ax$', since we may define 'A' by

$$Ax \equiv (y)(xRy \to xSy).$$

We may perhaps come to the conclusion that natural laws have another form than the one here described (*cf.* appendix *x): that they are logically still *stronger* than is here assumed; and that, if forced into a form like '$(x)Ax$', the predicate A becomes essentially non-observational (*cf.* notes *1 and *2 to the 'Third Note', reprinted in appendix *ix) although, of course, deductively testable. But in this case, our considerations remain valid *a fortiori*.

that since a is supposed to describe a law of nature, the various a_i are instantiation statements, and thus more probable than their negations which are potential falsifiers. (*Cf.* note *1 to section 28.) This objection however, relates to an inessential part of the argument. For whatever probability—short of unity—we attribute to A, the infinite product a will have zero probability (assuming independence, which will be discussed later on). Indeed, we have struck here a particularly trivial case of *the one-or-zero law of probability* (which we may also call, with an allusion to neuro-physiology, 'the all-or-nothing principle'). In this case it may be formulated thus: if a is the infinite product of $a_1, a_2, \ldots$, where $p(a_i) = p(a_j)$, and where every a_i is independent of all others, then the following holds:

(4) $p(a) = \lim\limits_{n \to \infty} p(a^n) = 0$, unless $p(a) = p(a^n) = 1$

But it is clear that $p(a) = 1$ is unacceptable (not only from my point of view but also from that of my inductivist opponents who clearly cannot accept the consequence that the probability of a universal law can never be increased by experience). For 'all swans are black' would have the probability 1 as well as 'all swans are white'—and similarly for all colours; so that 'there exists a black swan' and 'there exists a white swan', *etc.*, would all have zero probability, in spite of their intuitive logical weakness. In other words, $p(a) = 1$ would amount to asserting on purely logical grounds with probability 1 the emptiness of the universe.

Thus (4) establishes (1).

Although I believe that this argument (including the assumption of independence to be discussed below) is incontestable, there are a number of much weaker arguments which do not assume independence and which still lead to (1). For example we might argue as follows.

It was assumed in our derivation that for every k_i, it is logically possible that it has the property A, and alternatively, that it has the property non-A: this leads essentially to (4). But one might also assume, perhaps, that what we have to consider as our fundamental possibilities are not the possible properties of every individual in the universe of n individuals, but rather the possible proportions with which the properties A and non-A may occur within a sample of individuals. In a sample of n individuals, the possible proportions with which A may occur are: $0, 1/n, \ldots, n/n$. If we consider the occurrences of any of these proportions as our fundamental possibilities, and thus treat them as equi-probable ('Laplace's distribution'[2]) then (4) would have to be replaced by

(5) $$p(a^n) = 1/n; \text{ so that } \lim p(a^n) = 0.$$

Although from the point of view of a derivation of (1), formula (5) is much weaker than (4^c), it still allows us to derive (1)—and it allows us to do so without identifying the observed cases as the favourable ones or assuming that the number of observed cases is finite.

A very similar argument leading to (1) would be the following. We may consider the fact that every universal law a *entails* (and is therefore at most equally as probable as) a statistical hypothesis h of the form '$p(x, y) = 1$', and that the absolute probability of h may be calculated with the help of Laplace's distribution, with the result $p(h) = 0$. (Cf. appendix *ix, the *Third Note*, especially *13.) But since a entails h, this leads to $p(a) = 0$; that is to say, to (1).

To me, this proof appears the simplest and most convincing: it makes it possible to uphold (4) *and* (5), by assuming that (4) applies to a and (5) to h.

So far our considerations were based on the classical definition of probability. But we arrive at the same result if instead we adopt as our basis the logical interpretation of the formal calculus of probability. In this case, the problem becomes one of dependence or independence of statements.

If we again regard a as the logical product of the singular statements $a_1\ a_2, \ldots$, then the only reasonable assumption seems to be that, in the absence of any (other than tautological) information, we must consider all these singular statements as mutually *independent* of one another, so that a may be followed by a_j or by its negation, $\bar{a}_j$, with the probabilities

$$p(a_j, a_i) = p(a_j)$$
$$p(\bar{a}_j, a_i) = p(\bar{a}_j) = 1 - p(a_j).$$

Every other assumption would amount to postulating *ad hoc* a kind of after-effect; or in other words, to postulating that there is something like a casual connection between a_i and a_j. But this would obviously,

---

[2] It is the assumption underlying Laplace's derivation of his famous 'rule of succession'; this is why I call it 'Laplace's distribution'. It is an adequate assumption if our problem is one of *mere sampling*; it seems inadequate if we are concerned (as was Laplace) with a succession of individual events. See also appendix *ix, points 7 *ff.* of my 'Third Note'; and note 10 to appendix *viii.

be a non-logical, a synthetic assumption, to be formulated as a hypo-thesis. It thus cannot form part of a purely logical theory of probability.

The same point may be put a little differently thus: in the presence of some hypothesis, h say, we may of course have

(6) $$p(a_j, a_i h) > p(a_j, h)$$

For h may inform us of the existence of a kind of after-effect. Conse-quently, we should then have

(7) $$p(a_i a_j, h) > p(a_i, h)p(a_j, h),$$

since (7) is equivalent to (6). But in the absence of h, or if h is tau-tologous or, in other words, if we are concerned with absolute logical probabilities, (7) must be replaced by

(8) $$p(a_i a_j) = p(a_i)p(a_j)$$

which means that a_i and a_j are *independent*, and which is equivalent to

(9) $$p(a_j, a_i) = p(a_j).$$

But the assumption of mutual independence leads, together with $p(a_i) < 1$, as before to $p(a) = 0$; that is to say, to (1).

Thus (8), that is, the assumption of the mutual independence of the singular statements a_i leads to (1); and mainly for this reason, some authors have, directly or indirectly, rejected (8). The argument has been, invariably, that (8) must be false because if it were true, *we could not learn from experience*: empirical knowledge would be impossible. But this is incorrect: we may learn from experience even though $p(a) = p(a, b) = 0$; for example, $C(a, b)$—that is to say, the degree of corroboration of a by the tests b—may none the less increase with new tests. (*Cf.* appendix *ix). Thus this 'transcendental' argument fails to hit its target; at any rate, it does not hit my theory.[3]

[3] An argument which appeals to the fact that we possess knowledge or that we can learn from experience, and which concludes from this fact that knowledge or learning from experience must be possible, and further, that every theory which entails the impossibility of knowledge, or of learning from experience, must be false, may be called a 'transcendental argument'. (This is an allusion to Kant.) I believe that a transcendental argument may indeed be valid if it is used critically—against a theory which entails the impossibility of knowledge, or of learning from experience. But one must be very careful in using it. Empirical knowledge *in some sense* of the word 'knowledge', exists. But in other senses—for example in the sense of *certain* knowledge, or of *demonstrable* knowledge—it does not. And we must not assume, uncritically, that we have 'probable' knowledge—knowledge that is probable in the sense of the calculus of probability. It is indeed my contention that we do not have probable knowledge in this sense. For I believe that what we may call 'empirical knowledge', including 'scientific knowledge', consists of guesses, and that many of these guesses are not probable (or have a probability zero) even though they may be very well corroborated. See also my *Postscript*, sections *28 and *32.

But let us now consider the view that (8) is false, or in other words, that

$$p(a_i a_j) > p(a_i)p(a_j)$$

is valid, and consequently

$$p(a_j, a_i) > p(a_j),$$

and also the following:

$$(^+) \qquad p(a_n, a_1 a_2 \ldots a_{n-1}) > p(a_n)$$

This view asserts that once we have found some k_i to possess the property A, the probability increases that another k_j possesses the same property; and even more so if we have found the property in a number of cases. Or in Hume's terminology, $(^+)$ asserts '*that those instances*' (for example, k_n), '*of which we have had no experience, are likely to resemble those, of which we have had experience*'.

The quotation, except for the words 'are likely to', is taken from Hume's criticism of induction.[4] And Hume's criticism fully applies to $(^+)$, or its italicized verbal formulation. For, Hume argues, '*even after the observation of the frequent constant conjunction of objects, we have no reason to draw any inference concerning any object beyond those of which we have had experience*'.[5] If anybody should suggest that our experience entitles us to draw inferences from observed to unobserved objects, then, Hume says, 'I wou'd renew my question, *why from this experience we form any conclusion beyond those past instances, of which we have had experience*'. In other words, Hume points out that we get involved in an infinite regress if we appeal to experience in order to justify *any* conclusion concerning unobserved instances—*even mere probable conclusions*, as he adds in his *Abstract*. For there we read: 'It is evident that Adam, with all his science, would never have been able to *demonstrate* that the course of nature must continue uniformly the same. . . . Nay, I will go farther, and assert that he could not so much as prove by any *probable* arguments that the future must be conformable to the past. All probable arguments are built on the supposition that there is conformity betwixt the future and the past, and therefore can never prove it.'[6] Thus $(^+)$ is not justifiable by experience; yet in

---

[4] *Treatise of Human Nature*, 1739–40, book i, part iii, section vi (the italics are Hume's). See also my *Postscript*, note 1 to section *2 and note 2 to section *50.

[5] *loc. cit.*, section xii (the italics are Hume's). The next quotation is from *loc. cit.*, section vi.

[6] *Cf. An Abstract of a Book lately published entitled A Treatise of Human Nature*, 1740, ed. by J. M. Keynes and P. Sraffa, 1938, p. 15. *Cf.* note 2 to section 81. (The italics are Hume's.)

order to be logically valid, it would have to be of the character of a tautology, valid in every logically possible universe. But this is clearly not the case.

Thus (+), if true, would have the logical character of a *synthetic a priori principle* of induction, rather than of an analytic or logical assertion. But it does not quite suffice even as a principle of induction. For (+) may be true, and $p(a) = 0$ may be valid none the less. (An example of a theory that accepts (+) which, we have seen, must be synthetic, as *a priori* valid and which accepts at the same time $p(a) = 0$, is Carnap's.[7])

An effective probabilistic principle of induction would have to be even stronger than (+). It would have to allow us, at least, to conclude that for some fitting singular evidence b, we may obtain $p(a, b) > 1/2$, or in words, that a may be made, by accumulating evidence in its favour, more probable than its negation. But this is only possible if (1) is false, that is to say, if we have $p(a) > 0$.

A more direct disproof of (+) and a proof of (2) can be obtained from an argument which Jeffreys gives in his *Theory of Probability*, § 1.6.[8] Jeffreys discusses a formula which he numbers (3) and which in our symbolism amounts to the assertion that, provided $p(b_i, a) = 1$ for every $i \leqslant n$, so that $p(ab^n) = p(a)$, the following formula must hold:

(10) $p(a, b^n) = p(a)/p(b^n) = p(a)/p(b_1)p(b_2, b^1) \ldots p(b_n, b^{n-1})$

Discussing this formula, Jeffreys says (I am still using my symbols in place of his): 'Thus, with a sufficient number of verifications, one of three things must happen: (1) The probability of a on the information available exceeds 1. (2) it is always 0. (3) $p(b_n, b^{n-1})$ will tend to 1.' To this he adds that case (1) is impossible (trivially so), so that only (2) and (3) remain. Now I say that the assumption that case (3) holds universally, for some obscure logical reasons (and it would have to hold universally, and indeed *a priori*, if it were to be used in induction),

[7] Carnap's requirement that his 'lambda' (which I have shown to be the reciprocal of a dependence measure) must be finite entails our (+); *cf.* his *Continuum of Inductive Methods*, 1952. Nevertheless, Carnap accepts $p(a) = 0$, which according to Jeffreys would entail the impossibility of learning from experience. And yet, Carnap bases his demand that his 'lambda' must be finite, and thus that (+) is valid, on precisely the same transcendental argument to which Jeffreys appeals—that without it, we could not learn from experience. See his *Logical Foundations of Probability*, 1950, p. 565, and my contribution to the forthcoming Carnap volume of the *Library of Living Philosophers*, ed. by P. A. Schilpp, especially note 87.

[8] I translate Jeffreys's symbols into mine, omitting his H since nothing in the argument prevents us from taking it to be either tautological or at least irrelevant; in any case, my argument can easily be restated without omitting Jeffreys's H.

can be easily refuted. For the only condition needed for deriving (10), apart from $0 < p(b_i) < 1$, is that *there exists* some statement a such that $p(b^n, a) = 1$. But this condition can *always* be satisfied, for any sequence of statements b_i. For assume that the b_i are reports on penny tosses; then it is always possible to construct a universal law a which entails the reports of all the $n - 1$ observed penny tosses, and which allows us to predict all further penny tosses (though probably incorrectly).[9] Thus the required a always exists; and there always is also another law, a', yielding the same first $n - 1$ results but predicting, for the nth toss, the opposite result. It would be paradoxical, therefore, to accept Jeffreys's case (3), since for a sufficiently large n we would always obtain $p(b_n, b^{n-1})$ close to 1, and also (from another law, a') $p(\bar{b}_n, b^{n-1})$ close to 1. Accordingly, Jeffreys's argument, which is mathematically inescapable, can be used to prove his case (2), which happens to coincide with my own formula (2), as stated at the beginning of this appendix.[10]

We may sum up our criticism of ($^+$) as follows. Some people believe that, for purely logical reasons, the probability that the next thing we meet will be red increases in general with the number of red things seen in the past. But this is a belief in magic—in the magic of human language. For 'red' is merely a predicate; and there will always be predicates A and B which both apply to all the things so far observed, but lead to incompatible probabilistic predictions with respect to the next thing. These predicates may not occur in ordinary languages, but they can always be constructed. (Strangely enough, the magical belief here criticized is to be found among those who construct artificial model languages, rather than among the analysts of ordinary language.) By thus criticizing ($^+$) I am defending, of course, the principle of the (absolute logical) *independence* of the various a_n from any combination $a_i\, a_j \ldots$; that is to say, my criticism amounts to a defence of (4) and (1).

---

[9] Note that there is nothing in the conditions under which (10) is derived which would demand the b_i to be of the form '$B(k_i)$', with a common predicate 'B', and therefore nothing to prevent our assuming that $b_i = $ 'k_i is heads' and $b_j = $ 'k_j is tails'. Nevertheless, we can construct a predicate 'B' so that every b_i has the form '$B(k_i)$' : we may define B as 'having the property heads, or tails, respectively, if and only if the corresponding element of the sequence determined by the mathematical law a is 0, or is 1, respectively'. (It may be noted that a predicate like this can be defined only with respect to a universe of individuals which are *ordered*, or which may be *ordered*; but this is of course the only case that is of interest if we have in mind applications to problems of science. *Cf.* my Preface, 1958, and note 2 to section *49 of my *Postscript*.)

[10] Jeffreys himself draws the opposite conclusion: he adopts as valid the possibility stated in case (3).

There are further proofs of (1). One of them which is fundamentally due to an idea of Jeffreys and Wrinch[11] will be discussed more fully in appendix *viii. Its main idea may be put (with slight adjustments) as follows.

Let *e* be an *explicandum*, or more precisely, a set of singular facts or data which we wish to explain with the help of a universal law. There will be, in general, an infinite number of possible explanations—even an infinite number of explanations (mutually *exclusive*, given the data *e*) such that the sum of their probabilities (given *e*) cannot exceed unity. But this means that the probability of almost all of them must be zero—unless, indeed, we can order the possible laws in an infinite sequence, so that we can attribute to each a positive probability in such a way that their sum converges and does not exceed unity. And it means, further, that to laws which appear earlier in this sequence, a greater probability must be attributed (in general) than to laws which appear later in the sequence. We should therefore have to make sure that the following important *consistency condition* is satisfied:

Our method of ordering the laws must never place a law before another one if it is possible to prove that the probability of the latter is greater than that of the former.

Jeffreys and Wrinch had some intuitive reasons to believe that a method of ordering the laws satisfying this consistency condition may be found: they proposed to order the explanatory theories according to their decreasing simplicity ('simplicity postulate'), or according to their increasing complexity, measuring complexity by the number of the adjustable parameters of the law. But it can be shown (and it will be shown in appendix *viii) that this method of ordering, or any other possible method, violates the consistency condition.

Thus we obtain $p(a, e) = 0$ for all explanatory hypotheses, whatever the data *e* may be; that is to say, we obtain (2), and thereby indirectly (1).

(An interesting aspect of this last proof is that it is valid even in a finite universe, provided our explanatory hypotheses are formulated in a mathematical language which allows for an infinity of (mutually exclusive) hypotheses. For example, we may construct the following universe.[12] On a much extended chessboard, little discs or draught pieces are placed by somebody according to the following rule:

[11] *Philos. Magazine* **42**, 1921, pp. 369 *ff.*
[12] A similar example is used in appendix *viii, text to note 2.

there is a mathematically defined function, or curve, known to him but not to us, and the discs may be placed only in squares which lie on the curve; within the limits determined by this rule, they may be placed at random. Our task is to observe the placing of the discs, and to find an 'explanatory theory', that is to say, the unknown mathematical curve, if possible, or one very close to it. Clearly, there will be an infinity of possible solutions which are mathematically imcompatible, although some of them will be indistinguishable with respect to the discs placed on the board. Any of these theories may, of course, be 'refuted' by discs placed on the board after the theory was announced. Although the 'universe'—that of possible positions—may here be chosen to be a finite one, there will be nevertheless an infinity of mathematically incompatible explanatory theories. I am aware, of course, that instrumentalists or operationalists might say that the differences between any two theories determining the same squares would be 'meaningless'. But apart from the fact that *this example does not form part of my argument*—so that I need really not reply to this objection—the following should be noted. It will be possible, in many cases, to give 'meaning' to these 'meaningless' differences by making our mesh sufficiently fine, *i.e.* subdividing our squares.)

The detailed discussion of the fact that my consistency condition cannot be satisfied will be found in appendix *viii. I will now leave the problem of the validity of formulae (1) and (2), in order to proceed to the discussion of a formal problem arising from the fact that these formulae are valid, so that all universal theories, whatever their content, have zero probability.

There can be no doubt that the content or the logical strength of two universal theories can differ greatly. Take the two laws $a_1 =$ 'All planets move in circles' and $a_2 =$ 'All planets move in ellipses'. Owing to the fact that all circles are ellipses (with excentricity zero), a_1 entails a_2, but not *vice versa*. The content of a_1 is greater by far than the content of a_2. (There are, of course, other theories, and logically stronger ones, than a_1; for example, 'All planets move in concentric circles round the sun'.)

The fact that the content of a_1 exceeds that of a_2 is of the greatest significance for all our problems. For example, there are *tests* of a_1— that is to say, attempts to refute a_1 by discovering some deviation from circularity—which are not tests of a_2; but there could be no genuine test of a_2 which would not, at the same time, be an attempt to refute a_1. Thus a_1 can be more severely tested than a_2, it has the greater degree of testability; and if it stands up to its more severe tests, it will attain a higher degree of corroboration than a_2 can attain.

Similar relationships may hold between two theories, a_1 and a_2, even if a_1 does not logically entail a_2, but entails instead a theory to which a_2 is a very good approximation. (Thus a_1 may be Newton's dynamics and a_2 may be Kepler's laws which do not follow from Newton's theory, but merely 'follow with good approximation'; see also section *15 of my *Postscript*.) Here too, Newton's theory is better testable, because its content is greater.[13]

Now our proof of (1) shows that these differences in content and in testability cannot be expressed immediately in terms of the absolute logical probability of the theories a_1 and a_2, since $p(a_1) = p(a_2) = 0$. And if we define a measure of content, $C(a)$, by $C(a) = 1 - p(a)$, as suggested in the book, then we obtain, again, $C(a_1) = C(a_2)$, so that the differences in content which interest us here remain unexpressed by these measures. (Similarly, the difference between a self-contra-

---

[13] Whatever C. G. Hempel may mean by 'confirming evidence' of a theory, he clearly cannot mean the result of tests which corroborate the theory. For in his papers on the subject (*Journal of Symbolic Logic* 8, 1943, pp. 122 ff., and especially *Mind* 54, 1945, pp. 1 ff. and 97 ff.; 55, 1946, pp. 79 ff.), he states (*Mind* 54, pp. 102 ff.) among his conditions for adequacy the following condition (8.3): if e is confirming evidence of several hypotheses, say h_1 and h_2, then h_1 and h_2 and e must form a consistent set of statements.

But the most typical and interesting cases tell against this. Let h_1 and h_2 be Einstein's and Newton's theories of gravitation. They lead to incompatible results for strong gravitational fields and fast moving bodies, and therefore contradict each other. And yet, all the known evidence supporting Newton's theory is also evidence supporting Einstein's, and corroborates both. The situation is very similar for Newton's and Kepler's theories, or Newton's and Galileo's. (Also, any unsuccessful attempt to find a red or yellow swan corroborates both the following two theories which contradict each other in the presence of the statement 'there exists some swan': (i) 'All swans are white' and (ii) 'All swans are black'.)

Quite generally, let there be a hypothesis h, corroborated by the result e of severe tests, and let h_1, and h_2 be two incompatible theories each of which entails h. (h_1 may be ah, and h_2 may be $\bar{a}h$.) Then any test of h is one of both h_1 and h_2, since any successful refutation of h would refute both h_1 and h_2; and if e is the report of unsuccessful attempts to refute h, then e will corroborate both h_1 and h_2. (But we shall, of course, look for crucial tests between h_1 and h_2.) With 'verifications' and 'instantiations', it is, of course, otherwise. But these need not have anything to do with *tests*.

Yet quite apart from this criticism, it should be noted that in Hempel's model language identity cannot be expressed; see especially his p. 143 (line 5 from end of paper) and my second Preface, 1958. For a simple ('semantical') *definition of instantiaton*, see the last footnote of my note in *Mind*, 64, 1955, p. 391.

dictory statement $a\bar{a}$ and a universal theory a remains unexpressed since $p(a\bar{a}) = p(a) = 0$, and $C(a\bar{a}) = C(a) = 1$.[14])

All this does not mean that we cannot express the difference in content between a_1 and a_2 in terms of probability, at least in some cases. For example, the fact that a_1 entails a_2 but not *vice versa* would give rise to

$$p(a_1, a_2) = 0 \; ; \; p(a_2, a_1) = 1$$

even though we should have, at the same time, $p(a_1) = p(a_2) = 0$.

Thus we should have

$$p(a_1, a_2) < p(a_2, a_1)$$

which would be an indication of the greater content of a_1.

The fact that there are these differences in content and in absolute logical probability which cannot be expressed immediately by the corresponding measures may be expressed by saying that there is a *'fine structure'* of content, and of logical probability, which may allow us to differentiate between greater and smaller contents and absolute probabilities even in cases where the measures $C(a)$ and $p(a)$ are too coarse, and insensitive to the differences; that is, in cases where they yield equality. In order to express this fine structure, we may use the symbols ' $\succ$ ' ('is higher') and ' $\prec$ ' ('is lower'), in place of the ordinary symbols ' $>$ ' and ' $<$ '. (We may also use ' $\succeq$ ', or 'is higher or equally high', and ' $\preceq$ '.) The use of these symbols can be explained by the following rules:

---

[14] That a self-contradictory statement may have the same probability as a consistent synthetic statement is *unavoidable in any probability theory* if applied to some infinite universe of discourse: this is a simple consequence of the multiplication law which demands that $p(a_1 a_2 \ldots a_n)$ must tend to zero provided all the a_i are mutually independent. Thus the probability of tossing n successive heads is, according to *all* probability theories, $1/2^n$, which becomes zero if the number of throws becomes infinite.

A similar problem of probability theory is this. Put into an urn n balls marked with the numbers 1 to n, and mix them. What is the probability of drawing a ball marked with a prime number? The well-known solution of this problem, like that of the previous one, tends to zero when n tends to infinity; which means that the probability of drawing a ball marked with a divisible number becomes 1, for $n \to \infty$, even though there is an infinite number of balls with non-divisible numbers in the urn. This result must be the same in *any* adequate theory of probability. One must not, therefore, single out a particular theory of probability, such as the frequency theory, and criticize it as 'at least mildly paradoxical' because it yields this perfectly correct result. (A criticism of this kind will be found in W. Kneale's *Probability and Induction*, 1949, p. 156.) In view of our last 'problem of probability theory'—that of drawing numbered balls—Jeffreys's attack on those who speak of the 'probability distribution of prime numbers' seems to me equally unwarranted. (*Cf. Theory of Probability*, 2nd edition, p. 38, footnote.)

(1) '$C(a) \succ C(b)$' and thus its equivalent '$p(a) \prec p(b)$' may be used to state that the content of a is greater than that of b—*at least* in the sense of the fine structure of content. We shall thus assume that $C(a) \succ C(b)$ entails $C(a) \succeq C(b)$, and that this in turn entails $C(a) \geqslant C(b)$, that is to say, the falsity of $C(a) < C(b)$. None of the opposite entailments hold.

(2) $C(a) \succeq C(b)$ and $C(a) \preceq C(b)$ together entail $C(a) = C(b)$, but $C(a) = C(b)$ is compatible with $C(a) \succ C(b)$, or with $C(a) \prec C(b)$ and, of course, also with $C(a) \succeq C(b)$ and with $C(a) \preceq C(b)$.

(3) $C(a) > C(b)$ always entails $C(a) \succ C(b)$.

(4) Corresponding rules will hold for $p(a) \succ p(b)$, etc.

The problem now arises of determining the cases in which we may say that $C(a) \succ C(b)$ holds even though we have $C(a) = C(b)$. A number of cases are fairly clear; for example, unilateral entailment of b by a. More generally, I suggest the following rule:

If for all sufficiently large *finite* universes (that is, for all universes with more than N members, for some sufficiently large N), we have $C(a) > C(b)$, and thus, in accordance with rule (3), $C(a) \succ C(b)$, we retain $C(a) \succ C(b)$ for an infinite universe even if, for an infinite universe, we obtain $C(a) = C(b)$.

This rule seems to cover most cases of interest, although perhaps not all.[15]

The problem of $a_1 =$ 'All planets move in circles' and $a_2 =$ 'All planets move in ellipses' is clearly covered by our rule, and so is even the case of comparing a_1 and $a_3 =$ 'All planets move in ellipses with an excentricity other than zero'; for $p(a_3) > p(a_1)$ will hold in all sufficiently large finite universes (of possible observations, say) in the simple sense that there are more possibilities compatible with a_3 than with a_1.

The fine-structure of content and of probability here discussed not only affects the limits, 0 and 1, of the probability interval, but it affects in principle all probabilities between 0 and 1. For let a_1 and a_2

[15] Related problems are discussed in considerable detail in John Kemeny's very stimulating paper 'A Logical Measure Function', *Journal of Symb. Logic* 18, 1953, pp. 289 ff. Kemeny's model language is the second of three to which I allude in my second Preface. It is, in my opinion, by far the most interesting of the three. Yet as he shows on pp. 294, his language is such that infinitistic theorems—such as the principle that every number has a successor—must not be demonstrable in it. It thus cannot contain the usual system of arithmetic.

be universal laws with $p(a_2) = 0$ and $p(a_1) \prec p(a_2)$, as before; let b be not entailed by either a_1 or a_2 or their negations; and let $0 < p(b) = r < 1$. Then we have

$$p(a_1 \vee b) = p(a_2 \vee b) = r$$

and at the same time

$$p(a_1 \vee b) \prec p(a_2 \vee b).$$

Similarly we have

$$p(\bar{a}_1 b) = p(\bar{a}_2 b) = r$$

and at the same time

$$p(\bar{a}_1 b) \succ p(\bar{a}_2 b),$$

since $p(\bar{a}_1) \succ p(\bar{a}_2)$, although of course $p(\bar{a}_1) = p(\bar{a}_2) = 1$. Thus we may have for every b such that $p(b) = r$, a c_1 such that $p(c_1) = p(b)$ and $p(c_1) \prec p(b)$, and also a c_2 such that $p(c_2) = p(b)$ and $p(c_2) \succ p(b)$.

The situation here discussed is important for the treatment of the *simplicity or the dimension of a theory*. This problem will be further discussed in the next appendix.

377

Appendix *viii.* *Content, Simplicity, and Dimension.*

As indicated in the book,[1] I do not believe in hampering scientific language by preventing the scientist from using freely, whenever it is convenient, new ideas, predicates, 'occult' concepts, or anything else. For this reason, I cannot support the various recent attempts to introduce into the philosophy of science the method of artificial calculi or 'language systems'—systems supposed to be models of a simplified 'language of science'. I believe that these attempts have not only been useless so far, but that they have even contributed to the obscurity and confusion prevalent in the philosophy of science.

It has been briefly explained in section 38 and in appendix i that, had we (absolutely) atomic statements at our disposal—or what amounts to the same, (absolutely) *atomic predicates*—then we might introduce, as a measure of the *content* of a theory, the reciprocal of the minimum number of *atomic statements* needed for refuting that theory. For since the degree of content of a theory is the same as its degree of testability or refutability, the theory which is refutable by fewer atomic statements would also be the one which is the more easily refutable or testable, and thus the one with the greater content. (In brief, the smaller the number of atomic statements needed to compose a potential falsifier, the greater the content of the theory.)

But I do not want to operate either with the fiction of atomic statements, or with an artificial language system in which atomic statements are available to us. For it seems to me quite clear that there are no 'natural' atomic predicates available in science. To some older logicians, the predicates 'man' and 'mortal' seem to have presented themselves as examples of something like atomic predicates. Carnap uses 'blue' or 'warm' as examples—presumably because 'man' and 'mortal' are highly complex ideas which (some may think) can be defined in terms of simpler ideas such as 'blue' or 'warm'. Yet it is characteristic of scientific discussions that neither these nor any other

[1] See section 38, especially the text after note 2 and my appendix i; also my second Preface, 1958.

predicates are treated as (absolutely) atomic. Depending upon the problem under consideration, not only 'man' and 'mortal' but also 'blue' or 'warm' may be treated as highly complex; 'blue', say, as the colour of the sky, explicable in terms of atomic theory. Even the phenomenal term 'blue' may be treated, in certain contexts, as definable—as a character of visual images correlated with certain physiological stimuli. It is characteristic of scientific discussion that it proceeds freely; and the attempt to take away its freedom, by tying it down upon the Procrustean bed of a pre-established language system would, if successful, be the end of science.

For these reasons I rejected in advance the idea of using atomic statements for the purpose of measuring the degree of *content or simplicity* of a theory; and I suggested that we might use, instead, the idea of *relative-atomic statements*; and further, the idea of a *field of statements* which are relative-atomic with respect to a theory or a set of theories to the testing of which they are relevant; a field which could be interpreted as a *field of application* of the theory, or of the set of theories.

If we take as our example again, as in the preceding appendix, the two theories $a_1 =$ 'All planets move in circles' and $a_2 =$ 'All planets move in ellipses', then we can take as our field all the statements of the form 'At the time x the planet y was in the position z', which will be our relative atomic statements. And if we assume that we already know that the track of the planet is a plane curve, then we can take a graph-paper that represents the field and enter on it the various positions, marking in each case the time, and the name of the planet in question, so that each entry represents one of the relative-atomic statements. (We can, of course, make the representation three-dimensional, by marking the position with a pin whose length represents the time, measured from some assumed zero instant; and variations in the colour of the pinhead may be used to indicate the names of the various planets.)

It has been explained, mainly in sections 40 to 46, and in my old appendix i, how the minimum number of the relative-atomic statements needed to refute a certain theory could be used as a measure of the complexity of the theory. And it was shown that the *formal simplicity* of a theory might be measured by the *paucity of its parameters*, in so far as this paucity was not the result of a 'formal' rather than a 'material' reduction in the number of the parameters. (*Cf.* especially sections 40, 44, *f.*, and appendix i.)

379

Now all these comparisons of the simplicity of theories, or of their contents, will, clearly, amount to comparisons of the 'fine-structure' of their contents, in the sense explained in the preceding appendix, because their absolute probabilities will all be equal (that is, equal to zero). And I wish first to show that the number of the parameters of a theory (with respect to a field of application) can indeed be interpreted as measuring the fine-structure of its content.

What I have to show, to this end, is that *for a sufficiently large finite universe, the theory with the greater number of parameters will always be more probable (in the classical* sense) *than the theory with the smaller number of parameters.*

This can be shown as follows. In the case of a continuous geometrical field of applications, our universe of possible events, each described by a possible relative-atomic statement, is of course infinite. As shown in sections 38 *f.*, we can in this case compare two theories with respect to the *dimension*, rather than the *number*, of the possibilities which they leave open; that is, the possibilities which are favourable to them. The dimension of these possibilities turns out to be equal to the number of parameters. We now replace the infinite universe of relative-atomic statements by a *finite* (although very large) universe of relative-atomic statements, corresponding to the chessboard example in the preceding appendix.[2] That is to say, we assume that every relative-atomic statement refers to a little *square* with the side ε as the position of a planet rather than to a point of the plane, and that the possible positions do not overlap.[3] Somewhat differently from the example of the preceding appendix, we now replace the various *curves* which are the usual geometrical representations of our theories by 'quasi curves' (of a width approximately equal to ε); that is to say, by sets, or chains, of squares. As a result of all this, the number of the possible theories becomes finite.

We now consider the representation of a theory with *d* parameters which in the continuous case was represented by a *d*-dimensional continuum whose points (*d*-tuples) each represented a curve. We find that we can still use a similar representation, except that our *d*-dimen-

---

[2] *Cf.* appendix *vii, text to note 12.

[3] The assumption that the possible positions do not overlap is made in order to simplify the exposition. We could just as well assume that any two neighbouring squares partly overlap—say, for a quarter of their area; or we could replace the squares by overlapping circles (overlapping so as to enable us to cover the whole area with them). This last assumption would be a little closer to an interpretation of the 'positions' as the never absolutely sharp results of possible *measurements* of positions.

sional continuum will be replaced by a d-dimensional arrangement of d-dimensional 'cubes' (with the side ε). Each of these little cubes will now represent one 'quasi curve', and thus one of the possibilities favourable to the theory; and the d-dimensional arrangement will represent the set of all 'quasi-curves' compatible with, or favourable to, the theory.

But we can now say that the theory with fewer parameters—that is to say, the *set* of quasi curves which is represented by an arrangement of fewer dimensions—will not only have fewer dimensions, but will also contain a smaller number of 'cubes'; that is, of favourable possibilities.

Thus we are justified in applying the results of the preceding section: if a_1 has fewer parameters than a_2, we can assert that, in a sufficiently large but finite universe, we shall have

$$p(a_1) < p(a_2)$$

and therefore

(*) $$p(a_1) \prec p(a_2).$$

But formula (*) remains valid when we assume that ε tends to zero, which in the limit amounts to replacing the finite universe by an infinite one. We arrive, therefore, at the following *theorem*.

(1) If the number of parameters of a_1 is smaller than the number of parameters of a_2, then the assumption

$$p(a_1) > p(a_2)$$

contradicts the laws of the calculus of probability.

Writing '$d_F(a)$', or more simply '$d(a)$', for the dimension of the theory a (with respect to the field of application F) we can formulate our theorem as follows.

(1)   If $d(a_1) > d(a_2)$ then $p(a_1) \prec p(a_2)$;
consequently, '$p(a_1) > p(a_2)$' is incompatible with $d(a_1) < d(a_2)$.

This theorem (which is implied in what has been said in the body of the book) is in keeping with the following considerations. A theory a requires a minimum of $d(a) + 1$ relative-atomic statements for its refutation. Its *'weakest falsifiers'*, as we may call them, consist of a conjunction of $d(a) + 1$ relative-atomic statements. This means that if $n \leqslant d(a)$, then *no* conjunction of n relative-atomic statements is logically strong enough for deriving from them $\bar{a}$, that is, the negation of a. The strength or content of $\bar{a}$, accordingly, can be measured by $d(a) + 1$, since a will be stronger than any conjunction of $d(a)$ relative-atomic statements but certainly not stronger than some conjunctions of $d(a) + 1$ such statements. But from the probability rule

$$p(\bar{a}) = 1 - p(a)$$

we know that the probability of a theory a decreases with increasing probability of its negation $\bar{a}$, and *vice versa*, and that the same relations hold between the contents of a and of $\bar{a}$. From this we see, again, that $d(a_1) < d(a_2)$ means that the content of a_1 is greater than that of a_2, so that $d(a_1) < d(a_2)$ entails $p(a_1)) \prec p(a_2)$, and is thus *incompatible* with $p(a_1) < p(a_2)$. But this result is nothing but the theorem (1) derived above.

Our theorem has been derived by considering finite universes, and it is indeed quite independent of the transition to infinite universes. It is therefore independent of the formulae (1) and (2) of the preceding appendix, that is to say, of the fact that in an infinite universe, we have for any universal law a and any finite evidence e,

(2) $$p(a) = p(a, e) = 0.$$

We may therefore legitimately use (1) for another derivation of (2); and this can indeed be done, if we utilize an idea due to Dorothy Wrinch and Harold Jeffreys.

As briefly indicated in the preceding appendix,[4] Wrinch and Jeffreys observed that if we have an infinity of mutually incompatible or exclusive explanatory theories, the sum of the probabilities of these theories cannot exceed unity, so that almost all of these probabilities must be zero, unless we can order the theories in a sequence, and assign to each, as its probability, a value from a convergent sequence of fractions whose sum does not exceed 1. For example, we may make the following assignments: we may assign the value $1/2$ to the first theory, $1/2^2$ to the second, and, generally, $1/2^n$ to the nth. But we may also assign to each of the first 25 theories the value $1/50$, that is to say, $1/(2.25)$; to each of the next 100, say, the value $1/400$, that is to say, $1/(2^2.100)$ and so on.

However we may construct the order of the theories and however we may assign our probabilities to them, there will always be some *greatest* probability value, P say (such as $1/2$ in our first example, or

4 *Cf.* appendix *vii, text to footnote 11.

1/50), and this value P will be assigned to at most n theories (where n is a finite number, and $n.P < 1$). Each of these n theories to which the maximum probability P has been assigned, has a *dimension*. Let D be the largest dimension present among these n theories, and let a_1 be one of them, with $d(a_1) = D$. Then, clearly, none of the theories with dimensions greater than D will be among our n theories with maximum probability. Let a_2 be a theory with a dimension greater than D, so that $d(a_2) > D = d(a_1)$. Then the assignment leads to:

$$(-) \qquad\qquad d(a_1) < d(a_2); \text{ and } p(a_1) > p(a_2).$$

This result shows that our theorem (1) is violated. But an assignment of the kind described which leads to this result is unavoidable if we wish to avoid assigning the same probability—that is, zero—to all theories. Consequently our theorem (1) entails the assignment of zero probabilities to all theories.

Wrinch and Jeffreys themselves arrived at a very different result. They believed that the possibility of empirical knowledge required the possibility of raising the probability of a law by accumulating evidence in its favour. From this they concluded that (2) must be false, and further, that a legitimate method must exist of assigning non-zero probabilities to an infinite sequence of explanatory theories. Thus Wrinch and Jeffreys drew very strong positive conclusions from the 'transcendental' argument (as I called it in the preceding appendix).[5] Believing, as they did, that an increase in probability means an increase in knowledge (so that obtaining a high probability becomes an aim of science), they did not consider the possibility that *we may learn from experience more and more about universal laws without ever increasing their probability*; that we may test and corroborate some of them better and better, thereby increasing their *degree of corroboration* without altering their *probability* whose value remains zero.

Jeffreys and Wrinch never described the sequence of theories, and the assignment of probability values, in a sufficiently clear way. Their main idea, called the 'simplicity postulate',[6] was that the theories should be so ordered that their complexity, or number of parameters, increases, while the probabilities which they assign to them decrease;

[5] *Cf.* note 3 to appendix *vii.
[6] In his *Theory of Probability*, § 3.0, Jeffreys says of the 'simplicity postulate' that it 'is not . . . a separate postulate but an immediate application of rule 5'. But all that rule 5 contains, by way of reference to rule 4 (both rules are formulated in § 1.1) is a very vague form of the 'transcendental 'principle. Thus it does not affect our argument.

this, incidentally, would mean that *any two* theories of the sequence would violate our theorem (1). But this way of ordering cannot be carried through, as Jeffreys himself noticed. For there may be theories with *the same number* of parameters. Jeffreys himself gives as examples $y = ax$ and $y = ax^2$; and he says of them: 'laws involving the same number of parameters can be taken as having the same prior probability'.[7] But the number of laws having prior probability is infinite, for Jeffreys's own examples can be continued to infinity: $y = ax^3$, $y = ax^4, \ldots y = ax^n$, and so on, with $n \to \infty$. Thus for each number of parameters, the same problem would recur as for the whole sequence.

Moreover, Jeffreys himself recognizes, in the same § 3.0,[8] that a law, a_1, say, may be obtained from a law a_2 with one additional parameter, by assuming that parameter to be equal to zero; and that in this case, $p(a_1) < p(a_2)$, since a_1 is a *special case* of a_2, so that fewer possibilities belong to a_1.[9] Thus in this special case, he recognizes that a theory with fewer parameters will be less probable than one with more parameters—in agreement with our theorem (1). But he recognizes this fact only in this special case; and he does not comment at all on the fact that a contradiction may well arise between his simplicity postulate and this case. Altogether, he nowhere tries to show that the simplicity postulate is consistent with his axiom system; but in view of the special case mentioned (which of course follows from his axiom system) it should have been clear that a proof of consistency was urgently needed.

Our own considerations show that a consistency proof cannot be given, and that the 'simplicity postulate' must contradict every adequate axiom system for probability; for it must violate our theorem (1).

In concluding this appendix, I wish to attempt something like an explanation why Wrinch and Jeffreys may have regarded their 'simplicity postulate' as harmless—as unable to create trouble.

It should be kept in mind that they were the first to identify simplicity and paucity of parameters. (I do not simply identify

---

[7] *Theory of Probability*, § 3.0 (1st edition, p. 95; 2nd edition, p. 100).

[8] *Op. cit.*, 1st edition, p. 96; 2nd edition, p. 101.

[9] Jeffreys, *loc. cit.*, remarks that 'half the prior probability [of a_2] is concentrated at $\alpha_{m+1} = 0$', which seems to mean that $p(a_1) = p(a_2)/2$; but this rule may lead to contradictions if the number of parameters of a_2 is greater than 2.

these two: I distinguish between a formal and a material reduction in the number of the parameters—*cf.* sections 40, 44, 45—and intuitive simplicity thus becomes something like formal simplicity; but otherwise my theory of simplicity agrees with that of Wrinch and Jeffreys in this point.) They also saw clearly that simplicity is one of the things aimed at by scientists—that these prefer a simpler theory to a more complicated one, and that they therefore try the simplest theories first. On all these points, Wrinch and Jeffreys were right. They were also right in believing that there are comparatively few simple theories, and many complex ones whose numbers increase with the number of their parameters.

This last fact may have led them to believe that the complex theories were the less probable ones (since the available probability was somehow to be divided among the various theories). And since they also assumed that a high degree of probability was indicative of a high degree of knowledge and therefore was one of the scientist's aims, they may have thought that it was intuitively evident that the simpler (and therefore more desirable) theory was to be identified with the more probable (and therefore more desirable) theory; for otherwise the aims of the scientist would become inconsistent. Thus the simplicity postulate appeared to be necessary on intuitive grounds and therefore *a fortiori* consistent.

But once we realize that the scientist does not and cannot aim at a high degree of probability, and that the opposite impression is due to mistaking the intuitive idea of probability for another intuitive idea (here labelled 'degree of corroboration'),[10] it will also become clear to us that simplicity, or paucity of parameters, is linked with, and tends to increase with improbability rather than with probability. And it will also become clear to us that a high degree of simplicity is nevertheless linked with a high

---

[10] It is shown in point 8 of my 'Third Note', reprinted in appendix ∗ix, that if h is a statistical hypothesis asserting '$p(a, b) = 1$', then after n severe tests passed by the hypothesis h, its degree of corroboration will be $(n-1)/(n+1) = 1 - 2/(n+1)$. There is a striking similarity between this formula and Laplace's 'rule of succession' according to which the probability that h will pass its next test is $(n+1)/(n+2) = 1 - 1/(n+2)$. The numerical similarity of these results, together with the failure to distinguish between probability and corroboration, may explain why Laplace's and similar results were intuitively felt to be satisfactory. I believe Laplace's result to be mistaken because I believe that his assumptions (I have in mind what I call the 'Laplacean distribution') do not apply to the cases he has in mind; but these assumptions apply to other cases; they allow us to estimate the absolute probability of a report on a statistical sample. *Cf.* my 'Third Note' (appendix ∗ix).

degree of corroboration. For a high degree of testability or corroborability is the same as a high prior improbability or simplicity.

The problem of corroboration will be discussed in the next appendix.

*Appendix *ix.   Corroboration, the Weight of Evidence, and Statistical tests.*

The three notes reprinted in the present appendix were originally published in *The British Journal for the Philosophy of Science.*[1]

Even before my book was published, I felt that the problem of degree of corroboration was one of those problems which should be further investigated. By 'the problem of degree of corroboration' I mean the problem (i) of showing that there exists a measure (to be called degree of corroboration) of the *severity of tests* to which a theory has been subjected, and of the manner in which it has passed these tests, or failed them; and (ii) of showing that *this measure cannot be a probability*, or more precisely, that it does not satisfy the formal laws of the probability calculus.

An outline of the solution of both of these tasks—especially the second—was contained in my book. But I felt that a little more was needed. It was not quite enough to show the failure of the existing theories of probability—of Keynes and of Jeffreys, for example, or of Kaila, or of Reichenbach, none of whom could establish even their central doctrine: that a universal law, or a theory, could ever reach a probability $> 1/2$. (They even failed to establish that a universal law, or a theory, could ever have a probability other than zero.) What was needed was a perfectly general treatment. I therefore aimed at constructing a formal probability calculus which could be interpreted in various senses. I had in mind (i) the logical sense, outlined in my book as (absolute) logical probability of statements; (ii) the sense of relative logical probability of statements or propositions, as envisaged by Keynes: (iii) the sense of a calculus of relative frequencies in sequences; (iv) the sense of a calculus of a measure of ranges, or of predicates, classes, or sets.

The ultimate aim was, of course, to show that *degree of corroboration was not a probability*; that is to say that *it was not one of the possible interpretations of the probability calculus.* Yet I realized that the task of

[1] *B.J.P.S.* **5**, 1954, pp. 143 *ff.* (see also corrections on pp. 334 and 359); **7**, 1957, pp. 350 *ff.*, and **8**, 1958, pp. 294 *ff.*

constructing a formal calculus was not only needed for this purpose, but was interesting in itself.

This led to my paper in *Mind*, reprinted here as appendix *ii, and to other work extending over many years and aimed both at simplifying my axiom systems and at producing a probability calculus in which $p(a, b)$—the probability of a given b—could have definite values, rather than 0/0, even if $p(b)$ was equal to zero. The problem arises, of course, because the definition

$$p(a, b) = p(ab)/p(b)$$

breaks down if $p(b) = 0$.

A solution of this last problem was needed because I soon found that, in order to define $C(x, y)$—the degree of corroboration of the theory x by the evidence y—I had to operate with some converse $p(y, x)$, called by Fisher the '*likelihood* of x' (in the light of the evidence y, or given y; note that both, my 'corroboration' and Fisher's likelihood, are intended to measure the acceptability of the hypothesis x; it is thus x which is important, while y represents merely the changing empirical evidence or, as I prefer to say, the reports of *the results of tests*). Now I was convinced that, if x is a theory, $p(x) = 0$. I saw therefore that I had to construct a new probability calculus in which the likelihood, $p(y, x)$, could be a definite number, other than 0/0, even if x was a universal theory with $p(x) = 0$.

I will now briefly explain how the problem of $p(y, x)$—of the likelihood of x—arises.

If we are asked to give a criterion of the fact that the evidence y supports or corroborates or confirms a statement x, the most obvious reply is: 'that y *increases* the probability of x'. We can put this in symbols if we write '$Co(x, y)$' for 'x is supported or corroborated or confirmed by y'. We then can formulate our criterion as follows.

(1) $Co(x, y)$ if, and only if, $p(x, y) > p(x)$.

This formulation, however, has a defect. For if x is a universal theory and y some empirical evidence, then, as we have seen in the two preceding appendices,[2]

(2) $p(x) = 0 = p(x, y)$.

But from this it would follow that, for a theory x and evidence y,

[2] See, especially, appendix *vii, formulae (1) and (2), and appendix *viii, formula (2).

$Co(x, y)$ is always false; or in other words, that a universal law can never be supported, or corroborated, or confirmed by empirical evidence.

(This holds not only for an infinite universe but also for any extremely large universe, such as ours; for in this case, $p(x, y)$ and $p(x)$ will be both immeasurably small, and thus practically equal.)

This difficulty however can be got over, as follows. Whenever $p(x) \neq 0 \neq p(y)$, we have

(3) $p(x, y) > p(x)$ if, and only if, $p(y, x) > p(y)$,

so that we can transform (1) into

(4) $Co(x, y)$ if, and only if, $p(y, x) > p(y)$.

Now let x be again a universal law, and let y be empirical evidence which, say, follows from x. In this case, that is whenever y follows from x, we shall say, intuitively, that $p(y, x) = 1$. And since y is empirical so that $p(y)$ will certainly be less than 1, we find that (4) can be applied, and that the assertion $Co(x, y)$ will be true. That is to say, x may be corroborated by y if y follows from x, provided only that $p(y) < 1$. Thus (4) is intuitively perfectly satisfactory; but in order to operate freely with (4), we have to have a calculus of probability in which $p(y, x)$ is a definite number—in our case, 1—rather than 0/0, even if $p(x) = 0$. In order to achieve this, a generalization of the usual calculus has to be provided, as explained above.

Although I had realized this by the time my note in *Mind* appeared (*cf.* appendix *ii), the pressure of other work which I considered more urgent prevented me from completing my researches in this field. It was only in 1954 that I published my results concerning degree of corroboration, in the first of the notes here reprinted; and another six months elapsed before I published an axiom system of relative probability[3] (equivalent to, though less simple than, the one which will be found in appendix *iv) which satisfied the demand that $p(x, y)$ should be a definite number even if $p(y)$ was equal to zero. This paper provided the technical prerequisites for a satisfactory definition of likelihood and of degree of corroboration or confirmation.

My first note 'Degree of Confirmation', published in 1954 in the *B.J.P.S.*, contains a mathematical refutation of all those theories of induction which identify the degree to which a statement is supported or confirmed or corroborated by empirical tests with its degree of

[3] See *B.J.P.S.* **6**, 1955, pp. 56–57.

probability in the sense of the calculus of probability. The refutation consists in showing that if we identify degree of corroboration or confirmation with probability, we should be forced to adopt a number of highly paradoxical views, among them the following clearly self-contradictory assertion:

(*) There are cases in which x is strongly supported by z and y is strongly undermined by z while, at the same time, x is confirmed by z to a lesser degree than is y.

A simple example, showing that this devasting consequence would follow if we were to identify corroboration or confirmation with probability, will be found under point 6 of my first note.[4] In view of the brevity of that passage, I may perhaps explain this point here again.

Consider the next throw with a homogeneous die. Let x be the statement 'six will turn up'; let y be its negation, that is to say, let $y = \bar{x}$; and let z be the information 'an even number will turn up'. We have the following absolute probabilities:

$$p(x) = 1/6 ; \quad p(y) = 5/6 ; \quad p(z) = 1/2.$$

Moreover, we have the following relative probabilities:

$$p(x, z) = 1/3 ; \quad p(y, z) = 2/3.$$

We see that x is supported by the information z, for z raises the probability of x from $1/6$ to $2/6 = 1/3$. We also see that y is undermined by z, for z lowers the probability of y by the same amount from $5/6$ to $4/6 = 2/3$. Nevertheless, we have $p(x, z) < p(y, z)$. This example proves the following theorem:

(5) There exist statements, x, y, and z, which satisfy the formula,

$$p(x, z) > p(x) \ \& \ p(y, z) < p(y) \ \& \ p(x, z) < p(y, z).$$

Obviously, we may replace here '$p(y, z) < p(y)$' by the weaker '$p(y, z) \leqslant p(y)$'.

This theorem is, of course, far from being paradoxical. And the same holds for its corollary (6) which we obtain by substituting for

---

[4] As opposed to the example here given in the text, the example given under points 5 and 6 of my first note are the simplest possible examples, that is to say, they operate with the smallest possible number of equiprobable exclusive properties. This holds also for the example given in the footnote to point 5. (As far as point 5 is concerned, there seems to be an equivalent, though more complicated, example in Carnap's *Logical Foundations of Probability*, 1950, § 71; I have been unable to follow it, because of its complexity. As to my point 6, I have found neither there, not anywhere else an example corresponding to it.)

'$p(x, z) > p(x)$' and '$p(y, z) \leqslant p(y)$' the expressions '$Co(x, z)$' and '$\sim Co(y, z)$'—that is to say '$non\text{-}Co(y, z)$'—respectively, in accordance with formula (1) above:

(6) There exist statements x, y and z which satisfy the formula

$$Co(x, z) \ \& \ \sim Co(y, z) \ \& \ p(x, z) < p(y, z).$$

Like (5), theorem (6) expresses a fact we have established by our example: that x may be supported by z, and y undermined by z, and that nevertheless x, given z, may be more probable than y, given z.

There at once arises, however, a clear self-contradiction if we now identify in (6) degree of confirmation and probability. In other words, the formula

(**) $Co(x, z) \ \& \ \sim Co(y, z) \ \& \ C(x, z) < C(y, z)$

is clearly self-contradictory, and cannot, therefore, be satisfied by any set of statements.

Thus we have proved that the identification of degree of corroboration or confirmation with probability is absurd on both formal and intuitive grounds: it leads to self-contradiction.

Here 'degree of corroboration or confirmation' may be taken in a sense wider than the one I usually have in mind. While I usually take it to be a synonym for 'degree of severity of the tests which a theory has passed', it is here used merely as 'degree to which a statement x is supported by a statement y'.

If we look at this proof, then we see that it depends upon two assumptions only:

(a) Formula (1);

(b) The assumption that any assertion of the following form is *self-contradictory*:

(***) x has the property P (for example, the property 'warm') and y has not the property P and y has the property P in a higher degree than x (for example, y is warmer than x).

Every careful reader of my first note, and especially of the example given under point 6, will find that all this is clearly implied there, except perhaps the general formulation (***) of the contradictions (*) and (**). Admittedly it is put here in a more explicit form; but the purpose of my note was not so much to criticize as to give a definition of degree of corroboration.

The criticism contained in my note was directed against *all* those who identify, explicitly or implicitly, degree of corroboration, or of confirmation, or of acceptability, with probability; the philosophers I had in mind were especially Keynes, Jeffreys, Reichenbach, Kaila, Hosiasson and, more recently, Carnap.

As to Carnap, I wrote a critical footnote which, I believe, speaks for itself. It was motivated by the fact that Carnap, in stating adequacy criteria for degree of confirmation, speaks of the consensus of 'practically all modern theories of degree of confirmation', but does not mention my dissent, in spite of the fact that he introduced the English term 'degree of confirmation' as a translation of my term '*Grad der Bewährung*'. (*Cf.* the footnote before section 79, above.) Moreover, I wished to point out that his division of probability into probability$_1$ (= his degree of confirmation) and probability$_2$ (= statistical frequency) was insufficient: that there were at the very least two interpretations of the calculus of probability (the logical and the statistical) and that, *in addition*, there was my degree of corroboration *which was not a probability* (as has now been shown here, and as was shown in my note).

It seems that this ten-line footnote of mine has drawn more attention to itself than the rest of my note. It led to a discussion in the *B.J.P.S.*[5] in which Bar-Hillel asserted that my criticism of what he termed 'the current theory of confirmation' (*i.e.* Carnap's theory) was purely verbal and that all I had to say was anticipated by Carnap; and it led to a review of my paper in the *Journal of Symbolic Logic*[6] in which Kemeny summed up my note by the words: 'The principal thesis of this paper is that Carnap's proposed measures of degree of confirmation, or any other assignment of logical probability, are not suited to measure degrees of confirmation.'

This was certainly not my principal thesis. My note was a continuation of some work of mine published fifteen years before Carnap's book was written; and as far as criticism is concerned, the point at issue —the identification of corroboration or confirmation or acceptability with probability—though it is of course the main thesis of Carnap's book, is far from being an original thesis of Carnap's; for he is here merely following the tradition of Keynes, Jeffreys, Reichenbach, Kaila, Hosiasson, and others. Moreover, both Bar-Hillel and Kemeny

[5] See *B.J.P.S.* **6**, 1955, pp. 155 to 163; and **7**, 1956, pp. 243 to 256.
[6] See *J.S.L.* **20**, 1955, p. 304. The following is an error of fact in Kemeny's review: in line 16 from the bottom of the page, 'measure of support given by y to x' should read 'measure of the explanatory power of x with respect to y'.

suggest that this criticism, as far as it applies to Carnap's theory, is purely verbal, and that there is no reason why Carnap's theory should be given up; and I therefore feel compelled to say now quite clearly that Carnap's theory is self-contradictory, and that its contradictoriness is not a minor matter which can be easily repaired, but is due to mistakes in its logical foundations.

First, both the assumptions (a) and (b) which, as we have seen, suffice for the proof that degree of confirmation must not be identified with probability, are explicitly asserted in Carnap's theory: (a), that is to say our formula (1), can be found in Carnap's book as formula (4) on p. 464;[7] (b), that is to say (***), or the assumption that our (**) is self-contradictory, can be found on p. 73 of Carnap's book where he writes: 'If the property Warm and the relation Warmer were designated by . . . , say, 'P' and 'R', then '$Pa.{\sim}Pb.Rba$' would be self-contradictory.' But this is our (***). Of course, in a way it is quite irrelevant to my argument which shows the absurdity of the identification of C and p whether or not (a) and (b) are explicitly admitted in a book; but it so happens that in Carnap's book, they are.

Moreover, the contradiction here explained is crucial for Carnap: by accepting (1), or more precisely, by defining on pp. 463 f. 'x is confirmed by y' with the help of '$p(x, y) > p(x)$' (in our symbolism), Carnap shows that the intended meaning of 'degree of confirmation' (his '*explicandum*') is, *roughly*, the same as the one intended by myself. It is the intuitive idea of degree of support by empirical evidence. (Kemeny *loc. cit.* is mistaken when he suggests the opposite. In fact, 'a careful reading' of my paper—and, I should add, of Carnap's book—will *not* 'show that Popper and Carnap have two different *explicanda* in mind', but it will show that Carnap had inadvertently two different and incompatible '*explicanda*' in mind with his probability$_1$, one of them my C, the other my p; and it will show that I have repeatedly pointed out the dangers of this confusion—for example in the paper reviewed by Kemeny.) Therefore, any change of assumption (a) would be *ad hoc*. It is not my criticism that is 'purely verbal', but the attempts to rescue the 'current theory of confirmation'.

For further details, I must refer to the discussion in the pages of the *B.J.P.S.* I may say that I was a little disappointed both by this

---

[7] See also formula (6) on p. 464. Carnap's formula (4) on p. 464 is written as an equivalence, but this does not make any difference. Note that Carnap writes 't' for tautology; a usage which would allow us to write $p(x, t)$ instead of $p(x)$.

discussion and by Kemeny's review in the *Journal of Symbolic Logic*. From a rational point of view, the situation appears to me quite serious. In this post-rationalist age of ours, more and more books are written in symbolic languages, and it becomes more and more difficult to see why: what it is all about, and why it should be necessary, or advantageous, to allow oneself to be bored by volumes of symbolic trivialities. It almost seems as if the symbolism were becoming a value in itself, to be revered for its sublime 'exactness': a new expression of the old quest for certainty, a new symbolic ritual, a new substitute for religion. Yet the only possible value of this kind of thing—the only possible excuse for its dubious claim to exactness—seems to be this. Once a mistake, or a contradiction, is pin-pointed, there can be no verbal evasion: it can be proved, and that is that. (Frege did not try evasive manœuvres when he received Russell's criticism.) So if one has to put up with a lot of tiresome technicalities, and with a formalism of unnecessary complexity, one might at least hope to be compensated by the ready acceptance of a straight-forward proof of contradictoriness—a proof consisting of the simplest of counter-examples. It was disappointing to be met, instead, by merely verbal evasions, combined with the assertion that the criticism offered was 'merely verbal'.

Still, one must not be impatient. Since Aristotle, the riddle of induction has turned many philosophers to irrationalism—to scepticism or to mysticism. And although the philosophy of the identity of C and p seems to have weathered many a storm since Laplace, I still think that it will be abandoned one day. I really cannot bring myself to believe that the defenders of the faith will be satisfied for ever with mysticism and Hegelianism, upholding '$C = p$' as a self-evident axiom, or as the dazzling object of an inductive intuition. (I say 'dazzling' because it seems to be an object whose beholders are smitten with blindness when running into its logical contradictions.)

I may perhaps say here that I regard the doctrine that *degree of corroboration or acceptability cannot be a probability* as one of the most interesting findings of the philosophy of knowledge. It can be put very simply like this. A report of the result of testing a theory can be summed up by an appraisal. This can take the form of assigning some degree of corroboration to the theory. But it can never take the form of assigning to it a degree of probability; for *the probability of a statement (given some test statements) simply does not express an appraisal of the severity of the tests a theory has passed, or of the manner in which it has passed these tests.* The main reason for this is that the *content* of a theory—which is the same as its *improbability*—determines its *testability* and its *corroborability*.

I believe that these two ideas—that of *content* and of *degree of corroboration*—are the most important logical tools developed in my book.[8]

So much by way of introduction. In the three notes which follow here I have left the word 'confirmation' even where I should now only write 'corroboration'. I have also left '$P(x)$' where I now usually write '$p(x)$'. But I have corrected some misprints;[9] and I have added a few footnotes, preceded by stars, and also two new points, *13 and *14, to the end of the Third Note.

Degree of Confirmation.

1. The purpose of this note is to propose and to discuss a definition, in terms of probabilities, of *the degree to which a statement x is confirmed by a statement y.* (Obviously this may be taken to be identical with *the degree to which a statement y confirms a statement x.*) I shall denote this degree by the symbol '$C(x, y)$', to be pronounced 'the *degree of con-*

---

[8] As far as I am aware, the recognition of the significance of the *empirical content* or assertive power of a theory; the suggestion that this content increases with the class of the potential falsifiers of the theory—that is to say, the states of affairs which it forbids, or excludes (see sections 23, and 31); and the idea that content may be measured by the improbability of the theory, were not taken by me from any other source but were 'all my own work'. I was therefore surprised when I read in Carnap's *Introduction to Semantics*, 1942, p. 151, in connection with his definition of 'content': '. . .the assertive power of a sentence consists in its excluding certain states of affairs (Wittgenstein); the more it excludes, the more it asserts.' I wrote to Carnap, asking for details and reminding him of certain relevant passages in my book. In his reply he said that his reference to Wittgenstein was due to an error of memory, and that he actually had a passage from my book in mind; and he repeated this correction in his *Logical Foundations of Probability*, 1950, p. 406. I mention this here because in a number of papers published since 1942, the idea of content—in the sense of empirical or informative content—has been attributed, without definite reference, to Wittgenstein, or to Carnap, and sometimes to Wittgenstein and myself. But I should not like anybody to think that I have taken it without acknowledgment from Wittgenstein or anybody else: as a student of the history of ideas, I think that it is quite important to refer to one's sources. (See also my discussion in section 35 of the distinction between *logical content* and *empirical content*, with references to Carnap in footnotes 1 and 2.)

[9] I have also, of course, incorporated the corrections mentioned in *B.J.P.S.* **5**, pp. 334 and 359.

firmation of x by y'. In particular cases, x may be a hypothesis, h; and y may be some empirical evidence, e, in favour of h, or against h, or neutral with respect to h. But $C(x, y)$ will be applicable to less typical cases also.

The definition is to be in terms of probabilities. I shall make use of both, $P(x, y)$, *i.e.* the (relative) probability of x given y, and $P(x)$, *i.e.* the (absolute) probability of x.[1] But either of these two would be sufficient.

2. It is often assumed that the degree of confirmation of x by y must be the same as the (relative) probability of x given y, *i.e.* that $C(x, y) = P(x, y)$. My first task is to show the inadequacy of this view.

3. Consider two contingent statements, x and y. From the point of view of the confirmation of x by y, there will be two extreme cases: the complete support of x by y or the establishment of x by y, when x follows from y; and the complete undermining or refutation or disestablishment of x by y, when x follows from y. A third case of special importance is that of mutual independence or irrelevance, characterized by $P(xy) = P(x)P(y)$. Its value of $C(x, y)$ will lie below establishment and above disestablishment.

Between these three special cases—establishment, independence, and disestablishment—there will be intermediate cases: *partial support* (when y entails part of the content of x); for example, if our contingent y follows from x but not *vice versa*, then it is itself part of the content of x and thus entails part of the content of x, supporting x; and *partial undermining* of x by y (when y partially supports $\bar{x}$); for example, if y follows from $\bar{x}$. We shall say, then, that y supports x, or that it undermines x, whenever $P(xy)$, or $P(\bar{x}y)$, respectively, exceed their values for independence. (The three cases—support, undermining, independence—are easily seen to be exhaustive and exclusive in view of this definition.)

4. Consider now the conjecture that there are three statements, x_1, x_2, and y, such that (i) x_1 and x_2 are each independent of y (or

---

[1] '$P(x)$' may be defined, in terms of relative probability, by the definiens '$P(x, \overline{zz})$' or, more simply, '$P(x, \overline{xx})$'. (I use throughout 'xy' to denote the conjunction of x and y, and '$\bar{x}$' to denote the negation of x.) Since we have, generally, $P(x, y\overline{zz}) = P(x, y)$, and $P(x, yz) = P(xy, z)/P(y, z)$, we obtain $P(x, y) = P(xy)/P(y)$—a serviceable formula for defining relative probability in terms of absolute probability. (See my note in *Mind*, 1938, **47**, 275, f., where I identified absolute probability with what I called 'logical probability' in my *Logik der Forschung*, Vienna, 1935, esp. sects. 34 f. and 83, since the term 'logical probability' is better used for the 'logical interpretation' of both $P(x)$ and $P(x, y)$, as opposed to their 'statistical interpretation' which may be ignored here.)

undermined by y) while (ii) y supports their conjunction x_1x_2. Obviously, we should have to say in such a case that y confirms x_1x_2 to a higher degree than it confirms either x_1 or x_2; in symbols,

(4.1) $$C(x_1, y) < C(x_1x_2, y) > C(x_2, y)$$

But this would be incompatible with the view that $C(x, y)$ is a probability, i.e. with

(4.2) $$C(x, y) = P(x, y)$$

since for probabilities we have the generally valid formula

(4.3) $$P(x_1, y) \geqslant P(x_1x_2, y) \leqslant P(x_2, y)$$

which, in the presence of (4.1) contradicts (4.2). Thus we should have to drop (4.2). But in view of $0 \leqslant P(x, y) \leqslant 1$, (4.3) is an immediate consequence of the general multiplication principle for probabilities. Thus we should have to discard such a principle for the degree of confirmation. Moreover, it appears that we should have to drop the special addition principle also. For a consequence of this principle is, since $P(x, y) \geqslant 0$,

(4.4) $$P(x_1x_2 \text{ or } x_1\bar{x}_2, y) \geqslant P(x_1x_2, y)$$

But for $C(x, y)$, this could not remain valid, considering that the alternative, x_1x_2 or $x_1\bar{x}_2$, is equivalent to x_1, so that we obtain by substitution on the left-hand side of (4.1):

(4.5) $$C(x_1x_2 \text{ or } x_1\bar{x}_2, y) < C(x_1x_2, y).$$

In the presence of (4.4), (4.5) contradicts (4.2).[2]

5. These results depend upon the conjecture that statements x_1, x_2, and y exist such that (i) x_1 and x_2 are each independent of y (or undermined by y) while (ii) y supports x_1x_2. I shall prove this conjecture by an example.[3]

[2] In his *Logical Foundations of Probability*, Chicago, 1950, p. 285, Carnap uses the multiplication and addition principles as '*conventions on adequacy*' for degree of confirmation. The only argument he offers in favour of the adequacy of these principles is that 'they are generally accepted in practically all modern theories of probability₁', i.e. our $P(x, y)$ which Carnap identifies with 'degree of confirmation'. But the very term 'degree of confirmation' ('*Grad der Bewährung*') was introduced by me in sections 82 *f.* of my *Logik der Forschung* (a book to which Carnap sometimes refers), in order to show that both logical and statistical probability are *inadequate* to serve for a degree of confirmation, since confirmability must increase with testability, and thus with (absolute) logical improbability and content. (See below.)

[3] The example satisfies (i) for *independence* rather than *undermining*. (To obtain one for undermining, add amber as a fifth colour, and put $y =$ '*a* is amber or blue or yellow'.)

Take coloured counters, called 'a', 'b', . . . , with four exclusive and equally probable properties, blue, green, red, and yellow. Let x_1 be the statement 'a is blue or green'; $x_2 = $ 'a is blue or red'; $y = $ 'a is blue or yellow'. Then all our conditions are satisfied. (That y supports $x_1 x_2$ is obvious: y follows from $x_1 x_2$, and its presence raises the probability of $x_1 x_2$ to twice the value it has in the absence of y.)

6. But we may even construct a more striking example to show the inadequacy of identifying $C(x, y)$ and $P(x, y)$. We choose x_1 so that it is strongly supported by y, and x_2 so that it is strongly undermined by y. Thus we shall have to demand that $C(x_1, y) > C(x_2, y)$. But x_1 and x_2 can be so chosen that $P(x_1, y) < P\, x_2, y)$. The example is this: take $x_1 = $ 'a is blue'; $x_2 = $ 'a is not red'; and $y = $ 'a is not yellow'. Then $P(x_1) = 1/4; P(x_2) = 3/4;$ and $1/3 = P(x_1, y) < P(x_2, y) = 2/3$. That y supports x_1 and undermines x_2 is clear from these figures, and also from the fact that y follows from x_1 and also from x_2.[*1]

7. Why have $C(x, y)$ and $P(x, y)$ been confounded so persistently? Why has it not been seen that it is absurd to say that some evidence y of which x is completely independent can yet strongly 'confirm' x? And that y can strongly 'confirm' x, even if y undermines x? And this even if y is the total available evidence? I do not know the answer to these questions, but I can make a few suggestions. First, there is the powerful tendency to think that whatever may be called the 'likelihood' or 'probability' of a hypothesis must be a probability in the sense of the calculus of probabilities. In order to disentangle the various issues here involved, I distinguished twenty years ago what I then called the 'degree of confirmation' from both, the logical and the statistical probability. But unfortunately, the term 'degree of confirmation' was soon used by others as a new name for (logical) probability; perhaps under the influence of the mistaken view that science, unable to attain certainty, must aim at a kind of '*Ersatz*'—at the highest attainable probability.

Another suggestion is this. It seems that the phrase 'the degree of confirmation of x by y' was never turned round into 'the degree to which y confirms x', or '*the power of y to support x*'. Yet in this form it would have been quite obvious that, in a case in which y supports x_1 and undermines x_2, $C(x_1, y) < C(x_2, y)$ is absurd—although

[*1] This fact—that is to say, $p(y, x_1) = p(y, x_2) = 1$—means that Fisher's 'likelihood' of x_1, and thus of x_2, in the light of y, is maximal. See the introduction to the present appendix in which the argument here outlined in the text is elaborated.

$P(x_1, y) < P(x_2, y)$ may be quite in order, indicating, in such a case, that we had $P(x_1) < P(x_2)$ to start with. Furthermore, there seems to be a tendency to confuse measures *of* increase or decrease with the measures *that* increase and decrease (as shown by the history of the concepts of velocity, acceleration, and force). But the power of y to support x, it will be seen, is essentially a *measure of the increase or decrease* due to y, in the probability of x. (See also 9 (vii), below.)

8. It will perhaps be said, in reply to all this, that it is in any case legitimate to call $P(x, y)$ by any name, and also by the name 'degree of confirmation'. But the question before us is not a verbal one.

The degree of confirmation of a hypothesis x by empirical evidence y is supposed to be used for estimating the degree to which x is *backed by experience*. But $P(x, y)$ cannot serve this purpose, since (Px_1, y) may be higher than (Px_2, y) even though x_1 is undermined by y and z_2 supported by y, and since this is due to the fact that $P(x, y)$ depends very strongly upon $P(x)$, *i.e.* the absolute probability of x, which has nothing whatever to do with the empirical evidence.

Furthermore, the degree of confirmation is supposed to have an influence upon the question whether we should *accept*, or *choose*, a certain hypothesis x, if only tentatively; a high degree of confirmation is supposed to characterise a hypothesis as 'good' (or 'acceptable'), while a disconfirmed hypothesis is supposed to be 'bad'. But $P(x, y)$ cannot help here. *Science does not aim, primarily, at high probabilities. It aims at a high informative content, well backed by experience. But a hypothesis may be very probable simply because it tells us nothing, or very little.* A high degree of probability is therefore not an indication of 'goodness'—it may be merely a symptom of low informative content. On the other hand, $C(x, y)$ must, and can, be so defined that only hypotheses with a high informative content can reach high degrees of confirmation. The *confirmability* of x (*i.e.* the maximum degree of confirmation which a statement x can reach) should increase with $C(x)$, *i.e.* the measure of the content of x, which is equal to $P(\bar{x})$, and therefore to the *degree of testability* of x. Thus, while $P(x\bar{x}, y) = 1$, $C(x\bar{x}, y)$ should be zero.

9. A definition of $C(x, y)$ that satisfies all these and other *desiderata* indicated in my *Logik der Forschung*, and stronger ones besides, may be based upon $E(x, y)$, *i.e.* a non-additive measure of the *explanatory power of x with respect to y*, designed so as to have -1 and $+1$ as its lower and upper bounds. It is defined as follows.

(9.1) Let x be consistent,[4] and $P(y) \neq 0$; then we define,

$$E(x, y) = \frac{P(y, x) - P(y)}{P(y, x) + P(y)}$$

$E(x, y)$ may also be interpreted as a non-additive measure of the dependence of y upon x, or the support given to y by x (and *vice versa*). It satisfies the most important of our *desiderata*, but not all: for example, it violates (viii, *c*) below, and satisfies (iii) and (iv) only approximately in special cases. To remedy these defects, I propose to define $C(x, y)$ as follows.[*2]

(9.2) Let x be consistent and $P(y) \neq 0$; then we define,

$$C(x, y) = E(x, y)(1 + P(x)P(x, y))$$

This is less simple than, for example, $E(x, y)(1 + P(xy))$, which satisfies most of our *desiderata* but violates (iv), while for $C(x, y)$ the theorem holds that it satisfies all of the following *desiderata*:

(i) $C(x, y) \gtreqless 0$ respectively if and only if y supports x, or is independent of x, or undermines x.

(ii) $-1 = C(\bar{y}, y) \leqslant C(x, y) \leqslant C(x, x) \leqslant 1$

(iii) $0 \leqslant C(x, x) = C(x) = P(x) \leqslant 1$

Note that $C(x)$, and therefore $C(x, x)$, is an additive measure of the content of x, definable by $P(\bar{x})$, *i.e.* the absolute probability of x to be false, or the *a priori* likelihood of x to be *refuted*. Thus *confirmability* equals *refutability or testability*.[5]

---

[4] This condition may be dropped if we accept the general convention that $P(x, y) = 1$ whenever y is inconsistent.

[*2] The following is a somewhat simpler alternative definition which also satisfies all my adequacy conditions or *desiderata*. (I first stated it in the *B.J.P.S.* **5**, p. 359.)

(9.2*)
$$C(x,y) = \frac{P(y, x) - P(y)}{P(y, z) - P(xy) + P(y)}$$

Similarly, I now put

(10.1*)
$$C(x,y,z) = \frac{P(y, xz) - P(y, z)}{P(y, xz) - P(xy, z) + P(y, z)}$$

[5] See section 83 of my *L.d.F.*, which bears the title 'Confirmability, Testability, Logical Probability'. (Before 'logical', 'absolute' should be inserted, in agreement with the terminology of my note in *Mind*, *loc. cit.*)

(iv) If y entails x, then $C(x, y) = C(x, x) = C(x)$

(v) If y entails $\bar{x}$, then $C(x, y) = C(\bar{y}, y) = -1$

(vi) Let x have a high content—so that $C(x, y)$ approaches $E(x, y)$—and let y support x. (We may, for example, take y to be the total available empirical evidence.) *Then for any given y, $C(x, y)$ increases with the power of x to explain y* (i.e. to explain more and more of the content of y), and therefore with the *scientific interest* of x.

(vii) If $C(x) = C(y) \neq 1$, then $C(x, u) \gtreqless C(y, w)$ whenever $P(x, u) \gtreqless P(y, w)$.[*3]

(viii) If x entails y, then: (a) $C(x, y) \geqslant 0$; (b) for any given x, $C(x, y)$ and $C(y)$ increase together; and (c) for any given y, $C(x, y)$ and $P(x)$ increase together.[6]

(ix) If $\bar{x}$ is consistent and entails y, then: (a) $C(x, y) \leqslant 0$; (b) for any given x, $C(x, y)$ and $P(y)$ increase together; and (c) for any given y, $C(x, y)$ and $P(x)$ increase together.

10. All our considerations, without exception, may be relativized with respect to some initial information z; adding at the appropriate places phrases like 'in the presence of z, assuming $P(z, \overline{zz}) \neq 0$'. The relativized definition of the degree of confirmation becomes:

(10.1) $\qquad C(x, y, z) = E(x, y, z)(1 + P(x, z)P(x, yz))$

where

(10.2) $\qquad\qquad E(x, y, z) = \dfrac{P(y, xz) - P(y, z)}{P(y, xz) + P(y, z)}$

$E(x, y, z)$ is the explanatory power of x with respect to y, in the presence of z.[7]

11. There are, I believe, some intuitive *desiderata* which cannot be satisfied by any formal definition. For example, a theory is the better

[*3] The condition '$\neq 1$' was printed neither in the original text, nor in the published corrections.

[6] (vii) and (viii) contain the only important *desiderata* which are satisfied by $P(x, y)$.

[7] Let x_1 be Einstein's gravitational theory; x_2 Newton's; y the (interpreted) empirical evidence available today, including 'accepted' laws (it does not matter if none or one or both of the theories in question are included, provided our conditions for y are satisfied); and z a part of y, for example, a selection from the evidence available one year ago. Since we may assume that x_1 explains more of y than x_2, we obtain $C(x_1, y, z) \geqslant C(x_2, y, z)$ for every z, and $C(x_1, y, z) > C(x_2, y, z)$ for any suitable z containing some of the relevant initial conditions. This follows from (vi) —*even if we have to assume* that $P(x_1, yz) = P(x_2, yz) = P(x_1) = P(x_2) = 0$.

confirmed the more ingenious our unsuccessful attempts at its refutation have been. My definition incorporates something of this idea—if not as much as can be formalized. But one cannot completely formalize the idea of a sincere and ingenious attempt.[8]

The particular way in which $C(x, y)$ is here defined I consider unimportant. What may be important are the *desiderata*, and *the fact that they can be satisfied together*.

A Second Note on Degree of Confirmation

1 The suggestion has been made by Professor J. G. Kemeny[1] (with a reference to my definition of *content*), and independently by Dr. C. L. Hamblin[2] that the *content* of x, denoted by '$C(x)$', should be measured by $-\log_2 P(x)$ instead of $1 - P(x)$, as I originally suggested. (I am here using my own symbols.) If this suggestion is adopted, then my *desiderata*[3] for *degree of confirmation* of x by y, denoted by $C(x, y)$, have to be slightly amended: in (ii) and in (v), we must replace ± 1 by $\pm \infty$; and (iii) becomes:

---

[8] There are many ways of getting nearer to this idea. For example, we may put a premium on crucial experiments by defining

$$C_{a,\,b}(h) = (C(h, e_b) \prod_{i-1}^{n} C(h, c_i, e_a))^{1/(n+1)}$$

where $c_1, c_2, \ldots$, is the sequence of experiments made between the moments of time, t_a and t_b. We have $t_a < t_1 \leqslant t_i \leqslant t_n = t_b$. e_a and e_b are the total evidence (which may include laws) *accepted* at t_a and t_b. We postulate $P(c_i, e_b) = 1$ and (to ensure that only new experiments are counted) $P(c_i, e_a) \neq 1$ and $P(c_i, Uc_j) \neq 1$, whenever $j < i$. ('Uc_j' is the spatio-temporal universalization of c_j.)

*Today, I should be inclined to treat this question in a different way. We may, very simply, distinguish between the formula '$C(x, y)$' (or '$C(x, y, z)$') and the *applications* of this formula to what we mean, intuitively, by corroboration, or acceptability. Then it suffices to say that $C(x, y)$ must not be interpreted as degree of corroboration, and must not be applied to problems of acceptability, unless y represents the (total) results of sincere attempts to refute x. See also point *14 of my 'Third Note', below.

I have here put 'total' in brackets, because there is another possibility to be considered: we may confine our tests to a certain field of application F (*cf.* the old appendix i, and appendix *viii), we may thus relativize C, and write '$C_F(x,y)$'. The total corroboration of a theory may then be said, simply, to be the sum of its corroborations in its various (independent) fields of application.

[1] John G. Kemeny, *Journal of Symbolic Logic*, 1953, **18**, p. 297. (Kemeny's reference is to my *Logik der Forschung*.)

[2] C. L. Hamblin, 'Language and the Theory of Information', a thesis submitted to the University of London in May 1955 (unpublished); see p. 62. Dr. Hamblin produced this definition independently of Professor Kemeny's paper (to which he refers in his thesis).

[3] 'Degree of Confirmation', this *Journal*, 1954, **5**, 143 *sqq.*; see also p. 334.

(iii) $0 \leqslant C(x, xy) = C(x, x) = C(x) = -\log_2 P(x) \leqslant +\infty.$

The other *desiderata* remain as they were.

Dr. Hamblin suggests[4] that we define degree of confirmation by

(1) $C(x, y) = \log_2(P(xy)/P(x)P(y))$

which for finite systems, but not necessarily for infinite systems, is the same as

(2) $C(x, y) = \log_2(P(y, x)/P(y)),$

a formula which has the advantage of remaining determinate even if $P(x) = 0$, as may be the case if x is a universal theory. The corresponding relativized formula would be

(3) $C(x, y, z) = \log_2(P(y, xz)/P(y, z)).$

The definition (1) does not, however, satisfy my desideratum viii (c), as Dr. Hamblin observes; and the same holds for (2) and (3). *Desiderata* ix (b) and (c) are also not satisfied.

Now my desideratum viii (c) marks, in my opinion, the difference between a measure of explanatory power and one of confirmation. The former may be symmetrical in x and y, the latter not. For let y follow from x (and support x) and let a be unconfirmed by y. In this case it does not seem satisfactory to say that ax is always as well confirmed by y as is x. (But there does not seem to be any reason why ax and x should not have the same explanatory power with respect to y, since y is completely explained by both.) This is why I feel that viii(c) should not be dropped.

Thus I prefer to look upon (2) and (3) as highly adequate definitions of *explanatory power*—of $E(x, y)$ and $E(x, y, z)$—rather than of degree of confirmation. The latter may be defined, on the basis of explanatory power, in many different ways so as to satisfy viii(c). One way is as follows (I think that better ways may be found):

(4) $C(x, y) = E(x, y)/(1 + nP(x)P(\bar{x}, y))$

(5) $C(x, y, z) = E(x, y, z)/(1 + nP(x, z)P(\bar{x}, yz))$

Here we may choose $n \geqslant 1$. And if we wish viii(c) to have a marked effect, we can make n a large number.

In case x is a universal theory with $P(x) = 0$ and y is empirical evidence, the difference between E and C disappears, as in my original definitions, and as demanded by desideratum (vi). It also disappears if x

[4] C. L. Hamblin, *op. cit.*, p. 83. A similar suggestion (without, however, specifying 2 as basis of the logarithm) is made in Dr. I. J. Good's review of my 'Degree of Confirmation'; *cf. Mathememematical Review*, 1955, 16, 376.

follows from y. Thus at least some of the advantages of operating with a logarithmic measure remain: as explained by Hamblin, the concept defined by (1) becomes closely related to the fundamental idea of information theory. Good also comments on this point (see footnote 4).

The transition from the old to the new definitions is order-preserving. (This holds also for explanatory power, as Hamblin's observations imply.) Thus the difference is metrical only.

2. The definitions of explanatory power, and even more of degree of confirmation (or corroboration or acceptability or attestation, or whatever name may be chosen for it) give of course full weight to the '*weight of evidence*' (or the 'weight of an argument' as Keynes called it in his chapter vi).*¹ This becomes obvious with the new definitions, based upon Hamblin's suggestions, which seem to have considerable advantages if we are at all interested in metrical questions.

3. However, we must realize that the metric of our C will depend entirely upon the metric of P. *But there cannot be a satisfactory metric of P; that is to say, there cannot be a metric of logical probability which is based upon purely logical considerations.* To show this we consider the logical probability of any measurable physical property (non-discreet random variable) such as length, to take the simplest example. We make the assumption (favourable to our opponents) that we are given some logically necessary finite lower and upper limits, l and u, to its values. Assume that we are given a distribution function for the logical probability of this property; for example, a generalized equidistribution function between l and u. We may discover that an empirically desirable change of our theories leads to a non-linear correction of the *measure* of our physical property (based, say, on the Paris metre). Then 'logical probability' has to be corrected also; which shows that its metric depends upon our empirical knowledge, and that it cannot be defined *a priori*, in purely logical terms. In other words, the metric of the 'logical probability' of a measurable property would depend upon the metric of the measurable property itself; and since this latter is liable to correction on the basis of empirical theories, there can be no purely 'logical' measure of probability.

These difficulties can be largely, but not entirely, overcome by making use of our 'background knowledge' z. But they establish, I think, the significance of the topological approach to the

*¹ See the 'Third Note', below.

problem of both degree of confirmation and logical probability.[*2]

But, even if we were to discard all metric considerations, we should still adhere, I believe, to the concept of probability, as defined, implicitly, by the usual axiom systems for probability. These retain their full significance, exactly as pure metrical geometry retains its significance even though we may not be able to define a yardstick in terms of pure (metrical) geometry. This is especially important in view of the need to *identify logical independence with probabilistic independence* (special multiplication theorem). If we assume a language such as Kemeny's (which, however, breaks down for continuous properties) or a language with *relative-atomic* statements (as indicated in appendix 1 of my *Logic of Scientific Discovery*), then we shall have to postulate independence for the atomic, or relative-atomic, sentences (of course, as far as they are not 'logically dependent', in Kemeny's sense). *On the basis of a probabilistic theory of induction*, it then turns out that *we cannot learn* if we identify logical and probabilistic independence in the way here described; but *we can learn* very well in the sense of my C-functions; that is to say, we can corroborate our theories.

Two further points may be mentioned in this connection.

4. The first point is this. On the basis of my axiom systems for relative probability,[5] $P(x, y)$ can be considered as defined for any value of x and y, including such values for which $P(y) = 0$. More especially, in the logical interpretation of the system, whenever x follows from y, $P(x, y) = 1$, even if $P(y) = 0$. There is thus no reason to doubt that our definition works for languages containing both singular statements and universal laws, even if all the latter have zero probability, as is the case, for example, if we employ Kemeny's measure function m, by postulating $P(x) = m(x)$. (In the case of our definitions of E and C, there is no need whatever to depart from the assignment of equal weight to the 'models'; cf. Kemeny, *op. cit* p. 307.) On the contrary,

---

[*2] I now believe that I have solved these difficulties, as far as a system S (in the sense of appendix *iv) is concerned whose elements are *probability statements*; that is to say, as far as the logical metric of the *probability of probability* statements is concerned or, in other words, the logical metric of *secondary probabilities*. The method of the solution is described in my *Third Note*, points 7 *ff.*; see especially point *13.

As far as primary properties are concerned, I believe that the difficulties described here in the text are in no way exaggerated. (Of course, z may help, by pointing out, or assuming, that we are confronted, in a certain case, with a finite set of symmetrical or equal possibilities.)

[5] This *Journal*, **6**, p. 56 *sq*, (see also pp. 176 and 351). Simplified versions are given in *British Philosophy in the Mid-Century* (ed. by C. A. Mace), p. 191; and in my *Logic of Scientific Discovery*, appendix *iv.

any such departure must be considered as a deviation from a *logical* interpretation, since it would violate the equality of logical and probabilistic independence demanded in 3, above.)

5. The second point is this. Among the derived desiderata, the following are not satisfied by all definitions of '*x* is confirmed by *y*' which have been proposed by other authors. It might therefore be mentioned separately as a tenth desideratum:[6]

(x) If *x* is confirmed or corroborated or supported by *y* so that $C(x, y) > 0$, then (a) $\bar{x}$ is always undermined by *y*, i.e. $C(\bar{x}, y) < 0$, and (b) *x* is always undermined by $\bar{y}$, i.e. $C(x, \bar{y}) < 0$.

It seems to me clear that this desideratum is an indispensable adequacy condition, and that any proposed definition which does not satisfy it is intuitively paradoxical.

A Third Note on Degree of Corroboration or Confirmation

In this note I wish to make a number of comments on the problem of the *weight of evidence*, and on *statistical tests*.

1. The theory of corroboration or 'confirmation' proposed in my two previous notes on 'Degree of Confirmation'[1] is able to solve with ease the so-called *problem of the weight of evidence*.

This problem was first raised by Peirce, and discussed in some detail by Keynes who usually spoke of the 'weight of an argument' or of the 'amount of evidence'. The term 'the weight of evidence' is taken from J. M. Keynes and from I. J. Good.[2] Considerations of the 'weight of evidence' lead, within the subjective theory of probability, to paradoxes which, in my opinion, are insoluble within the framework of this theory.

2. By the subjective theory of probability, or the subjective interpretation of the calculus of probability, I mean a theory that interprets probability as a measure of our ignorance, or of our partial knowledge,

---

[6] Compare the remark in this *Journal*, 1954, 5, end of the first paragraph on p. 144.

[1] This *Journal*, 1954, 5, 143, 324, and 359; and 1957, 7, 350. See also 1955, 6, and 1956, 7, 244, 249. To the first paragraph of my 'Second Note', a reference should be added to a paper by R. Carnap and Y. Bar-Hillel, 'Semantic Information', this *Journal*, 1953, 4, 147 *sqq.* Moreover, the first sentence of note 1 on p. 351 should read, '*Op. cit.*, p. 83', rather than as at present, because the reference is to Dr. Hamblin's thesis. *(This last correction has been made in the version printed in this book.)

[2] *Cf.* C. S. Peirce, *Collected Papers*, 1932, Vol. 2, p. 421 (first published 1878); J. M. Keynes, *A Treatise on Probability*, 1921, pp. 71 to 78 (see also 312 *sq.*, 'the amount of evidence', and the *Index*); I. J. Good, *Probability and the Weight of Evidence*, 1950, pp. 62 *f.* See also C. I. Lewis, *An Analysis of Knowledge and Valuation*, 1946, pp. 292 *sq.*; and R. Carnap, *Logical Foundations of Probability*, 1950, pp. 554 *sq.*

or, say, of the degree of the rationality of our beliefs, in the light of the evidence available to us.

(I may mention, in parentheses, that the more customary term, 'degree of rational belief', may be a symptom of a slight confusion, since what is intended is 'degree of rationality of a belief'. The confusion arises as follows. Probability is first explained as a measure of the strength or intensity of a belief or conviction—measurable, say, by our readiness to accept odds in betting. Next it is realized that the intensity of our belief often depends, in fact, upon our wishes or fears rather than upon rational arguments; thus, by a slight change, probability is then interpreted as the intensity, or the degree, of a belief *in so far as it is rationally justifiable*. But at this stage, the reference to the intensity of a belief, or to its degree, clearly becomes redundant; and 'degree of belief' should therefore be replaced by 'degree of the rationality of a belief'. These remarks should not be taken to mean that I am prepared to accept *any* form of the subjective interpretation; see point 12, below, and chapter *ii of my *Postscript: After Twenty Years*.)

3. In order to save space, I shall explain the problem of the weight of evidence merely by giving one instance of the paradoxes to which I referred above. It may be called the '*paradox of ideal evidence*'.

Let z be a certain penny, and let a be the statement 'the nth (as yet unobserved) toss of z will yield heads'. Within the subjective theory, it may be assumed that the absolute (or prior) probability of the statement a is equal to $1/2$, that is to say,

$$(1) \qquad\qquad P(a) = 1/2$$

Now let e be some *statistical evidence*; that is to say, a *statistical report*, based upon the observation of thousands or perhaps millions of tosses of z; and let this evidence e be *ideally favourable* to the hypothesis that z is strictly symmetrical—that it is a 'good' penny, with equidistribution. (Note that here e is not the full, detailed report about the results of each of these tosses—this report we might assume to have been lost—but only a *statistical abstract* from the full report; for example, e may be the statement, 'among a million of observed tosses of z, heads occurred in $500,000 \pm 20$ cases'. It will be seen, from point 8, below, that an evidence e' with $500,000 \pm 1,350$ cases would still be ideal, if my functions C and E are adopted; indeed, from the point of view of these functions, e is ideal precisely because it entails e'.) We then have no other option concerning $P(a, e)$ than to assume that

(2) $$P(a, e) = 1/2$$

This means that the probability of tossing heads remains unchanged, in the light of the evidence e; for we now have

(3) $$P(a) = P(a, e).$$

But this formula has to be interpreted as asserting that e is, on the whole, (absolutely) *irrelevant* information with respect to a.

Now this is a little startling; for it means, more explicitly, that our so-called *'degree of rational belief'* in the hypothesis, a, *ought to be completely unaffected by the accumulated evidential knowledge, e*; that the absence of any statistical evidence concerning z justifies precisely the same 'degree of rational belief' as the weighty evidence of millions of observations which, *prima facie*, support or confirm or strengthen our belief.

4. I do not think that this paradox can be solved within the framework of the subjective theory, for the following reason.

The *fundamental postulate of the subjective theory* is the postulate that degrees of the rationality of beliefs in the light of evidence exhibit a *linear order*: that they can be measured, like degrees of temperature, on a one-dimensional scale. But from Peirce to Good, all attempts to solve the problem of the weight of evidence within the framework of the subjective theory proceed by introducing, in addition to probability, *another measure of the rationality of belief in the light of evidence*. Whether this new measure is called 'another dimension of probability', or 'degree of reliability in the light of the evidence', or 'weight of evidence' is quite irrelevant. Relevant is only the implicit admission that it is not possible to attribute linear order to degrees of the rationality of beliefs in the light of the evidence: that there may be *more than one way in which evidence may affect the rationality of a belief*. This admission is sufficient to overthrow the fundamental postulate on which the subjective theory is based.

Thus the naïve belief that there really are intrinsically different kinds of entities, some to be called, perhaps, 'degree of the rationality of belief' and others 'degree of reliability' or of 'evidential support', is no more able to rescue the subjective theory than the equally naïve belief that these various measures 'explicate' different *'explicanda'*; for the claim that there exists an *'explicandum'* here—such as 'degree of rational belief'—capable of 'explication' in terms of probability stands or falls with what I have called the 'fundamental postulate'.

5. All these difficulties disappear as soon as we interpret our

probabilities *objectively*. (It does not matter, in the context of the present paper, whether the objective interpretation is a *purely* statistical interpretation or a propensity interpretation.[3]) According to the objective interpretation, we have to introduce b, the statement of the conditions of the experiment (the conditions defining the sequence of the experiments from which we take our example); for instance, b may be the information: 'the toss in question will be a toss of z, randomized by spinning'. Moreover, we have to introduce the *objective* probabilistic hypothesis h, that is to say, the hypothesis, '$P(a, b) = 1/2$'.[4]

From the point of view of the objective theory, what we are mainly interested in is the hypothesis h, that is to say, the statement

$$'P(a, b) = 1/2'.$$

6. If we now consider the ideally favourable statistical evidence e which led to the 'paradox of ideal evidence', it is quite obvious that from the point of view of the objective theory, e is to be considered as evidence bearing upon h rather than evidence bearing upon a: it is ideally favourable to h, and quite neutral to a. Upon the assumption that the various tosses are *independent, or random*, the objective theory yields for *any* statistical evidence e quite naturally $P(a, be) = P(a, b)$; thus e is indeed irrelevant to a, in the presence of b.

Since e is evidence in favour of h, our problem turns, as a matter of course, into the question of how the evidence e corroborates h (or 'confirms' h). The answer is that if e is ideally favourable evidence, then both $E(h, e)$ and $C(h, e)$, *i.e.* the degree of corroboration of h, given e, will approach 1, if the size of the sample upon which e is based goes to infinity.[5] Thus ideal evidence produces a correspondingly ideal behaviour of E and C. Consequently, no paradox arises; and we may

[3] For the 'propensity interpretation' of probability, see my papers 'Three Views Concerning Human Knowledge'; 'Philosophy of Science: A Personal Report', and 'The Propensity Interpretation of Probability and the Quantum Theory', published, respectively, in *Contemporary British Philosophy*, ed. by H. D. Lewis; in *British Philosophy in the Mid-Century*, ed. by C. A. Mace; and in *Proceedings of the Ninth Symposium of the Colston Research Society*, 1957 (*The Colston Papers*, **9**), ed. by S. Körner.

[4] Note that 'b' may be interpreted, alternatively, not as a name of a statement but as a name of the sequence of tosses—in which case we should have to interpret 'a' as a name of a class of events rather than as a name of a statement; but 'h' remains the name of a statement in any case.

[5] Both E and C are defined in my first note. It is sufficient here to remember that $E(h\ e) = (P(e, h) - (P(e))/(P(e, h) + P(e))$, and that C approaches E in most cases of importance. In this *Journal*, 1954, **5**, 324, I suggested that we define
$$C(x, y, z) = (P(y, xz) - P(y, z))/(P(y, xz) - P(xy, z) + P(y, z)).$$
From this we obtain $C(x, y)$ by assuming z (the 'background knowledge') to be tautological.

quite naturally measure *the weight of the evidence e with respect to the hypothesis h* by either $E(h, e)$, or $C(h, e)$, or else—keeping more closely to Keynes' idea—by the absolute values of either of these functions.

7. If, as in our case, h is a statistical hypothesis, and e the report of the results of statistical tests of h, then $C(h, e)$ is a measure of the degree to which these tests corroborate h, exactly as in the case of a non-statistical hypothesis.

It should be mentioned, however, that as opposed to the case of a non-statistical hypothesis, it might sometimes be quite easy to estimate the numerical values of $E(h, e)$ and even of $C(h, e)$, if h is a statistical hypothesis.[6] (In 8 I will briefly indicate how such numerical calculations might proceed in simple cases, including, of course, the case of $h = $ '$p(a, b) = 1$'.)

The expression

(4) $$P(e, h) - P(e)$$

is crucial for the functions $E(h, e)$ and $C(h, e)$; indeed, these functions are nothing but two different ways of 'normalizing' the expression (4); they thus increase and decrease with (4). This means that in order to find a *good* test-statement e—one which, if true, is highly favourable to h—we must construct a statistical report e such that (i) e makes $P(e, h)$ —which is Fisher's 'likelihood' of h given e—large, *i.e.* nearly equal to 1, and such that (ii) e makes $P(e)$ small, *i.e.* nearly equal to 0. Having constructed a test statement e of this kind, we must submit e itself to empirical tests. (That is to say, we must *try* to find evidence refuting e.)

Now let h be the statement

(5) $$P(a, b) = r$$

and let e be the statement 'In a sample which has the size n and which satisfies the condition b (or which is taken at random from the population b), a is satisfied in $n(r \pm \delta)$ of the instances'.[*1] Then we may put, especially for small values of δ,

(6) $$P(e) \approx 2\delta. \text{[*2]}$$

[6] It is quite likely that in numerically calculable cases, the logarithmic functions suggested by Hamblin and Good (see my 'Second Note') will turn out to be improvements upon the functions which I originally suggested. Moreover, it should be noted that from a numerical point of view (but not from the theoretical point of view underlying our *desiderata*) my functions and the 'degree of factual support' of Kemeny and Oppenheim will in most cases lead to similar results.

[*1] It is here assumed that if the size of the sample is n, the frequency within this sample can be determined at best with an imprecision of $\pm 1/2n$; so that for a finite n, we have $\delta \geqslant 1/2n$. (For large samples, this simply leads to $\delta > 0$.)

[*2] Formula (6) is a direct consequence of the fact that the informative content of a

We may even put $P(e) = 2\delta$; for this would mean that we assign equal probabilities—and therefore, the probabilities $1/n$—to each of the possible proportions, $1/n$, $2/n$, ... n/n, with which a property a may occur in a sample of the size n. It follows that we should have to assign the probability, $P(e) = (2d + 1)/n$, to a statistical report e informing us that $m \pm d$ members of a population of the size n have the property a; so that, by putting $\delta = (d + 1/2)/n$, we obtain $P(e) = 2\delta$. (The equidistribution here described is the one which Laplace assumes in the derivation of his rule of succession. It is adequate for assessing the *absolute* probability, $P(e)$, if e is a *statistical report about a sample*. But it is inadequate for assessing the relative probability $P(e, h)$ of the same report, given a hypothesis h according to which the sample is the product of an n times repeated experiment whose possible results occur each with a certain probability. For in this case, it is adequate to assume a combinatoric, *i.e.* a Bernoullian rather than a Laplacean, distribution.) We see from (6) that, if we wish to make $P(e)$ small, we have to make δ small.

On the other hand, $P(e, h)$—the likelihood of h—will be close to 1 *either* if δ is comparatively large (roughly, if $\delta \approx 1/2$) *or*—in case δ is small —if n, the sample size, is a large number. We therefore find that $P(e, h) - P(e)$, and thus our functions E and C, can only be large if δ is small and n large; or in other words, if e is a *statistical report asserting a good fit in a large sample*.

Thus the test-statement e will be the better the greater its precision (which will be inverse to 2δ) and consequently its refutability or content, and the larger the sample size n, that is to say, the statistical material required for testing e. And the test-statement e so constructed may then be confronted with the results of actual observations.

We see that accumulating statistical evidence will, if favourable, increase E and C. Accordingly, E or C may be taken as measures of the weight of the evidence in favour of h; or else, their absolute values may be taken as measuring the weight of the evidence with respect to h.

8. Since the numerical value of $P(e, h)$ can be determined with the help of the binomial theorem (or of Laplace's integral), and since especially

---

statement increases with its precision, so that its absolute logical probability increases with its degree of imprecision; see sections 34 to 37. (To this we have to add the fact that in the case of a statistical sample, the degree of imprecision and the probability have the same minima and maxima, o and 1.)

for a small δ we can, by (6), put $P(e)$ equal to 2δ, it is possible to calculate $P(e, h) - P(e)$ numerically, and also E.

Moreover, we can calculate for any given n a value $\delta = P(e)/2$ for which $P(e, h) - P(e)$ would become a maximum. (For $n = 1,000,000$, we obtain $\delta = 0.0018$.) Similarly, we can calculate another value, of $\delta = P(e)/2$, for which E would become a maximum. (For the same n, we obtain $\delta = 0.00135$, and $E(h, e) = 0.9946$.)

For a universal law h such that $h = 'P(a, b) = 1'$ which has passed n severe tests, *all* of them with the result a, we obtain, first, $C(h, e) = E(h, e)$, in view of $P(h) = 0$; and further, evaluating $P(e)$ with the help of the Laplacean distribution and $d = 0$, we obtain $C(h, e) = (n - 1)/(n + 1) = 1 - 2/(n + 1)$. It should be remembered, however, that non-statistical scientific theories have as a rule a form totally different from that of the h here described; moreover, if they are forced into this form, then any instance a, and therefore the 'evidence' e, would become essentially non-observational.*³

9. One may see from all this that the testing of a statistical hypothesis is deductive—as is that of all other hypotheses: first a test-statement is constructed in such a way that it follows (or almost follows) from the hypothesis, although its content or testability is high; and afterwards it is confronted with experience.

It is interesting to note that if e were chosen so as to be a full report of our observations—say, a full report about a long sequence of tosses, head, head, tail, . . ., etc., a sequence of one thousand elements—then e would be useless as evidence for a statistical hypothesis; for *any* actual sequence of the length n has the same probability as any other sequence

---

*³ One might, however, speak of the degree of corroboration of a theory *with respect to a field of application*, in the sense of appendices i and *viii; and one might then use the method of calculation here described. But since this method ignores the fine-structure of content and probability, it is very crude, as far as non-statistical theories are concerned. Thus in these cases, we may rely upon the comparative method explained in footnote 7 to the 'First Note' above. It should be stressed that by formulating a theory in the form '$(x)Ax$', we are in general forced to make 'A' a highly complex and non-observational predicate. (See also appendix *vii, especially footnote 1.)

I believe that it is of some interest to mention here that the method developed in the text allows us to obtain *numerical results*—that is, numerical degrees of corroboration—in all cases envisaged either by Laplace or by those modern logicians who introduce artificial language systems, in the vain hope of obtaining in this way an *a priori* metric for the probability of their predicates, believing as they do that this is needed in order to get numerical results. Yet I get numerical degrees of corroboration in many cases far beyond those language systems, since measurable predicates do not create any new problem for our method. (And it is a great advantage that we do not have to introduce a metric for the logical probability of any of the '*predicates*' dealt with; see my criticism in point 3 of the 'Second Note', above. See also my second Preface, 1958.)

(given *h*). Thus we should arrive at the same value for $P(e, h)$, and thus for E and C—*viz.* $E = C = O$—whether *e* contains, say *only* heads, or whether it contains exactly half heads and half tails. This shows that we cannot use, as evidence for or against *h*, our *total* observational knowledge, but that we must extract, from our observational knowledge, such *statistical* statements as can be compared with statements which either follow from *h*, or which have at least a high probability, given *h*. Thus if *e* consists of the complete results of a long sequence of tosses, then *e* is, *in this form*, completely useless as a test-statement of a statistical hypothesis. But a logically weaker statement of the *average frequency* of heads, extracted from the same *e*, could be used. For a probabilistic hypothesis can explain only *statistically interpreted* findings, and it can therefore be tested and corroborated only by statistical abstracts—and not, for example, by the 'total available evidence', if this consists of a full observation report; not even if its various statistical interpretations may be used as excellent and weighty test-statements.*4

Thus our analysis shows that statistical methods are essentially hypothetical-deductive, and that they proceed by the elimination of inadequate hypotheses—as do all other methods of science.

10. If δ is very small, and therefore also $P(e)$—which is possible only for large samples—then we have, in view of (6),

(7) $$P(e, h) \approx P(e, h) - P(e).$$

In this case, and only in this case, it will therefore be possible to accept Fisher's likelihood function as an adequate measure of degree of corroboration. We can interpret, *vice versa*, our measure of degree of

---

*4 This point is of considerable interest in connection with the problem of the numerical value of the absolute probabilities needed for the determination of $C(x, y)$, *i.e.* the problem discussed under point 3 of the 'Second Note', and also in the present note (see especially footnote *1). Had we to determine the absolute probability of the 'total available evidence' consisting of the conjunction of a large number of observational reports, then we should have to know the absolute probability (or 'width') of each of these reports, in order to form their product, under the assumption (discussed in appendix *vii above) of the absolute independence of these reports. But in order to determine the absolute probability of a statistical abstract, we do not have to make any assumptions concerning either the absolute probability of the observational reports or their independence. For it is clear, even without assuming a Laplacean distribution, that (6) must hold for small values of δ, simply because the *content* of *e* must be always a measure of its *precision* (*cf.* section 36), and thus absolute probability must be measured by the *width* of *e*, which is 2δ. The Laplacean distribution, then, may be accepted merely as the simplest equiprobability assumption leading to (6). It may be mentioned, in this context, that the Laplacean distribution may be said to be based upon a *universe of samples*, rather than a universe of things or events. The universe of samples chosen depends, of course, upon the hypothesis to be tested. It is within each universe of samples that an assumption of equiprobability leads to a Laplacean (or 'rectangular') distribution.

corroboration as *a generalization of Fisher's likelihood function*; a generalization which covers cases, such as a comparatively large δ, in which Fisher's likelihood function would become clearly inadequate. For the likelihood of h in the light of the statistical evidence e should certainly not reach a value close to its maximum merely because (or partly because) the available statistical evidence e was lacking in precision.

It is unsatisfactory, not to say paradoxical, that statistical evidence e, based upon a million tosses and δ=0.00135, may result in numerically *the same lik.lihood*—i.e. $P(e, h) = 0.9930$—as would statistical evidence e', based on only a hundred tosses and δ = 0.135.*[5] (But it is quite satisfactory to find that $E(h, e) = 0.9946$ while $E(h, e') = 0.7606$.)

11. It should be noticed that the absolute logical probability of a universal law h—that is, $P(h)$—will be in general zero, in an infinite universe. For this reason, $P(e, h)$—that is, the likelihood of h—will become indefinite, in most systems of probability, since in most systems $P(e, h)$ is defined as $P(eh)/P(h) = 0/0$. We therefore need a formal calculus of probability which yields definite values for $P(e, h)$ even if $P(h) = 0$, and which will always and unambiguously yield $P(e, h) = 1$ whenever e follows (or 'almost follows') from h. A system answering these demands was published by me some time ago.[7]

12. Our $E(h, e)$ may be adequately interpreted as a measure of the explanatory power of h with respect to e, even if e is not a report of genuine and sincere attempts to refute h. But our $C(h, e)$ can be adequately interpreted as degree of corroboration of h—or of the rationality of our belief in h, in the light of tests—only if e consists of reports of the outcome of sincere attempts to refute h, rather than of attempts to verify h.

As hinted in the preceding sentence, I suggest that, while it is a mistake to think that probability may be interpreted as a measure of the rationality of our beliefs (this interpretation is excluded by the

---

*[5] Fisher's 'likelihood' turns out to be in many cases intuitively unsatisfactory. Let x be 'the next throw with this die is a six'. Then the likelihood of x in the light of the evidence y will be 1, and thus at its maximal value, if we take y to mean, for example, 'the next throw is even', or 'the next throw is a number > 4' or even 'the next throw is a number other than 2'. (The values of $C(x, y)$ are satisfactory, it seems: they are, respectively, 3/8; 4/7; and 1/10.)

[7] This *Journal*, 1955, **6**; see esp. 56 sq. A simplified form of this axiom system may be found in my papers 'Philosophy of Science: A Personal Report' (p. 191) and 'The Propensity Interpretation', etc., referred to in note 3 above. (In the latter paper, p. 67, note 3, the last occurrence of '$<$' should be replaced by '$\neq$', and in (B) and (C) a new line should commence after the second arrows.)

paradox of perfect evidence), degree of corroboration may be so interpreted.[8] As to the calculus of probability, it has a very large number of different interpretations.[9] Although 'degree of rational belief' is not among them, there is a *logical interpretation* which takes probability as a generalization of deducibility. But this probability logic has little to do with our hypothetical estimates of chances or of odds; for the probability statements in which we express these estimates are always hypothetical appraisals of the *objective possibilities* inherent in the particular situation—in the objective conditions of the situation, for example in the experimental set-up. These hypothetical estimates (which are *not derivable* from anything else, but freely conjectured, although they may be suggested by symmetry considerations, or by statistical material) can in many important cases be submitted to statistical tests. They are never estimates of our own nescience: the opposite view, as Poincaré saw so clearly, is the consequence of a (possibly unconscious) determinist view of the world.[10]

From this point of view, a 'rational gambler' always tries to estimate the *objective odds*. The odds which he is ready to accept do not represent a measure of his 'degree of belief' (as is usually assumed), but they are, rather, the object of his belief. He believes that there are, objectively, such odds: he believes in a probabilistic hypothesis h. If we wish to measure, behaviouristically, the degree of his belief (in these odds or in anything else) then we might have to find out, perhaps, what proportion of his fortune he is ready to risk on a one-to-one bet that his belief—his estimate of the odds—was correct, provided that this can be ascertained.

As to degree of corroboration, it is nothing but a measure of the degree to which a hypothesis h has been tested, and of the degree to which it has stood up to tests. It must not be interpreted, therefore, as a degree of the rationality of our belief in the *truth* of h; indeed, we know that $C(h, e) = 0$ whenever h is logically true. Rather, it is a measure of the rationality of *accepting*, tentatively, a problematic guess, knowing that it is a guess—but one that has undergone searching examinations.

*13. The foregoing twelve points constitute the 'Third Note', as published in the *B.J.P.S.* Two further remarks may be added, in order

---

[8] *Cf.* this *Journal*, 1955, **55** (the title of section).

[9] *Cf.* my note in *Mind*, 1938, **47**, 275 sq.

[10] *Cf.* H. Poincaré, *Science and Method*, 1914, IV, I. (This chapter was first published in *La Revue du mois*, 1907, **3**, pp. 257–276, and in *The Monist*, 1912, **22**, pp. 31–52.)

to make more explicit some of the more formal considerations which are implicit in this note.

The first problem I have in mind is, again, that of the *metric* of logical probability (*cf.* the second note, point 3), and its relation to the distinction between what I am going to call primary and secondary probability statements. My thesis is that on the secondary level, Laplace's and Bernoulli's distribution provide us with a *metric*.

We may operate with a system $S_1 = \{\, a, b, c, a_1, b_1, c_1, \dots \,\}$ of elements (in the sense of our system of postulates in appendix *iv). These elements will give rise to probability statements of the form '$p(a, b) = r$'. We may call them 'primary probability statements'. These primary probability statements may now be considered as the elements of a secondary system of elements, $S_2 = \{\, e, f, g, h, \dots \,\}$; so that '$e$', '$f$', etc., are now names of statements of the form '$p(a, b) = r$'.

Now Bernoulli's theorem tells us, roughly, the following: let h be '$p(a, b) = r$'; then if h is true, it is extremely probable that in a long sequence of repetitions of the conditions b, the frequency of the occurrence of a will be equal to r, or very nearly so. Let '$\delta_r(a)_n$' denote the statement that a will occur in a long sequence of n repetitions with a frequency $r \pm \delta$. Then Bernoulli's theorem says that the probability of $\delta_r(a)_n$ will approach 1, with growing n, given h, *i.e.* given that $p(a, b) = r$. (It also says that this probability will approach 0, given that $p(a, b) = s$, whenever s falls outside $r \pm \delta$; which is important for the refutation of probabilistic hypotheses.)

Now this means that we may write Bernoulli's theorem in the form of a (secondary) statement of *relative* probability about elements g and h of S_2; that is to say, we can write it in the form

$$\lim_{n \to \infty} p(g, h) = 1$$

where $g = \delta_r(a)_n$ and where h is the information that $p(a, b) = r$; that is to say, h is a primary probability statement and g is a primary statement of *relative frequency*.

These considerations show that we have to admit, in S_2, *frequency statements* such as g, that is to say, $\delta_r(a)_n$, and probabilistic assumptions, or hypothetical probabilistic estimates, such as h. It seems for this reason proper, in the interest of a homogeneous S_2, to identify *all* the probability statements which form the elements of S_2, with *frequency statements*, or in other words, to assume, for the *primary* probability statements $e, f, g, h, \dots$ which form the elements of S_2, some kind of *frequency interpretation of probability*. At the same time, we may assume the *logical interpretation of probability* for the probability statements of the form

$$P(g, h) = r$$

that is to say, for the *secondary* probability statements which make assertions about the degree of probability of the primary probability statements, g and h.

Although we may not have a logical (or absolute) metric of the primary probability statements, that is to say, although we may have no idea of the value of $p(a)$ or of $p(b)$, we may have a logical or absolute metric of the secondary probability statements: this is provided by the Laplacean distribution, according to which $P(g)$, the absolute probability of g, that is to say of $\delta,(a)_n$, equals 2δ, whether g is empirically observed, or a hypothesis; so that the typical probabilistic hypothesis, h, gets $P(h) = 0$, because h has the form '$p(a, b) = r$', with $\delta = 0$. Since Bernoulli's methods allow us to calculate the value of the relative probability $P(g, h)$, by pure mathematical analysis, we may consider the relative probabilities $P(g, h)$ as likewise determined on purely logical grounds. It therefore seems entirely adequate to adopt, on the secondary level, the logical interpretation of the formal calculus of probability.

To sum up, we may consider the methods of Bernoulli and Laplace as directed towards the establishment of a purely logical metric of probabilities on the secondary level, independently of whether or not there exists a logical metric of probabilities on the primary level. Bernoulli's methods determine thereby the logical metric of relative probabilities (secondary likelihood of primary hypotheses, in the main), and Laplace's the logical metric of absolute probabilities (of statistical reports upon samples, in the main).

Their efforts were, no doubt, directed to a large extent towards establishing a probabilistic theory of induction; they certainly tended to identify C with p. I need not say that I believe they were mistaken in this: statistical theories are, like all other theories, hypothetico-deductive. And statistical hypotheses are tested, like all other hypotheses, by attempts to falsify them—by attempts to reduce their secondary likelihood to zero, or to almost zero. Their 'degree of corroboration', C, is of interest only if it is the result of such tests; for nothing is easier than to select statistical evidence so that it is *favourable* to a statistical hypothesis—if we wish to do so.

*14. It might well be asked at the end of all this whether I have not, inadvertently, changed my creed. For it may seem that there is nothing to prevent us from calling $C(h, e)$ 'the inductive probability of h, given e' or—if this is felt to be misleading, in view of the fact that C does not obey the laws of the probability calculus—'the degree of the rationality of our belief in h, given e'. A benevolent inductivist critic might even congratulate me on having solved, with my C function, the age-old problem of induction *in a positive sense*—on having finally established, with my C function, the validity of inductive reasoning.

My reply would be as follows. I do not object to calling $C(h, e)$ by any name whatsoever, suitable or unsuitable: I am quite indifferent to terminology, so long as it does not mislead us. Nor do I object— so long as it does not mislead us—to an extension (inadvertent or otherwise) of the meaning of 'induction'. But I must insist that $C(h, e)$ can be interpreted as degree of corroboration only if e is *a report on the severest tests we have been able to design.* It is this point that marks the difference between the attitude of the inductivist, or verificationist, and my own attitude. The inductivist or verificationist wants *affirmation* for his hypothesis. He hopes to make it *firmer* by his evidence e and he looks out for '*firmness*'—for '*confirmation*'. At best, he may realize that we must not be biased in our selection of e: that we must not ignore unfavourable cases; and that e must comprise reports on our *total* observational knowledge, whether favourable or unfavourable. (Note that the inductivist's requirement that e must comprise our *total* observational knowledge cannot be represented in any formalism. It is a non-formal requirement, a condition of adequacy which must be satisfied if we wish to *interpret* $p(h, e)$ as *degree of our imperfect knowledge* of h.)

In opposition to this inductivist attitude, I assert that $C(h, e)$ must not be interpreted as the degree of corroboration of h by e, unless e reports the results of *our sincere efforts to overthrow h.* The requirement of sincerity cannot be formalized—no more than the inductivist requirement that e must represent our total observational knowledge. Yet if e is *not* a report about the results of our sincere attempts to overthrow h, then we shall simply deceive ourselves if we think we can interpret $C(h, e)$ as degree of corroboration, or anything like it.

My benevolent critic might reply that he can still see no reason why my C function should not be regarded as a positive solution to the classical problem of induction. For my reply, he might say, should be perfectly acceptable to the classical inductivist, seeing that it merely consists in an exposition of the so-called 'method of eliminative induction'—an inductive method which was well known to Bacon, Whewell, and Mill, and which is not yet forgotten even by some of the probability theorists of induction (though my critic may well admit that the latter were unable to incorporate it effectively into their theories).

My reaction to this reply would be regret at my continued failure to explain my main point with sufficient clarity. For the sole purpose of the elimination advocated by all these inductivists was to *establish as firmly as possible the surviving theory* which, they thought, must be the *true* one (or perhaps only a *highly probable* one, in so far as we may not have fully succeeded in eliminating every theory except the true one).

As against this, I do not think that we can ever seriously reduce, by elimination, the number of the competing theories, since this number remains always infinite. What we do—or should do—is *hold on to the most improbable of the surviving theories* which is the one that can be most severely tested. We tentatively '*accept*' this theory— but only in the sense that we select it as worthy to be subjected to further criticism, and to the severest tests we can design.

On the positive side, we may be entitled to add that the surviving theory is the best theory—and the best tested theory—of which we know.

*Appendix *x.   Universals, Dispositions, and Natural or Physical Necessity.*

(1) The fundamental doctrine which underlies all theories of induction is *the doctrine of the primacy of repetitions*. Keeping Hume's attitude in mind, we may distinguish two variants of this doctrine. The first (which Hume criticized) may be called the doctrine of the logical primacy of repetitions. According to this doctrine, repeated instances furnish a kind of *justification* for the acceptance of a universal law. (The idea of repetition is linked here, as a rule, with that of probability.) The second (which Hume upheld) may be called the doctrine of the temporal (and psychological) primacy of repetitions. According to this second doctrine, repetitions, even though they should fail to furnish any kind of *justification* for a universal law and for the expectations and beliefs which it entails, nevertheless induce and *arouse* these expectations and beliefs in us, as a matter of fact—however little 'justified' or 'rational' this fact (or these beliefs) may be.

Both variants of this doctrine of the primacy of repetitions, the stronger doctrine of their logical primacy and the weaker doctrine of their temporal (or causal or psychological) primacy, are untenable. This may be shown with the help of two entirely different arguments.

My first argument against the primacy of repetitions is the following. All the repetitions which we experience are *approximate repetitions*; and by saying that a repetition is approximate I mean that the repetition B of an event A is not identical with A, or indistinguishable from A, but only *more or less similar* to A. But if repetition is thus based upon mere similarity, it must share one of the main characteristics of similarity; that is, its relativity. Two things which are similar are

420

always similar *in certain respects*. The point may be illustrated by a simple diagram.

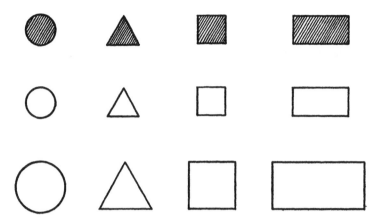

If we look at this diagram, we find that some of the figures are similar with respect to shading (hatching) or to its absence; others are similar with respect to shape; and others are similar with respect to size. The table might be extended like this.

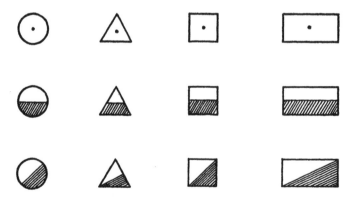

One can easily see that there is no end to the possible kinds of similarity.

These diagrams show that things may be similar in *different respects*, and that any two things which are from one point of view similar may be dissimilar from another point of view. Generally, similarity, and with it repetition, always presuppose the adoption of *a point of view*:

some similarities or repetitions will strike us if we are interested in one problem, and others if we are interested in another problem. But if similarity and repetition presuppose the adoption of a point of view, or an interest, or an expectation, it is logically necessary that points of view, or interests, or expectations, are logically prior, as well as temporally (or causally or psychologically) prior, to repetition. But this result destroys both the doctrines of the logical and of the temporal primacy of repetitions.[1]

The remark may be added that for any given finite group or set of things, however variously they may be chosen, we can, with a little ingenuity, find always points of view such that all the things belonging to that set are similar (or partially equal) if considered from one of these points of view; which means that anything can be said to be a 'repetition' of anything, if only we adopt the appropriate point of view. This shows how naïve it is to look upon repetition as something ultimate, or given. The point here made is closely related to the fact (mentioned in appendix *vii, footnote 9; cf. the property B) that we can find, for any given finite sequence of noughts and ones, a mathematical rule or 'law' for constructing an infinite sequence such that it commences with the given finite sequence.

I now come to my second argument against the primacy of repetitions. It is this. There are laws and theories of a character altogether different from 'All swans are white', even though they may be formulated in a similar way. Take ancient atomic theory. Admittedly, it may be expressed (in one of its simplest forms) as 'All material bodies are composed of corpuscles'. Yet clearly, the 'all'-form is comparatively unimportant in the case of this law. What I mean is this. The problem of showing that one single physical body—say, a piece of· iron—is composed of atoms or 'corpuscles' is at least as difficult as that of showing that all swans are white. Our assertions transcend, in both cases, all observational experience. It is the same with almost all scientific theories. We cannot show, directly, even of one physical body that, in the absence of forces, it moves along a straight line; or that it attracts, and is attracted by, one other physical body in accordance with the inverse square law. All these theories describe what we may call *structural properties of the world*; and they all transcend all

[1] Some illustrations of this argument, so far as it is directed against the doctrine of temporal primacy of repetitions (that is, against Hume) may be found in sections iv and v of my paper 'Philosophy of Science: A Personal Report', in *British Philosophy in the Mid-Century*, ed. by C. A. Mace, 1957.

possible experience. The difficulty with these structural theories is not so much to establish the universality of the law from repeated instances as to establish that it holds even for one single instance.

This difficulty has been seen by many inductivists. Most of those who saw it tried, like Berkeley, to make a sharp distnction between pure observational generalizations and more 'abstract' or 'occult' theories, such as the corpuscular theory, or Newton's theory; and they tried, as a rule, to resolve the problem by saying, as did Berkeley, that abstract theories are not genuine assertions about the world, but that they are *nothing but instruments*—instruments for the prediction of observable phenomena. I have called this view '*instrumentalism*', and I have criticized it in some detail elsewhere.[2] Here I will only say that I reject instrumentalism, and I will give only one reason for rejecting it: that it does not solve the problem of the 'abstract' or 'occult' or 'structural' properties. For such properties do not only occur in the 'abstract' theories which Berkeley and his successors had in mind. They are mentioned all the time, by everybody, and in ordinary speech. Almost every statement we make transcends experience. There is no sharp dividing line between an 'empirical language' and a 'theoretical language': *we are theorizing all the time*, even when we make the most trivial singular statement. With this remark, we have arrived at the main problem which I intend to examine in this appendix.

(2) Admittedly, if we say 'All swans are white', then the whiteness we predicate is an observable property; and to this extent, a singular statement such as 'This swan here is white' may be said to be based on observation. Yet it transcends experience—not because of the word 'white', but because of the word 'swan'. For by calling something a 'swan', we attribute to it properties which go far beyond mere observation—almost as far as when we assert that it is composed of 'corpuscles'.

Thus not only the more abstract explanatory theories transcend experience, but even the most ordinary singular statements. For even ordinary singular statements are always *interpretations of 'the facts' in the light of theories*. (And the same holds even for 'the facts' of the case. They contain *universals*; and universals always entail a *law-like* behaviour.)

---

[2] *Cf.* my papers 'A Note on Berkeley as a Precursor of Mach', *B.J.P.S.* **4**, 1953, and 'Three Views Concerning Human Knowledge' in *Contemporary British Philosophy* iii, ed. by H. D. Lewis, 1956. See also sections *11 to *15 of my *Postscript*.

I explained briefly at the end of section 25 how it is that the use of universals such as 'glass' or 'water', in a statement like 'here is a glass of water', necessarily transcends experience. It is due to the fact that words like 'glass' or 'water' are used to characterize the *law-like behaviour* of certain things; which may be expressed by calling them 'dispositional words'. Now since every law transcends experience—which is merely another way of saying that it is not verifiable—every predicate expressing law-like behaviour transcends experience also: this is why the statement 'this container contains water' is a testable but non-verifiable hypothesis, transcending experience.[3] It is for this reason impossible to 'constitute' any true universal term (as Carnap tried to do) that is to say, to define it in purely experimental or observational terms—or to 'reduce' it to purely experiential or observational terms: since *all universals are dispositional*, they cannot be reduced to experience. We must introduce them as undefined terms, except those which we may define in terms of other non-experiential universals (such as 'water' if we choose to define it as 'a compound of two atoms of hydrogen and one of oxygen').

(3) That *all* universals are dispositional is often overlooked, owing to the fact that universals can be dispositional in varying degrees. Thus 'soluble' or 'breakable' are clearly dispositional in a higher degree than 'dissolved' or 'broken'. But it is sometimes not realized that even 'dissolved' and 'broken' are dispositional. A chemist would not say that sugar or salt has *dissolved* in water if he did not expect that he could get the sugar or the salt back, by evaporating the water. Thus 'dissolved' denotes a dispositional state. And as to 'broken', we need only consider how we proceed *if we are in doubt* whether or not a thing is broken—something we have dropped, perhaps, or say, a bone in our body: we test the behaviour of the thing in question, trying to find out whether it does not show a certain undue mobility. Thus 'broken', like 'dissolved', describes dispositions to behave in a certain regular or law-like manner. Similarly, we say of a surface that it is red, or white, if it has the disposition to reflect red, or white, light, and consequently the

---

[3] Since it is a singular statement, it is less incorrect to speak here of a symmetry between non-verifiability and non-falsifiability than in a case of universal statements; for in order to falsify it, we have to accept another singular statement, similarly non-verifiable, as true. But even here, a certain asymmetry remains. For quite generally in assuming the truth, or the falsity, of some test-statement, we can only establish the *falsity* of the statement under test, but not its truth. The reason is that the latter entails an infinite number of test statements. See also section 29 of the book, and section *22 of my *Postscript*.

disposition to look in daylight red, or white. In general, the dispositional character of any universal property will become clear if we consider what tests we should undertake if we are in doubt whether or not the property is present in some particular case.

Thus the attempt to distinguish between dispositional and non-dispositional predicates is mistaken, just as is the attempt to distinguish between theoretical terms (or languages) and non-theoretical or empirical or observational or factual or ordinary terms (or languages). It is, perhaps, somewhat like this. What people have learnt before reaching a certain critical age, they are inclined to regard as factual, or 'ordinary', and what they hear later, as theoretical or perhaps as 'merely instrumental'. (The critical age seems to depend on the psychological type.)

(4) Universal laws transcend experience, if only because they are universal and thus transcend any finite number of their observable instances; and singular statements transcend experience because the universal terms which normally occur in them entail dispositions to behave in a law-like manner, so that they entail universal laws (of some lower order of universality, as a rule). Accordingly, universal laws transcend experience in at least two ways: because of their universality, and because of the occurrence of universal or dispositional terms in them. And they transcend experience in a higher degree if the dispositional terms which occur in them are dispositional in a higher degree or more abstract. There are layers of higher and higher degrees of universality, and thus of transcendence. (In section *15 of the *Postscript*, an attempt is made to explain the sense in which these are also layers of what may be called 'depth'.)

It is of course because of this transcendence that scientific laws or theories are non-verifiable, and that *testability* or *refutability* is the only thing that distinguishes them, in general, from metaphysical theories.

If it is asked why we use these transcendent universal laws instead of keeping more closely to 'experience', two kinds of answer may be given.

(a) Because we need them: because there is no such thing as 'pure experience', but only experience interpreted in the light of expectations or theories which are 'transcendent'.

(b) Because a theorist is a man who *wishes to explain* experiences, and because explanation involves the use of explanatory hypotheses which (in order to be independently testable; see section *15 of the *Postscript*) must transcend what we hope to explain.

The reason given under (a) is a pragmatic or instrumentalist one, and although I believe that it is true, I do not think that it is comparable in importance with the reason given under (b); for even if a programme of eliminating explanatory theories for practical purposes (say, for prediction) were to succeed, the aim of the theorist would be un-affected.[4]

(5) That theories transcend experience in the sense here indicated was asserted in many places in the book. At the same time, theories were described as strictly universal statements.

A most penetrating criticism of the view that theories, or laws of nature, can be adequately expressed by a universal statement, such as 'All planets move in ellipses', has been advanced by William Kneale. I have found Kneale's criticism difficult to understand. Even now I am not entirely sure whether I understand him properly; but I hope I do.[5]

I believe that Kneale's point can be put as follows. Although universal statements are *entailed* by statements of natural law, the latter are logically stronger than the former. They do not only assert 'All planets move in ellipses', but rather something like 'All planets move *necessarily* in ellipses.' Kneale calls a statement of this kind a 'principle of necessitation'. I do not think that he succeeds in making quite clear what the difference is between a universal statement and a 'principle of necessitation'. He speaks of 'the need for a more precise formulation of the

---

[4] That it is possible to do without theories is asserted by Carnap, *Logical Foundations of Probability*, p. 574 f. Yet there is no reason whatever for the belief that Carnap's analysis, even if it were otherwise defensible, could be legitimately transferred from his model language to 'the language of science'; see my *Preface*, 1958. In two very interesting articles W. Craig has discussed certain reduction programmes. (See *Journal of Symb. Logic* **18**, 1953, pp. 30 f., and *Philosophical Review* **65**, 1956, pp. 38 ff.) To his own excellent critical comments on his method of eliminating 'auxiliary' (or 'trans-cendent') ideas, the following might be added. (i) He achieves the elimination of explanatory theories, essentially, by promoting infinitely many theorems to the rank of axioms (or by replacing the definition of 'theorem' by a new definition of 'axiom' which is co-extensive with it as far as the 'purified' sub-language goes). (ii) In the actual construction of the purified system, he is of course *guided by our knowledge of the theories* to be eliminated. (iii) The purified system is no longer an explanatory system, and no longer testable in the sense in which explanatory systems may be testable whose testa-bility is, essentially, related to their informative *content* and *depth*. (One might well say that the axioms of the purified system have zero depth in the sense of section *15 of my *Postscript*.)

[5] *Cf.* William Kneale, *Probability and Induction*, 1949. One of my minor difficulties in understanding Kneale's criticism was connected with the fact that he gives in some places very good outlines of some of my views, while in others he seems to miss my point completely. (See for example note 17, below.)

notions of contingency and necessity'.[6] But a little later, one reads to one's surprise: 'In fact, the word "necessity" is the least troublesome of those with which we have to deal in this part of philosophy.'[7] Admittedly, between these two passages, Kneale tries to persuade us that 'the sense of this distinction'—the distinction between contingency and necessity—'can be easily understood from examples'.[8] But I found his examples perplexing. Always assuming that I have succeeded in my endeavours to understand Kneale, I must say that his positive theory of natural laws seems to me utterly unacceptable. Yet his criticism seems to me most valuable.

(6) I am now going to explain, with the help of an example, what I believe to be essentially Kneale's criticism of the view that a characterization of laws of nature as universal statements is *logically sufficient* and also *intuitively adequate*.

Consider some extinct animal, say the moa, a huge bird whose bones abound in some New Zealand swamps. (I have there dug for them myself.) We decide to use the name 'moa' as a universal name (rather than as a proper name; cf. section 14) of a certain biological structure; but we ought to admit that it is of course quite possible—and even quite credible—that no moas have ever existed in the universe, or will ever exist, apart from those which once lived in New Zealand; and we will assume that this credible view is correct.

Now let us assume that the biological structure of the moa organism is of such a kind that under very favourable conditions, a moa might easily live sixty years or longer. Let us further assume that the conditions met by the moa in New Zealand were far from ideal (owing, perhaps, to the presence of some virus), and that no moa ever reached the age of fifty. In this case, the strictly universal statement 'All moas die before reaching the age of fifty' will be true; for according to our assumption, there never is, was, or will be a moa in the universe more than fifty years of age. At the same time, this universal statement will not be a law of nature; for according to our assumptions, it would be *possible* for a moa to live longer, and it is only due to *accidental or contingent*

---

[6] *Op. cit.*, p. 32.
[7] *Op. cit.*, p. 80.
[8] *Op. cit.*, p. 32. One of the difficulties is that Kneale at times seems to accept Leibniz's view ('A truth is necessary when its negation implies a contradiction; and when it is not necessary, it is called contingent.' *Die philosophischen Schriften*, ed. by Gerhardt, 3, p. 400; see also 7, p. 390 *ff.*), while at other times he seems to use 'necessary' in a wider sense.

conditions—such as the co-presence of a certain virus—that in fact no moa did live longer.

The example shows that there may be *true, strictly universal statements* which have an accidental character rather than the character of true universal laws of nature. Accordingly, the characterization of laws of nature as strictly universal statements is logically insufficient and intuitively inadequate.

(7) The example may also indicate in what sense natural laws may be described as 'principles of necessity' or 'principles of impossibility', as Kneale suggests. For according to our assumptions—assumptions which are perfectly reasonable—it would be *possible*, under favourable conditions, for a moa to reach a greater age than any moa has actually reached. But were there a natural law limiting the age of any moa-like organism to fifty years, *then it would not be possible* for any moa to live longer than this. Thus natural laws set certain limits to what is possible.

I think that all this is intuitively acceptable; in fact, when I said, in several places in my book, that natural laws *forbid* certain events to happen, or that they have the character of *prohibitions*, I gave expression to the same intuitive idea. And I think that it is quite possible and perhaps even useful to speak of 'natural necessity' or of 'physical necessity', in order to describe this character of natural laws, and of their logical consequences.

(8) But I think it is a mistake to underrate the differences between this natural or physical necessity, and other kinds of necessity, for example, logical necessity. We may, roughly, describe as logically necessary what would hold in any conceivable world. But although Newton's inverse square law may conceivably be a true law of nature in some world, and to that extent naturally necessary in that world, a world in which it is not valid is perfectly *conceivable*.

Kneale has criticized this kind of argument by pointing out that Goldbach's conjecture (according to which any even number greater than two is the sum of two primes) may *conceivably* be true, or conceivably be false, even though it may well be demonstrable (or refutable), and in this sense mathematically or logically necessary (or impossible); and he argues that 'the conceivability of the contradictory is not to be taken as a disproof of necessity in mathematics'.[9] But if so, why, he asks, 'should it be supposed to furnish . . . a disproof in natural

---

[9] *Op. cit.*, p. 80.

science?'[10] Now I think that this argument lays too much stress on the *word* 'conceivable'; moreover, it operates with a sense of 'conceivable' different from the one intended: once we have a proof of Goldbach's theorem, we may say that this proof establishes precisely that an even number (greater than two) which is not the sum of two primes is inconceivable—in the sense that it leads to inconsistent results: to the assertion, among others, that $0 = 1$, which is 'inconceivable'. In another sense, $0 = 1$ may be quite conceivable: it may even be used, like any other mathematically false statement, as an assumption in an indirect proof. Indeed, an indirect proof may well be put thus: '*Conceive* that *a* is true. Then we should have to admit that *b* is true. But we know that *b* is absurd. Thus it is *inconceivable* that *a* is true.' It is clear that although this use of 'conceivable' and 'inconceivable' is a little vague and ambiguous, it would be misleading to say that this way of arguing must be invalid since the truth of *a* cannot be inconceivable, considering that we did start by conceiving, precisely, the truth of *a*.

Thus 'inconceivable' in logic and mathematics is simply another word for 'leading to an obvious contradiction'. *Logically* possible or 'conceivable' is everything that does not lead to an obvious contradiction, and logically impossible or 'inconceivable' is everything that does. When Kneale says that the contradictory of a theorem may be 'conceivable', he uses the word in another sense—and in a very good sense too.

(9) Thus an assumption is logically possible if it is not self-contradictory; it is physically possible if it does not contradict the laws of nature. The two meanings of 'possible' have enough in common to explain why we use the same word; but to gloss over their difference can only lead to confusion.

Compared with logical tautologies, laws of nature have a contingent, an accidental character. This is clearly recognized by Leibniz who teaches (*cf. Philos. Schriften*, Gerhardt, **7**, p. 390) that the world is the work of God, in a sense somewhat similar to that in which a sonnet, or a rondeau, or a sonata, or a fugue, is the work of an artist. The artist may freely choose a certain *form*, voluntarily restricting his freedom by this choice: he imposes certain principles of impossibility upon his creation, for example upon his rhythm, and, to a lesser extent, his words which, as compared to the rhythm, may appear contingent, accidental. But this does not mean that his choice of

[10] *Ibid.*

form, or of rhythm, was not contingent also. For another form or rhythm could have been chosen.

Similarly with natural laws. They restrict the (logically) possible choice of singular facts. They are thus principles of impossibility with respect to these singular facts; and the singular facts seem highly contingent as compared with the natural laws. But the natural laws, though necessary as compared with singular facts, are contingent as compared with logical tautologies. For there may be *structurally different worlds*—worlds with different natural laws.

Thus natural necessity or impossibility is like musical necessity or impossibility. It is like the impossibility of a four-beat rhythm in a classical minuet, or the impossibility of ending it on a diminished seventh or some other dissonance. It imposes *structural* principles upon the world. But it still leaves a great deal of freedom to the more contingent singular facts—the initial conditions.

If we compare the situation in music with that of our example of the moa, we can say: there is no musical law prohibiting the writing of a minuet in G flat minor, but it is nevertheless quite possible that no minuet has ever been, or will ever be, written in this unusual key. Thus we can say that musically necessary laws may be distinguished from true universal statements about the historical facts of musical composition.

(10) The opposite view—the view that natural laws are in no sense contingent—seems to be the one which Kneale is advancing, if I understand him well. To me it seems quite as mistaken as the view which he justly criticizes—the view that laws of nature are nothing but true universal statements.

Kneale's view that laws of nature are necessary in the same sense in which logical tautologies are necessary may perhaps be expressed in religious terms thus: God may have faced the choice between creating a physical world and not creating a physical world, but once this choice was made, He was no longer free to choose the form, or the structure of the world; for since this structure—the regularities of nature, described by the laws of nature—is necessarily what it is, all He could freely choose were the initial conditions.

It seems to me that Descartes held a view very similar to this. According to him, all the laws of nature follow with necessity from the one analytic principle (the essential definition of 'body') according to which 'to be a body' means the same as 'to be extended'; which is taken to imply that two *different* bodies cannot take up the same extension, or space. (Indeed, this principle is similar to Kneale's standard example—'that nothing which is red is also green'.[11]) But it is by going beyond these 'truisms' (as Kneale calls them, stressing their similarity to logical tautologies[12]) that, beginning with Newton, physical theory has reached a depth of insight utterly beyond the Cartesian approach.

It seems to me that the doctrine that the laws of nature are *in no sense contingent* is a particularly severe form of a view which I have described and criticized elsewhere as 'essentialism'.[13] For it entails the doctrine of the existence of *ultimate explanations*; that is to say, of the existence of explanatory theories which in their turn are neither in need of any further explanation nor capable of being further explained. For should we succeed in the task of reducing all the laws of nature to the true 'principles of necessitation'—to truisms, such as that two essentially extended things cannot take up the same extension, or that nothing which is red is also green—further explanation would become both unnecessary and impossible.

I see no reason to believe that the doctrine of the existence of ultimate explanations is true, and many reasons to believe that it is false. The more we learn about theories, or laws of nature, the less do they remind us of Cartesian self-explanatory truisms or of essentialist definitions. It is not truisms which science unveils. Rather, it is part of the greatness and the beauty of science that we can learn, through our own critical investigations, that the world is utterly different from what we ever imagined—until our imagination was fired by the refutation of our earlier theories. There does not seem any reason to think that this process will come to an end.[14]

All this receives the strongest support from our considerations about content and (absolute) logical probability. If laws of nature are not merely strictly universal statements, they must be *logically stronger* than the corresponding universal statements, since the latter must be deducible from them. But the *logical necessity* of a, as we have seen (at the end of appendix *v) can be defined by the definiens

[11] *Cf.* Kneale, *op. cit.*, p. 32; see also, for example, p. 80.
[12] *Op. cit.*, p. 33.
[13] *Cf.* my *Poverty of Historicism*, section 10; *The Open Society*, chapter 3, section vi; chapter 11; 'Three Views Concerning Human Knowledge' (*Contemporary British Philosophy* 3) and my *Postscript*, for example sections *15 and *31.
[14] *Cf.* my *Postscript*, especially section *15.

$$p(a) = p(a, \bar{a}) = 1.$$

For universal statements a, on the other hand, we obtain (*cf.* the same appendix and appendices *vii and *viii):

$$p(a) = p(a, \bar{a}) = 0;$$

and the same must hold for any logically stronger statement. Accordingly, a law of nature is, by its great content, as far removed from a logically necessary statement as a consistent statement can be; and it is much nearer, in its logical import, to a 'merely accidentally' universal statement than to a logical truism.

(11) The upshot of this discussion is that I am prepared to accept Kneale's criticism in so far as I am prepared to accept the view that there exists a category of statements, the laws of nature, which are logically stronger than the corresponding universal statements. This doctrine is, in my opinion, incompatible with any theory of induction. To my own methodology it makes little or no difference, however. On the contrary, it is quite clear that a proposed or conjectured principle which declares the impossibility of certain events would have to be tested by trying to show that these events are possible; that is to say, by trying to bring them about. But this is precisely the method of testing which I advocate.

Thus from the point of view here adopted, no change whatever is needed, as far as methodology is concerned. The change is entirely on an ontological, a metaphysical level. It may be described by saying that if we conjecture that a is a natural law, we conjecture that a expresses a *structural property of our world*; a property which prevents the occurrence of certain logically possible singular events, or states of affairs of a certain kind—very much as explained in sections 21 to 23 of the book, and also in sections 79, 83, and 85.

(12) As Tarski has shown, it is possible to explain *logical necessity* in terms of universality: a statement may be said to be logically necessary if and only if it is deducible (for example, by particularization) from a '*universally valid*' statement function; that is to say, from a statement function which is *satisfied by every model*.[15] (This means, true in all possible worlds.)

I think that we may explain by the same method what we mean by *natural necessity*; for we may adopt the following definition.

[15] *Cf.* my 'Note on Tarski's Definition of Truth', *Mind* **64**, 1955, especially p. 391.

A statement may be said to be naturally or physically necessary if, and only if, it is deducible from a statement function which is satisfied in all worlds that differ from our world, if at all, only with respect to initial conditions.

We can never *know*, of course, whether a supposed law is a genuine law or whether it only looks like a law but depends, in fact, upon certain special initial conditions prevailing in our region of the universe. (*Cf.* section 79.) We cannot, therefore ever find out of any given non-logical statement that it is in fact naturally necessary: the conjecture that it is remains a conjecture for ever (not merely because we cannot search our whole world in order to ensure that no counter instance exists, but for the even stronger reason that we cannot search all worlds that differ from ours with respect to initial conditions.) But although our proposed definition excludes the possibility of obtaining *a positive criterion* of natural necessity, we can in practice apply our definition of natural necessity in a *negative* way: by finding initial conditions under which the supposed law turns out to be invalid, we can show that it was not necessary; that is to say, not a law of nature. Thus the proposed definition fits our methodology very well indeed.

The proposed definition would, of course, make all laws of nature, together with all their logical consequences, *naturally or physically necessary.*[16]

It will be seen at once that the proposed definition is in perfect agreement with the results reached in our discussion of the moa example (*cf.* points 6 and 7 above): it was precisely because we thought that moas would live longer under different conditions—under more favourable ones—that we felt that a true universal statement about their actual maximal age was of an accidental character.

(13) We now introduce the symbol '*N*' as a name of the class of statements which are necessarily true, in the sense of natural or physical necessity; that is to say, true whatever the initial conditions may be.

With the help of '*N*', we can define '$a \vec{N} b$' (or in words, 'If a then necessarily b') by the following somewhat obvious definition:

(D) $a \vec{N} b$ is true if, and only if, $a \rightarrow b \in N$.

In words, perhaps: 'If a then necessarily b' holds if, and only if, 'If a then b' is necessarily true. Here '$a \rightarrow b$' is, of course, the name of an

---

[16] Incidentally, logically necessary statements would (simply because they follow from any statement) become physically necessary also; but this does not matter, of course.

ordinary conditional with the antecedent a and the consequent b. If it were our intention to define logical entailment or 'strict implication', then we could also use (D), but we should have to interpret 'N' as 'logically necessary' (rather than as 'naturally or physically necessary').

Owing to the definition (D), we can say of '$a \xrightarrow{N} b$' that it is the name of a statement with the following properties.

(A) $a \xrightarrow{N} b$ is not always true if a is false, in contradistinction to $a \rightarrow b$.

(B) $a \xrightarrow{N} b$ is not always true if b is true, in contradistinction to $a \rightarrow b$.

(A') $a \xrightarrow{N} b$ is always true if a is impossible or necessarily false, or if its negation, $\bar{a}$, is necessarily true (whether by logical or by physical necessity).

(B') $a \xrightarrow{N} b$ is always true if b is necessarily true (whether by logical or physical necessity).

Here a and b may be either statements or statement functions.

$a \xrightarrow{N} b$ may be called a 'necessary conditional' or a 'nomic conditional'. It expresses, I believe, what some authors have called 'subjunctive conditionals', or 'counterfactual conditionals'. (It seems, however, that other authors—for example Kneale—meant something else by a 'counterfactual conditional': they took this name to imply that a is, in fact, false.[17] I do not think that this usage is to be recommended.)

A little reflection will show that the class N of naturally necessary statements comprises not only the class of all those statements which, like true universal laws of nature, can be intuitively described as being unaffected by changes of initial conditions, but also all those statements which follow from true universal laws of nature, or from the true structural theories about the world. There will be statements among

---

[17] In my 'Note on Natural Laws and so-called Contrary-to-Fact Conditionals' (*Mind* 58, N.S., 1949, pp. 62–66) I used the term 'subjunctive conditional' for what I here call 'necessary' or 'nomic conditional'; and I explained repeatedly that these subjunctive conditionals must be deducible from natural laws. It is therefore difficult to understand how Kneale (*Analysis* 10, 1950, p. 122) could attribute to me even tentatively the view that a subjunctive conditional or a 'contrary to fact conditional' was of the form '$\sim\phi(a).(\phi(a) \supset \psi(a))$'. I wonder whether Kneale realized that this expression of his was only a complicated way of saying '$\sim\phi(a)$'; for who would ever think of asserting that '$\sim\phi(a)$' was deducible from the law '$(x)(\phi(x) \supset \psi(x))$'?

these that describe a definite set of initial conditions; for example, statements of the form 'if in this phial under ordinary room temperature and a pressure of 1000 g per cm^2, hydrogen and oxygen are mixed . . . then . . .'. If conditional statements of this kind are deducible from true laws of nature, then their truth will be also invariant with respect to all changes of initial conditions: either the initial conditions described in the antecedent will be satisfied, in which case the consequent will be true (and therefore the whole conditional); or the initial conditions described in the antecedent will not be satisfied and therefore factually untrue ('counter-factual'). In this case the conditional will be true as vacuously satisfied. Thus the much discussed vacuous satisfaction plays its proper part to ensure that the statements deducible from naturally necessary laws are also 'naturally necessary' in the sense of our definition.

Indeed, we could have defined N simply as the class of natural laws and their logical consequences. But there is perhaps a slight advantage in defining N with the help of the idea of initial conditions (of a simultaneity class of singular statements). If we define N as, for example, the class of statements which are true in all worlds that differ from our world (if at all) only with respect to initial conditions, then we avoid the use of subjunctive (or counter-factual) wording, such as 'which would remain true even if different initial conditions held (in our world) than those which actually do hold'.

Nevertheless, the phrase 'all worlds which differ (if at all) from our world only with respect to the initial conditions' undoubtedly contains implicitly the idea of laws of nature. What we mean is 'all worlds which have the same structure—or the same natural laws—as our own world'. In so far as our *definiens* contains implicitly the idea of laws of nature, (D) may be said to be circular. But all definitions must be circular *in this sense*—precisely as all derivations (as opposed to proofs),[18] for example, all syllogisms, are circular: the conclusion must be contained in the premises. Our definition is not, however, circular in a more technical sense. Its *definiens* operates with a perfectly clear intuitive idea—that of varying the initial conditions of our world; for example, the distances of the planets, their masses, and the mass of the sun. It interprets the result of such changes as the construction of a kind of 'model' of our world (a model or 'copy' which does not need to be faithful with respect to the initial conditions); and it then imitates the

---

[18] The distinction between derivation and proof is dealt with in my paper 'New Foundations for Logic', *Mind* **56**, 1947, pp. 193 *f*.

well-known device of calling those statements 'necessary' which are true in (the universe of) *all* these models (*i.e.* for all *logically possible* initial conditions).

(14) My present treatment of this problem differs, intuitively, from a version previously published.[19] I think that it is a considerable improvement, and I gladly acknowledge that I owe this improvement, in a considerable measure, to Kneale's criticism. Nevertheless, from a more technical (rather than an intuitive) point of view the changes are slight. For in that paper, I operate (a) with the idea of natural laws, (b) with the idea of conditionals which *follow* from natural laws; but (a) and (b) together have the same extension as *N*, as we have seen. (c) I suggest that 'subjunctive conditionals' are those that follow from (a), *i.e.* are just those of the class (b). And (d) I suggest (in the last paragraph) that we may have to introduce the supposition that all logically possible initial conditions (and therefore all events and processes which are compatible with the laws) are somewhere, at some time, realized in the world; which is a somewhat clumsy way of saying more or less what I am saying now with the help of the idea of all worlds that differ (if at all) from our world only with respect to the initial conditions.[20]

My position of 1949 might indeed be formulated with the help of the following statement. Although our world may not comprise all logically possible worlds, since worlds of another structure—with different laws—may be logically possible, it comprises all physically possible worlds, in the sense that all physically possible initial conditions are realized in it—somewhere, at some time. My present view is that it is only too obvious that this metaphysical assumption may possibly be true—in both senses of 'possible'—but that we are much better off without it.

Yet once this metaphysical assumption is adopted, my older and my present views become (except for purely terminological differences) equivalent, as far as *the status of laws* is concerned. Thus my older view

[19] *Cf.* 'A Note on Natural Laws and So-Called Contrary-to Fact Conditionals', *Mind* 58, N.S., 1949, pp. 62–66. See also my *Poverty of Historicism*, 1957 (first published 1945), the footnote on p. 123.

[20] I call my older formulation 'clumsy' because it amounts to introducing the assumption that somewhere moas have once lived, or will one day live, under ideal conditions; which seems to me a bit far-fetched. I prefer now to replace this supposition by another – that among the 'models' of our world – which are not supposed to be real, but logical constructions as it were—there will be at least one in which moas live under ideal conditions. And this, indeed, seems to me not only admissible, but obvious. Apart from terminological changes, this seems to me to be the only change in my position, as compared with my note in *Mind* of 1949. But I think that it is an important change.

is, if anything, more 'metaphysical' (or less 'positivistic') than my present view, even though it does not make use of the *word* 'necessary' in describing the status of laws.

(15) To a student of method who opposes the doctrine of induction and adheres to the theory of falsification, there is not much difference between the view that universal laws are nothing but strictly universal statements and the view that they are 'necessary': in both cases, we can only test our conjecture by attempted refutations.

To the inductivist, there is a crucial difference here: he ought to reject the idea of 'necessary' laws, since these, being logically stronger, must be even less accessible to induction than mere universal statements.

Yet inductivists do not in fact always reason in this way. On the contrary, some seem to think that a statement asserting that laws of nature are necessary may somehow be used to justify induction—perhaps somewhat on the lines of a 'principle of the uniformity of nature'.

But it is obvious that no principle of this kind could ever justify induction. None could make inductive conclusions valid or even probable.

It is quite true, of course, that a statement like 'there exist laws of nature' might be appealed to if we wished to justify our search for laws of nature.[21] But in the context of this remark of mine, 'justify' has a sense very different from the one it has in the context of the question whether we can justify induction. In the latter case, we wish to establish certain statements—the induced generalizations. In the former case, we merely wish to justify an activity, the search for laws. Moreover, even though this activity may, in some sense, be justified by the knowledge that true laws exist—that there are structural regularities in the world —it could be so justified even without that knowledge: the hope that there may be some food somewhere certainly 'justifies' the search for it—especially if we are starving—even if this hope is far removed from knowledge. Thus we can say that, although the knowledge that true laws exist would add something to the justification of our

[21] *Cf.* Wittgenstein's *Tractatus*, 6.36: 'If there were a law of causality, it might run: "There are natural laws". But that can clearly not be said; it shows itself.' In my opinion, what shows itself, if anything, is that this clearly *can* be said: it *has* been said by Wittgenstein, for example. What can clearly not be done is to *verify* the statement that there are natural laws (or even to falsify it). But the fact that a statement is not verifiable (or even that it is not falsifiable) does not mean that it is meaningless, or that it can not be understood, or that it 'can clearly not be said', as Wittgenstein believed.

search for laws, this search is justified, even if we lack knowledge, by our curiosity, and by the mere hope that we may succeed.

Moreover, the distinction between 'necessary' laws and strictly universal statements does not seem to be relevant to this problem: whether necessary or not, the knowledge that laws exist would add something to the 'justification' of our search, without being needed for this kind of 'justification'.

(16) I believe, however, that the idea that there are necessary laws of nature, in the sense of natural or physical necessity explained under point (12), is metaphysically or ontologically important, and of great intuitive significance in connection with our attempts to understand the world. And although it is impossible to establish this metaphysical idea either on empirical grounds (because it is not falsifiable) or on other grounds, I believe that it is true, as I indicated in sections 79, and 83 to 85. Yet I am now trying to go beyond what I said in these sections by emphasizing the peculiar ontological status of universal laws (for example, by speaking of their 'necessity', or their 'structural character'), and also by emphasizing the fact that the metaphysical character or the irrefutability of the assertion that laws of nature exist need not prevent us from discussing this assertion rationally—that is to say, critically. (See my *Postscript*, especially sections *6, *7, *15, and *120.)

Nevertheless, I regard, unlike Kneale, 'necessary' as a mere word —as a label useful for distinguishing *the universality of laws* from 'accidental' universality. Of course, any other label would do just as well, for there is not much connection here with logical necessity. I largely agree with the spirit of Wittgenstein's paraphrase of Hume: 'A necessity for one thing to happen because another has happened does not exist. There is only logical necessity.'[22] Only in one way is $a \xrightarrow{N} b$ connected with logical necessity: the necessary link between a and b is neither to be found in a nor in b, but in the fact that the corresponding ordinary conditional (or 'material implication', $a \rightarrow b$ without 'N') follows *with logical necessity* from a law of nature—that it is necessary, relative to a law of nature.[23] And it may be said that a law of nature is necessary in its turn because it is logically derivable from, or explicable by, a law of a still higher degree of universality, or of greater

[22] *Cf. Tractatus*, 6.3637.
[23] I pointed this out in *Aristotelian Society Supplementary Volume* 22, 1948, pp. 141 to 154, section 3; see especially p. 148. In this paper I briefly sketched a programme which I have largely carried out since.

'depth'. (See my *Postscript*, section *15.) One might suppose that it is this logically necessary dependence upon true statements of higher universality, conjectured to exist, which suggested in the first instance the idea of 'necessary connection' between cause and effect.[24]

(17) So far as I can understand the modern discussions of 'subjunctive conditionals' or 'contrary-to-fact conditionals' or 'counterfactual conditionals', they seems to have arisen mainly out of the problem situation created by the inherent difficulties of inductivism or positivism or operationalism or phenomenalism.

The phenomenalist, for instance, wishes to translate statements about physical objects into statements about observations. For example, 'There is a flower-pot on the window sill' should be translatable into something like 'If anybody in an appropriate place looks in the appropriate direction, he will see what he has learned to call a flower-pot'. The simplest objection (but by no means the most important one) to regarding the second statement as a translation of the first is to point out that while the second will be (vacuously) true when nobody is looking at the window sill, it would be absurd to say that whenever nobody is looking at some window sill, there must be a flower-pot on it. The phenomenalist is tempted to reply to this that the argument depends on the truth-table definition of the conditional (or of 'material implication'), and that we have to realize the need for a different interpretation of the conditional—a *modal* interpretation which makes allowance for the fact that what we mean is something like 'If anybody looks, or if anybody were looking, then he will see, or would see, a flower-pot.'[25]

One might think that our $a \underset{N}{\rightarrow} b$ could provide the desired modal conditional, and in a way it does do this. Indeed, it does it as well as one can possibly expect. Nevertheless, our original objection stands, because we know that if $\bar{a}$ is necessary—that is, if $\bar{a} \varepsilon N$—then $a \underset{N}{\rightarrow} b$ holds for every b. This means that, if for some reason or other the place where a flower-pot is (or is not) situated is such that it is physically *impossible* for anybody to look at it, then 'If anybody looks, or if anybody were looking, at that place, then he will, or would,

---

[24] *Cf.* my paper quoted in the foregoing footnote.

[25] It was R. B. Braithwaite who replied along similar lines as these to my objection of vacuous satisfaction after a paper he read on phenomenalism in Professor Susan Stebbing's seminar, in the spring of 1936. It was the first time that I heard, in a context like this, of what is nowadays called a 'subjunctive conditional'. For a criticism of phenomenalist 'reduction programmes', see note 4 and text, above.

see a flower-pot' will be true, merely because nobody *can* look at it. But this means that the phenomenalist modal translation of 'At the place x is a flower-pot' will be true for all those places x which, for some physical reason or other, nobody *can* look at. (Thus there is a flower-pot—or whatever else you like—in the centre of the sun.) But this is absurd.

For this reason, and for many other reasons, I do not think that there is any chance of rescuing phenomenalism by this method.

As to the doctrine of operationalism—which demands that scientific terms, such as length, or solubility, should be defined in terms of the appropriate experimental procedure—it can be shown quite easily that all so-called operational definitions will be circular. I may show this briefly in the case of 'soluble'.[26]

The experiments by which we test whether a substance such as sugar is *soluble in water* involve such tests as the recovery of dissolved sugar from the solution (say, by evaporation of the water; *cf.* point 3 above). Clearly, it is necessary to identify the recovered substance, that is to say, to find out whether it has the same properties as sugar. Among these properties, *solubility in water* is one. Thus in order to define 'x is soluble in water' by the standard operational test, we should at least have to say something like this:

'x is *soluble in water* if and only if (a) when x is put into water then it (necessarily) disappears, and (b) when after the water evaporates, a substance is (necessarily) recovered which, again, is *soluble in water*.'

The fundamental reason for the circularity of this kind of definition is very simple: experiments are never conclusive; and they must, in their turn, be testable by further experiments.

Operationalists seem to have believed that once the problem of subjunctive conditionals was solved (so that the vacuous satisfaction of the defining conditional could be avoided) there would be no further obstacle in the way of operational definitions of dispositional terms. It seems that the great interest shown in the so-called problem of subjunctive (or counter-factual) conditionals was mainly due to this

---

[26] The argument is contained in a paper which I contributed in January 1955 to the still unpublished Carnap volume of the *Library of Living Philosophers*, ed. by P. A. Schilpp. As to the circularity of the operational definition of length, this may be seen from the following facts: (a) the 'operational' definition of *length* involves *temperature* corrections, and (b) the (usual) operational definition of *temperature* involves measurements of *length*.

belief. But I think I have shown that even if we have solved the problem of logically analysing subjunctive (or 'nomic') conditionals, we cannot hope to define dispositional terms, or universal terms, operationally. For universals, or dispositional terms, transcend experience, as explained here under points 1 and 2, and in section 25 of the book.

Appendix *xi. *On the Use and Misuse of Imaginary Experiments,*
Especially in Quantum Theory.

The criticisms presented in the later parts of this appendix are
logical in character. My point is not to refute certain arguments, some
of which, for all I know, may have long been discarded by their
originators. I try, rather, to show that certain *methods of argument* are
inadmissible—methods which have been used, without being
challenged, for many years in the discussions about the interpretation
of quantum theory. It is, in the main, the *apologetic use* of imaginary
experiments which I am criticizing here, rather than any particular
theory in whose defence these experiments were propounded.[1] Least
of all do I wish to create the impression that I am doubting the
fruitfulness of imaginary experiments.

(1) One of the most important imaginary experiments in the
history of natural philosophy, and one of the simplest and most
ingenious arguments in the history of rational thought about our
universe, is contained in Galileo's criticism of Aristotle's theory of
motion.[2] It disproves the Aristotelian supposition that the natural
velocity of a heavier body is greater than that of a lighter body. 'If
we take two moving bodies', Galileo's spokesman argues, 'such that
their natural velocities are unequal, it is manifest that if we join them
together, the slower and the faster one, then the latter will be partly
retarded by the slower one, and the slower partly sped up by the
faster one'. Thus 'if a big stone moves, for example, with a velocity
of eight steps and a smaller one with a velocity of four, then, after
being joined together, the composite system will move with a
velocity of less than eight steps. But the two stones joined together
make a stone bigger than the first one which moved with a velocity of

---

[1] More especially, I do not wish to criticize the quantum theory here, or any
particular interpretation of it.

[2] Galileo himself proudly says of his argument (he puts the words into the mouth of
Simplicio) 'In truth, your argument has proceeded exceedingly well.' *Cf. Dialogues
Concerning Two New Sciences*, 1638, First Day, p. 109 (p. 66 of vol. xiii, 1855, of the
Opere Complete; p. 64 and 62 of the English edition of Crew and Salvio, 1914).

eight steps. *Thus the composite body (although bigger than the first alone) will nevertheless move more slowly than the first alone*; which is contrary to your supposition.'[3] And since this Aristotelian supposition was the one from which the argument started, it is now refuted: it is shown to be absurd.

I see in Galileo's imaginary experiment a perfect model for the best use of imaginary experiments. It is the *critical use*. I do not wish to suggest, however, that there is no other way of using them. There is, especially, a *heuristic* use which is very valuable. But there are less valuable uses also.

An old example of what I call the heuristic use of imaginary experiments is one that forms the heuristic basis of atomism. We imagine that we take a piece of gold, or some other substance, and cut it into smaller and smaller parts 'until we arrive at parts so small that they cannot be any longer subdivided': a thought experiment used in order to explain 'indivisible atoms'. Heuristic imaginary experiments have become particularly important in thermodynamics (Carnot's cycle); and they have lately become somewhat fashionable owing to their use in relativity and in quantum theory. One of the best examples of this kind is Einstein's experiment of the accelerated lift: it illustrates the local equivalence of acceleration and gravity, and it suggests that light rays in a gravitational field may proceed on curved paths. This use is important and legitimate.

The main purpose of this note is to issue a warning against what may be called *the apologetic use of imaginary experiments*. This use goes back, I think, to the discussion of the behaviour of measuring rods and clocks from the point of view of special relativity. At first these experiments were used in an illustrative and expository way—a perfectly legitimate usage. But later, and in the discussion of quantum theory, they were also used, at times, as arguments, both in a critical and in a defensive or apologetic mood. (In this development, an important part was played by Heisenberg's imaginary microscope through which one could observe electrons; see points 9 and 10 below.)

Now the use of imaginary experiments in critical argumentation is, undoubtedly, legitimate: it amounts to an attempt to show that certain possibilities were overlooked by the author of a theory. Clearly, it must also be legitimate to counter such critical objections,

---

[3] *Op. cit.*, p. 107 (1638); p. 65 (1855); p. 63 (1914).

for example, by showing that the proposed imaginary experiment is in principle impossible, and that here, at least no possibility was overlooked.[4] An imaginary experiment designed in a critical spirit— designed in order to criticize a theory by showing that certain possibilities have been overlooked—is usually permissible, but great care must be taken with the reply: in a reconstruction of the controversial experiment, undertaken in defence of the theory, it is, more particularly, important *not to introduce any idealizations* or other special assumptions unless they are favourable to an opponent, or unless any opponent who uses the imaginary experiment in question would have to accept them.

(2) More generally, I think that the argumentative use of imaginary experiments is legitimate only if the views of the opponent in the argument are stated with clarity, and if the rule is adhered to that *the idealizations made must be concessions to the opponent, or at least acceptable to the opponent.* For example, in the case of Carnot's cycle all idealizations introduced increase the efficiency of the machine, so that the opponent to the theory—who asserts that a heat machine can produce mechanical work without transferring heat from a higher temperature to a lower temperature—must agree that they are concessions. Idealizations, clearly, become impermissible for the purpose of critical argumentation whenever this rule is violated.

(3) This rule may be applied, for example, to the discussion initiated by the imaginary experiment of Einstein, Podolski, and Rosen. (Their argument is briefly re-stated by Einstein in a letter here reproduced in appendix *xii; and this discussion is further commented upon in my *Postscript*, section *109.) Einstein, Podolski, and Rosen attempt, in their critical argument, to make use of idealizations acceptable to Bohr; and in his reply, Bohr does not challenge the legitimacy of their idealizations. They introduce (*cf.* section *109 and appendix *xii) two particles, A and B, which interact in such a way that by measuring the position (or momentum) of B, the theory allows us to calculate the position (or momentum) of A which has meanwhile moved far away and cannot be any longer disturbed by the measurement of B. Thus A's momentum (or position) cannot become blurred—or 'smeared', to use a term of Schrödinger's—as

---

[4] For example, my own experiment of section 77 has been shown to be in principle impossible (from the quantum-theoretical point of view) by Einstein in his letter printed in appendix *xii; see note *3 to section 77.

Heisenberg would have it.[5] Bohr, in his reply, operates with the idea that measurement of a position can be achieved only by 'some instrument *rigidly fixed to the support which defines the space frame of reference*' while measurement of the momentum would be done by a *movable* 'diaphragm' whose 'momentum . . . is measured before as well as after the passing of the particle'.[6] Bohr operates with the argument that in choosing one of these two systems of reference 'we . . . cut ourselves off from any . . . possibility' of using the other, in connection with the same physical system under investigation. He suggests, if I understand him properly, that though A is not interfered with, its co-ordinates may become smeared by the smearing of the *frame of reference*.

(4) Bohr's reply seems to me unacceptable for at least three different reasons.

First, prior to the proposed imaginary experiment of Einstein, Podolski, and Rosen, the reason given for the smearing of the position or momentum of a system was that by measuring it, we had interfered with the system. It seems to me that Bohr, surreptitiously, dropped this argument, and replaced it by saying (more or less clearly) that the reason is that we interfere with our frame of reference, or with the system of co-ordinates, rather than with the physical system. This is too big a change to be allowed to pass unnoticed. It would have to be explicitly acknowledged that the older position was refuted by the imaginary experiment; and it would have to be shown why this does not destroy the principle on which it was built.

We must not forget, in this connection, what the imaginary experiment of Einstein, Podolski, and Rosen, was intended to show. It was intended merely to refute certain *interpretations* of the *indeterminacy* formulae; it was certainly not intended to refute these formulae. In a sense, Bohr's reply, though not explicitly, acknowledged that the imaginary experiment succeeded in its purpose, for Bohr merely tried to defend the indeterminacy relations as such: he gave up the view that the

---

[5] Heisenberg thought, of course, of the smearing of one particle only, the one which is being measured. Einstein, Podolski, and Rosen show that it must be extended to another particle—one with which the measured particle had interacted at some time, perhaps years ago. But if so, how can we avoid having everything 'smeared'—the whole world—by one single observation? The answer is, presumably, that owing to the 'reduction of the wave packet', the observation does destroy the old *picture* of the system, and at the same time creates a new one. Thus the interference is not with the world, but merely with our way of representing it. This situation is illustrated, as will be seen, by Bohr's reply which follows in the text.

[6] Bohr, *Physical Review* 48, 1935, pp. 696–702. The quotations are from pp. 700 and 699. (The italics are mine.)

measurement would interfere with the system A which it was supposed to smear. Moreover, the argument of Einstein, Podolski, and Rosen could be carried a little further by the assumption that we measure the position A (accidentally) at the same instant of time at which we measure the momentum of B. We then obtain, *for that instant of time,* positions and momenta of both A and B. (Admittedly, the momentum of A and the position of B will have been destroyed or smeared by these measurements.) But this is sufficient to establish the point which Einstein, Podolski, and Rosen wanted to make: that it is incorrect to interpret the indeterminacy formulae as asserting that the system cannot have both a sharp position and a sharp momentum at the same time—even though it must be admitted that we cannot *predict* both at the same time. (For an interpretation which takes account of all this, see my *Postscript.*)

Secondly, Bohr's argument that we have 'cut ourselves off' from the other frame of reference seems to be *ad hoc.* For it is clearly possible to measure the momentum spectroscopically (either in a direct manner, or by using the Doppler effect), and the spectroscope will be rigidly fixed to the same frame as the first 'instrument'. (The fact that the spectroscope absorbs the particle B is irrelevant to the argument which concerns the fate of A.) Thus an arrangement with a movable frame of reference cannot be accepted as an essential part of the experiment.

Thirdly, Bohr does not explain here how to measure the momentum of B with the help of his movable diaphragm. In a later paper of his, a method of doing this is described; but this method seems to me again impermissible.[7] For the method described by Bohr consists in measuring (twice) the *position* of a 'diaphragm with a slit . . . suspended by weak springs from a solid yoke';[8] and since the measurement of the momentum with an arrangement of this kind depends on position measurements, it does not support Bohr's argument against Einstein, Podolski, and Rosen; nor does it succeed otherwise. For in this way we cannot get the momentum 'accurately before as well as after the passing' of B:[9] the first of these measurements of momentum (since it utilizes a *position* measurement) will interfere with the momentum of the diaphragm; it thus will be retrospective only, and will not be of

[7] See Bohr, in *Albert Einstein: Philosopher-Scientist,* ed. by P. A. Schilpp, 1949; see especially the diagram on p. 220.

[8] *Op. cit.,* p. 219.

[9] Bohr, *Physical Review* **48**, 1935, p. 699.

any use for calculating the momentum of the diaphragm at the time immediately before the interaction with B.

It does not seem, therefore, that Bohr in his reply adhered to the principle of making only such idealizations or special assumptions which favour his opponents (quite apart from the fact that it is far from clear what he wanted to contest).

(5) This shows that there is a grave danger, in connection with imaginary experiments of this kind, of carrying the analysis just as far as it serves one's purpose, and no further; a danger which can be avoided only if the above principles are strictly adhered to.

There are three similar cases which I wish to refer to, because I find them instructive.

(6) In order to meet a critical imaginary experiment of Einstein's, based upon his famous formula $E = mc^2$, Bohr had recourse to arguments from Einstein's gravitational theory (that is to say, from general relativity).[10] But $E = mc^2$ can be derived from special relativity, and even from non-relativistic arguments. In any case, in assuming $E = mc^2$, we certainly do not assume the validity of Einstein's theory of gravitation. If, therefore, as Bohr suggests, we must assume certain characteristic formulae of Einstein's gravitational theory in order to rescue the consistency of quantum theory (in the presence of $E = mc^2$), then this amounts to the strange assertion that quantum theory contradicts Newton's gravitational theory, and further to the still stranger assertion that the validity of Einstein's gravitational theory (or at least the characteristic formulae used, which are part of the theory of the gravitational field) can be derived from quantum theory. I do not think that even those who are prepared to accept this result will be happy about it.

Thus we have again an imaginary experiment which makes extravagant assumptions, with an apologetic purpose.

(7) David Bohm's reply to the experiment of Einstein, Podolski, and Rosen seems to me also highly unsatisfactory.[11] He believes that he has to show that Einstein's particle A which has run far away from B and from the measuring apparatus does nevertheless become smeared

[10] Bohr, in *Albert Einstein: Philosopher-Scientist*, ed. by P. A. Schilpp; the case is discussed on pp. 225–228. Dr. J. Agassi has drawn my attention to the invalidity of the argument.

[11] See D. Bohm, *Phys. Rev.* **85**, 1951, pp. 166 *ff*., 180 *ff*; see especially pp. 186 *f*. (I understand that Bohm does not any longer uphold some of the views expressed in the papers here criticized. But it seems to me that at least part of my argument may still be applicable to his later theories.)

in its position (or momentum) when the momentum (or position) of B is measured, and he tries, to this end, to show that A, in spite of having run away, is still interfered with in an unpredictable way. In this way he tries to show that his own theory agrees with Heisenberg's interpretation of the indeterminacy relations. But he does not succeed. This becomes manifest if we consider that the ideas of Einstein, Podolski, and Rosen allow us, by a slight extension of their experiment, to determine simultaneously positions *and* momenta of both A and B—although the result of this determination will have *predictive* significance only for the position of the one particle and the momentum of the other. For as explained under point (4) above, we may measure the position of B, and somebody far away may measure the momentum of A accidentally at the same instant, or at any rate before any smearing effect of our measurement of B could possibly reach A. Yet this is all that is needed to show that Bohm's attempt to save Heisenberg's idea of our interference with A is misplaced.

Bohm's reply to this objection is implicit in his assertion that the smearing effect proceeds with super-luminal velocity, or perhaps even instantaneously (*cf.* Heisenberg's super-luminal velocity commented on in section 76), an assumption to be supported by the further assumption that this effect cannot be used to transmit signals. But what *does* happen if the two measurements are carried out simultaneously? Does the particle you are supposed to observe in your Heisenberg microscope begin to dance under your very eyes? And if it does, is this not a signal? (This particular smearing effect of Bohm's, like the 'reduction of the wave packet', is not part of his formalism, but of its interpretation.)

(8) A similar example is a reply of Bohm's to another critical imaginary experiment proposed by Einstein (who thereby revived Pauli's criticism of de Broglie's pilot wave theory).[12]

Einstein proposes to consider a macroscopic 'particle' (it may be quite a big thing, say a billiard ball) moving with a certain constant velocity to and fro between two parallel walls by which it is elastically reflected. Einstein shows that this system can be represented in Schrödinger's theory by a standing wave; and he shows further that the pilot wave theory of de Broglie, or Bohm's so-called 'causal inter-

---

[12] See A. Einstein in *Scientific Papers Presented to Max Born*, 1953, pp. 33 *ff*; see especially p. 39.

pretation of quantum theory' leads to the paradoxial result (first pointed out by Pauli) that the velocity of the particle (or billiard ball) vanishes; or in other words, our original assumption that the particle moves with some arbitrarily chosen velocity leads in this theory, for every chosen velocity, to the conclusion that the velocity is zero, and that it does not move.

Bohm accepts this conclusion, and replies on the following lines: 'The example considered by Einstein', he writes, 'is that of a particle *moving freely* between perfectly reflecting and smooth walls.'[13] (We need not go into the finer peculiarities of the arrangement.) 'Now, in the causal interpretation of the quantum theory'—that is, in Bohm's interpretation—'... the particle is *at rest*', Bohm writes; and he goes on to say that, if we wish to *observe* the particle, we shall 'trigger' a process which will make the particle move.[14] But this argument about observation, whatever its merits, is no longer interesting. What is interesting is that Bohm's interpretation paralyses the freely moving particle: his argument amounts to the assertion that it cannot move between these two walls, as long as it is unobserved. For the assumption that it so *moves* leads Bohm to the conclusion that it is at *rest*, until triggered off by an observation. This paralysing effect is noted by Bohm, but simply not discussed. Instead, he proceeds to the assertion that though the *particle* does not move, our *observations* will show it to us moving (but this was not the point at issue); and further, to the construction of an entirely new imaginary experiment describing how our observation—the radar signal or photon used to observe the velocity of the particle—could trigger off the desired movement. But first, this again was not the problem. And secondly, Bohm fails to explain how the triggering photon could reveal to us the particle in its full, proper speed, rather than in a state of acceleration towards its proper speed. For this seems to demand that the particle (which may be as fast and as heavy as we like) acquires and reveals its full speed during the exceedingly short time of its interaction with the triggering photon. All these are *ad hoc* assumptions which few of his opponents will accept.

But we may elaborate Einstein's imaginary experiment by operating with two particles (or billiard balls) of which the one moves to and fro between the left wall and the centre of the box while the other moves

---

[13] D. Bohm, in the same volume, p. 13; the italics are mine.
[14] *Op. cit.*, p. 14; see also the second footnote on that page.

between the right wall and the centre; in the centre, the particles collide elastically with one another. This example leads again to standing waves, and thus to the disappearance of the velocity; and the Pauli-Einstein criticism of the theory remains unchanged. But Bohm's triggering effect now becomes even more precarious. For let us assume we observe the left particle by shooting at it a triggering photon from the left. This will (according to Bohm) overthrow the balance of forces which keep the particle at rest; and the particle will start moving—presumably from left to right. But although we triggered only the left particle, the right particle will have to start simultaneously, and in the opposite direction. It is asking much of a physicist to acquiesce in the possibility of all these processes—all assumed *ad hoc*, in order to avoid the consequences of the argument of Pauli and Einstein.

Einstein might have answered Bohm as follows, I think.

In the case considered, our physical system was a big macroscopic ball. No reason has been given why in such a case the usual classical view of measurement should be inapplicable. And this is a view that conforms, after all, as well with experience as one can desire.

But leaving measurement aside, is it seriously asserted that an oscillating ball (or two oscillating balls in a symmetric arrangement here described) simply *cannot* exist while unobserved? Or, what amounts to the same, is it seriously asserted that the assumption that it does move, or oscillate, while unobserved, must lead to the conclusion that it does not? And what happens if, once our observation has set the ball in motion, it is then no longer asymmetrically interfered with so that the system again becomes stationary? Does the particle then stop as suddenly as it started? And is its energy transformed into field energy? Or is the process irreversible?

Even assuming that all these questions may be answered somehow, they illustrate, I think, the significance of Pauli's and of Einstein's criticism, and of the critical use of imaginary experiments, especially the experiment of Einstein, Podolski, and Rosen. And I think that they also illustrate the danger of an apologetic use of imaginary experiments.

(9) So far I have discussed the problem of *pairs of particles*, introduced into the discussion by Einstein, Podolski, and Rosen. I now turn to some of the older imaginary experiments with single particles, such as Heisenberg's famous *imaginary microscope* through which one could 'observe' electrons, and 'measure' either their positions or their

momenta. Few imaginary experiments have exerted a greater influence on thought about physics than this one.

With the help of his imaginary experiment, Heisenberg tried to establish various points of which I may mention three: (a) the *interpretation* of the Heisenberg indeterminacy formulae as stating the existence of *insuperable barriers to the precision of our measurements*; (b) the *disturbance* of the measured object by the process of measurement, *whether of position or of momentum*; and (c) *the impossibility of testing the spatio–temporal 'path'* of the particle. I believe that Heisenberg's *arguments* tending to establish these points are all clearly invalid, whatever the merits of the three points in themselves may be. The reason is that Heisenberg's discussion *fails to establish that measurements of position and of momentum are symmetrical*; symmetrical, that is, with respect to the disturbance of the measured object by the process of measurement. For Heisenberg *does* show with the help of his experiment that in order to measure the *position* of the electron we should have to use light of a high frequency, that is to say, high energy photons, which means that we transfer an unknown momentum to the electron and thus *disturb* it, by giving it a severe knock, as it were. But Heisenberg *does not* show that the situation is analogous if we wish to measure the *momentum* of the electron, rather than its position. For in this case, Heisenberg says, we must observe it with a low frequency light—so low that we may assume that *we do not disturb the electron's momentum by our observation*. The resulting observation (though revealing the momentum) will fail to reveal the electron's position, which will thus remain indeterminate.

Now consider this last argument. There is no assertion here that we have *disturbed* (or 'smeared') the electron's position. For Heisenberg merely asserts that we have *failed to disclose it*. In fact, his argument implies that we have not disturbed the system at all (or only so slightly that we can neglect the disturbance): we have used photons of so low an energy level that there simply was not enough energy available to disturb the electron. Thus *the two cases—the measurement of position and that of momentum—are far from analogous or symmetrical*, according to Heisenberg's argument. This fact is veiled, however, by the customary talk (positivist or operationalist or instrumentalist talk) about the 'results of measurement' whose uncertainty is admittedly symmetrical with respect to position and momentum. Yet in countless discussions of the experiment, beginning with Heisenberg's own, it is always assumed that his argument establishes the *symmetry of the disturbances*. (In the formalism, the symmetry between position and momentum is complete, of course, but this does not mean that it is accounted for by Heisenberg's imaginary experiment.) Thus it is assumed—quite wrongly—that *we disturb the electron's position* if we measure its momentum with Heisenberg's microscope, and that this 'smearing' effect has been established by Heisenberg's discussion of his imaginary experiment.

My own imaginary experiment of section 77 was largely based on this asymmetry in Heisenberg's experiment. (*Cf.* note *1 to appendix xi.) Yet my experiment is invalid just because the asymmetry invalidates Heisenberg's whole discussion of measurement: only measurements resulting from *physical selection* (as I call it) can be used to illustrate Heisenberg's *formulae*, and a physical selection, as I quite correctly pointed out in the book, must always satisfy the 'scatter relations'. (Physical selection *does* disturb the system.)

Were Heisenberg's 'measurements' possible we could even check the momentum of an electron between two position measurements without disturbing it, which would also allow us—contrary to point (c), above—to check (part of) its spatio-temporal 'path' which is calculable from these two position measurements.

That the inadequacy of Heisenberg's argument has remained unnoticed for so long is no doubt due to the fact that the indeterminacy *formulae* clearly follow from the formalism of the quantum theory (the wave equation), and that the symmetry between position (q) and momentum (p) is also implicit in this formalism. This may explain why many physicists have failed to scrutinize Heisenberg's imaginary experiment with sufficient care: they did not take it seriously, I suppose, but merely as an illustration of a derivable formula. My point is that it is a bad illustration—just because it fails to account for the symmetry between position and momentum. And being a bad illustration, it is quite inadequate as a basis for interpreting these formulae—let alone the whole quantum theory.

(10) The immense influence of Heisenberg's imaginary experiment is, I am convinced, due to the fact that he managed to convey through it a new metaphysical picture of the physical world, whilst at the same time disclaiming metaphysics. (He thus ministered to a curiously ambivalent obsession of our post-rationalist age: its preoccupation with killing the Father—that is, Metaphysics—while keeping Him

inviolate, in some other form, and beyond all criticism. With some quantum physicists it sometimes looks almost as if the father was Einstein.) The metaphysical picture of the world, somehow conveyed through Heisenberg's discussion of his experiment although never really implied in it, is this. The *thing in itself* is unknowable: we can only know its appearances which are to be understood (as pointed out by Kant) as resulting from the thing in itself, and from our own perceiving apparatus. Thus the appearances result from a kind of interaction between the things in themselves and ourselves. This is why one thing may appear to us in different forms, according to our different ways of perceiving it—of observing it, and of interacting with it. We try to catch, as it were, the thing in itself, but we never succeed: we only find appearances in our traps. We can set either a classical *particle trap* or a classical *wave trap* ('classical' because we can build them and set them like a classical mouse trap); and in the process of triggering off the trap, and thus interacting with it, the thing is induced to assume the appearance of a particle or a wave. There is a symmetry between these two appearances, or between the two ways of trapping the thing. Moreover, we not only, by setting the trap, have to supply a stimulus for the thing in order to induce it to assume one of its two classical physical appearances, but we also have to bait the trap with energy—the energy needed for a classical physical realization or materialization of the unknowable thing in itself. In this way, we preserve the conservation laws.

This is the metaphysical picture conveyed by Heisenberg, and perhaps also by Bohr.

Now I am far from objecting to metaphysics of this kind (though I am not much attracted by this particular blend of positivism and transcendentalism). Nor do I object to its being conveyed to us through metaphors. What I do object to is the almost unconscious dissemination of this metaphysical picture, often combined with anti-metaphysical disclaimers. For I think that it should not be allowed to sink in unnoticed, and thus uncriticized.

It is interesting, I think, that much of David Bohm's work seems to be inspired by the same metaphysics. One might even describe his work as a valiant attempt to construct a physical theory that shall make this metaphysics clear and explicit. This is admirable. But I wonder whether this particular metaphysical idea is good enough, and really worth the trouble, considering that it cannot be supported (as we have seen) by Heisenberg's imaginary experiment which is the intuitive source of it all.

There seems to me a fairly obvious connection between Bohr's 'principle of complementarity' and this metaphysical view of an unknowable reality—a view that suggests the 'renunciation' (to use a favourite term of Bohr's) of our aspirations to knowledge, and the restriction of our physical studies to appearances and their interrelations. But I will not discuss this obvious connection here. Instead, I will confine myself to the discussion of certain arguments in favour of complementarity which have been based upon further imaginary experiments.

(11) In connection with this 'principle of complementarity' (discussed more fully in my *Postscript*; *cf.* also my paper 'Three Views Concerning Human Knowledge', *Contemporary British Philosophy* iii, edited by H. D. Lewis, 1956) Bohr has analysed a large number of subtle imaginary experiments in a similarly apologetic vein. Since Bohr's formulations of the principle of complementarity are vague and difficult to discuss, I shall have recourse to a well known and in many respects excellent book, *Anschauliche Quantentheorie*, by P. Jordan (and a book in which, incidentally, my *Logik der Forschung* was briefly discussed).[15]

Jordan gives a formulation of (part of) the contents of the principle of complementarity that brings it into the closest relation to the problem of the *dualism between particles and waves*. He puts it as follows. 'Any one experiment which would bring forth, *at the same time*, both the wave properties and the particle properties of light would not only contradict the classical theories (we have got used to contradictions of this kind), but would, over and above this, be absurd in a logical and mathematical sense.'[16]

Jordan illustrates this principle by the famous two-slit experiment. (See my old appendix v.) 'Assume that there is a source of light from which monochromatic light falls upon a black screen with two [parallel] slits which are close to each other. Now assume, *on the one hand*, that the slits and their distance are sufficiently small (as compared with the wave length of the light) to obtain interference fringes on a photographic plate which records the light that passes the two slits; and *on the other hand*, that some experimental arrangement would make it

[15] Jordan, *Anschauliche Quantentheorie*, 1936, p. 282.
[16] *Op. cit.*, p. 115.

possible to find out, of a single photon, which of the two slits it has passed through.'[17]

Jordan asserts 'that these two assumptions contain a contradiction'.[18]

I am not going to contest this, although the contradiction would not be a logical or mathematical absurdity (as he suggests in one of the previous quotations); rather, the two assumptions would, together, merely contradict the formalism of the quantum theory. Yet I wish to contest a different point. Jordan uses this experiment to illustrate his formulation of the contents of the principle of complementarity. But the very experiment by which he illustrates this principle may be shown to refute it.

For consider Jordan's description of the two-slit experiment, omitting at first his last assumption (the one introduced by the words *'on the other hand'*). Here we obtain interference fringes on the photographic plate. Thus this is an experiment which 'brings forth the wave properties of the light'. Now let us assume that the intensity of the light is sufficiently low to obtain on the plate distinct hits of the photons; or in other words, so low that the fringes are analysable as due to the density distribution of the single photon hits. Then we have here 'one experiment' that 'brings forth, *at the same time*, both the wave properties and the particle properties of light'—at least some of them. That is to say, it does precisely what according to Jordan must be 'absurd in a logical and mathematical sense'.

Admittedly, were we able, in addition, to find out through which of the slits a certain photon has passed, then we should be able to determine its path; and we might then say that this (presumably impossible) experiment would bring forth the particle properties of the photon even more strongly. I grant all this; but it is quite irrelevant. For what Jordan's principle asserted was not that *some* experiments which might seem at first sight possible turn out to be impossible—which is trivial—but that there are *no* experiments whatever which 'bring forth, *at the same time*, both the wave properties and the corpuscle properties of light'. And this assertion, we have shown, is simply false: it is refuted by *almost all* typical quantum mechanical experiments.

But what then did Jordan wish to assert? Perhaps that there was no experiment which would bring forth *all* the wave properties and *all* the particle properties of light? Clearly this cannot have been his

[17] *Op. cit.*, p. 115 *f.* (The italics are Jordan's.)
[18] *Op. cit.*, p. 116.

intention since even an experiment is impossible which would bring forth, at the same time, *all* the wave properties—even if we drop the demand that it should bring forth any of the particle properties. (And the same holds the other way round.)

What is so disturbing in this argumentation of Jordan's is its arbitrariness. From what has been said it is obvious that there must be some wave properties and some particle properties which no experiment can combine. This fact is first generalized by Jordan, and formulated as a principle (whose formulation by Jordan, at any rate, we have refuted). And then it is illustrated by an imaginary experiment which Jordan shows to be impossible. Yet as we have seen, that part of the experiment which everybody admits to be possible actually refutes the principle, at least in Jordan's formulation.

But let us look a little more closely at the other half of the imaginary experiment—the one introduced by the words 'on the other hand'. If we make arrangements to determine the slit through which the particle has passed, then, it is said, we destroy the fringes. Good. But do we destroy the wave properties? Take the simplest arrangement: we close one of the slits. If we do so, there still remain many signs of the wave character of light. (Even with one single slit we obtain a wave-like density distribution.) But it is now admitted, by our opponents, that the particle properties exhibit themselves very fully, since we can now trace the path of the particle.

(12) From a rational point of view, all these arguments are inadmissible. I do not doubt that there is an interesting intuitive idea behind Bohr's principle of complementarity. But neither he nor any other member of his school has been able to explain it, even to those critics who, like Einstein, tried hard for years to understand it.[19]

My impression is that it may well be the metaphysical idea described above under point (10). I may be wrong; but whatever it is, I feel that Bohr owes us a better explanation.

[19] *Cf. Albert Einstein: Philosopher-Scientist*, ed. by P. A. Schilpp, 1949, p. 674.

Appendix *xii. *The Experiment of Einstein, Podolski, and Rosen.*

A Letter from Albert Einstein, 1935.

The letter from Albert Einstein here printed in translation briefly and decisivley disposes of my imaginary experiment of section 77 of the book (it also refers to a slightly different version contained in an unpublished paper), and it goes on to describe with admirable clarity and brevity the imaginary experiment of Einstein, Podolski, and Rosen (also described under point 3 of my appendix *xi).

Between these two points, a few remarks will be found on the relation of theory and experiment in general, and upon the influence of positivistic ideas upon the interpretation of quantum theory.

The two last paragraphs of the letter also deal with a problem discussed in my book (and in my *Postscript*)—the problem of subjective probabilities, and of drawing statistical conclusions from nescience. In this I still disagree with Einstein: I believe that we draw these probabilistic conclusions from conjectures about equidistribution (often very natural conjectures, and for this reason perhaps not always consciously made), and therefore from probabilistic premises.

Einstein's literary executors requested that, if a translation of the letter were to be published, the original text should be published at the same time. This suggested to me the idea of reproducing Einstein's letter in his own handwriting.

Old Lyme, 11. IX. 35.

Dear Mr. Popper,

I have looked at your paper, and I largely [*weitgehend*] agree.[x] Only I do not believe in the possibility of producing a 'super-pure case' which would allow us to predict position *and* momentum (colour) of a photon with 'inadmissible' precision. The means proposed by you (a screen with a fast shutter in conjunction with a

[x] Main point: The ψ-function characterizes a statistical aggregate of systems rather than one single system. This is also the result of the considerations expounded below. This view makes it unnecessary to distinguish, more particularly, between "pure" and "non-pure" cases.

selective set of glass filters) I hold to be ineffective in principle, for the reason that I firmly believe that a filter of this kind would act in such a way as to 'smear' the position, just like a spectroscopic grid.

My argument is as follows. Consider a short light signal (precise position). In order to see more easily the effects of an absorbing filter, I assume that the signal is analysed into a larger number of quasi-monochromatic wave-trains W_n. Let the absorbing set [of filters] cut out all the colours W_n except one, W_1. Now this wave-group will have a considerable spatial extension ('smearing' of its position) because it is quasi-monochromatic; and this means that the filter will necessarily 'smear' the position.

Altogether I really do not at all like the now fashionable [*modische*] 'positivistic' tendency of clinging to what is observable. I regard it as trivial that one cannot, in the range of atomic magnitudes, make predictions with any desired degree of precision, and I think (like you, by the way) that theory cannot be fabricated out of the results of observation, but that it can only be invented.

I have no copies here of the paper which I wrote jointly with Mr. Rosen and Mr. Podolski, but I can tell you briefly what it is all about.

The question may be asked whether, from the point of view of today's quantum theory, the statistical character of our experimental findings is *merely the result of interfering with a system from without, which comprises measuring it*, while the systems in themselves—described by a ψ-function—behave in a deterministic fashion. Heisenberg flirts [*liebäugelt*] with this interpretation, without adopting it consistently. But one can also put the question thus: should we regard the ψ-function whose time-dependent changes are, according to Schrödinger's equation, deterministic, as a *complete* description of physical reality, and should we therefore regard the (insufficiently known) interference with the system from without as alone responsible for the fact that our predictions have a merely statistical character?

The answer at which we arrive is that the ψ-function should not be regarded as a complete description of the physical state of a system.

We consider a composite system, consisting of the partial systems A and B which interact for a short time only.

We assume that we know the ψ-function of the composite

system *before* the interaction—a collision of two free particles, for example—has taken place. Then Schrödinger's equation will give us the ψ-function of the composite system *after* the interaction.

Assume that now (after the interaction) an optimal [*vollständige*] measurement is carried out upon the partial system A, which may be done in various ways, however, depending upon the variables which one wants to measure precisely—for example, the momentum *or* the position co-ordinate. Quantum mechanics will then give us the ψ-function for the partial system B, and it will give us *various ψ-functions that differ, according to the kind of measurement which we have chosen to carry out upon A.*

Now it is unreasonable to assume that the physical state of B may depend upon some measurement carried out upon a system A which by now is separated from B [so that it no longer interacts with B]; and this means that two different ψ-functions belong to one and the same physical state of B. Since a *complete* description of a physical state must necessarily be an *unambiguous* description (apart from superficialities such as units, choice of the co-ordinates etc.), it is therefore not possible to regard the ψ-function as the *complete* description of the state of the system.

An orthodox quantum theorist will say, of course, that there is no such thing as a complete description and that there can be only a statistical description of an *aggregate* of systems, rather than a description of *one single* system. But first of all, he ought to *say* so clearly; and secondly, I do not believe that we shall have to be satisfied for ever with so loose and flimsy a description of nature.

It should be noticed that some of the precise predictions which I can obtain for the system B (according to the freely chosen way of measuring A) may well be related to each other in the same way as are measurements of momentum and of position. One can therefore hardly avoid the conclusion that the system B has indeed a definite momentum and a definite position co-ordinate. For if, upon freely choosing to do so [that is, without interfering with it], I am able to predict something, then this something must exist in reality.

A [method of] description which, like the one now in use, is statistical in principle, can only be a passing phase, in my opinion.

I wish to say again* that I do not believe that you are right in your thesis that it is impossible to derive statistical conclusions from

* This is an allusion to a previous letter. K. R. P.

a deterministic theory. Only think of classical statistical mechanics (gas theory, or the theory of Brownian movement). Example: a material point moves with constant velocity in a closed circle; I can calculate the probability of finding it at a given time within a given part of the periphery. What is essential is merely this: that I do not know the initial state, or that I do not know it precisely!

With kind regards,
Yours,
A. Einstein.

Reduced Facsimile

Old Lyme. 11.IX. 35.

Lieber Herr Popper!

Ich habe Ihre Abhandlung angesehen und stimme weitgehend überein.* Nur glaube ich nicht an die Herstellbarkeit eines „reinen Falles", der es erlauben würde, Ort und Impuls (Farbe) eines Lichtquants mit „unzulässiger" Genauigkeit zu prognostizieren. Ihr Mittel (Blende und Momentan-Verschlussklappe in Verbindung mit selektiv durchlässigem Gläsersatz) halte ich aus dem Grunde für prinzipiell unwirksam, weil ich bestimmt glaube, dass ein solches Filter „ortsverschmierend" wirkt wie etwa ein Beugungsgitter.

Meine Begründung ist folgende. Denken Sie an ein kurzes Lichtsignal (genauer Ort). Um die Wirksamkeit des Absorptionsfilters bequem zu übersehen, denke ich mir dieses rein formal in eine grosse Anzahl von quasi-monochromatischen Wellenzügen W_n zerlegt. Der Absorptionssatz wirke auf alle W_n (Farben) zerstörend bis auf W_i. Diese Wellengruppe hat aber eine erhebliche Ausdehnung, weil sie quasi-monochromatisch ist (Ortsverschmierung); d. h. das Filter wirkt notwendig „ortsverschmierend". —

Mir gefällt das ganze modische „positivistische" Kleben am Beobachtbaren überhaupt nicht. Ich

---

* Hauptsache: Die ψ-Funktion charakterisiert eine (statistische) System-Gesamtheit, nicht ein Einzelsystem. Dies ist auch das Ergebnis der weiter unten dargelegten Betrachtung. Diese Auffassung macht es auch überflüssig, zwischen „reinen" und „nicht-reinen" Fällen besonders zu unterscheiden.

halte es für trivial, dass man auf atomistischem
Gebiete nicht beliebig genau prognostizieren kann
und denke, dass Theorie nicht aus Beobachtungs-
resultaten fabriziert sondern nur erfunden
werden kann (wie Sie übrigens auch).–

Ich habe keine Exemplare meiner mit den
Herrn Rosen und Podolski zusammen verfassten
Arbeit hier, kann Ihnen aber kurz sagen, um was
es sich handelt.

~~Was Sie schon hervorgehoben haben.~~

Man kann sich fragen, ob der statistische Charakter
unserer experimentellen Befunde gemäss der heutigen
Quantentheorie erst durch die fremden Eingriffe inklusive
Messungen veranlasst wird, während die Systeme als
solche – durch eine ψ-Funktion beschrieben sich an
sich deterministisch verhalten. Heisenberg liebäugelt
mit einer solchen Auffassung, ohne sie konsequent
zu vertreten. Man kann auch so fragen: Ist die ψ-Funktion,
die sich nach der Schrödingergleichung zeitlich deterministisch
verändert, nicht als (vollständige) Beschreibung der physikalischen Realität
aufzufassen, wobei lediglich der fremde (uns genau bekannte) Eingriff durch
Beobachtung dafür verantwortlich ist, dass die Prognosen
nur statistischen Charakter haben?

Wir kommen zu dem Ergebnis, dass die ψ-Funktion
nicht als vollständige Beschreibung des physikalischen
Zustandes eines Systems aufgefasst werden kann.

Wir betrachten ein Gesamtsystem, das aus den Teilsystemen
A und B besteht, die nur während einer beschränkten
Zeit in Wechselwirkung miteinander stehen.

Die ψ - Funktion des Gesamtsystems vor der Wechselwirkung z. B. Zusammenstoss (zweier freier Teilchen) sei bekannt. Die Schrödinger-gleichung liefert dann die ψ Funktion des Gesamtsystems nach der Wechselwirkung. (nach der Wechselwirkung)

Es werde nun am Teilsystem A eine (vollständige) Messung ausgeführt, was aber in verschiedener Weise möglich ist, je nach den Variabeln, die man (genau) misst (z. B. Impuls oder Koordinate). Die Quanten-Mechanik liefert dann die ψ Funktion für das Teilsystem B, und zwar verschieden, je nach der Wahl der Messung, die man an A ausgeführt hat.

Da es aber ungereimt ist, anzunehmen, dass der physikalische Zustand von B davon abhängig sei, was für eine Messung ich an dem von ihm getrennten System A vornehme, so heisst dies, dass zu demselben physikalischen Zustande von B zwei verschiedene ψ - Funktionen gehören. Da eine vollständige Beschreibung eines physikalischen Zustandes notwendig eine eindeutige Beschreibung (abgesehen von Äusserlichkeiten wie Einheiten, Koordinatenwahl etc.) sein muss, so kann die ψ - Funktion nicht als die vollständige Beschreibung des Zustandes aufgefasst werden.

Natürlich wird ein orthodoxer Quantentheoretiker sagen, es gebe eben keine vollständige Beschreibung bezw. nur die statistische Beschreibung einer Systemgesamtheit und nicht eines Systems. Aber erstens will er dies sagen (und zweitens glaube ich nicht, dass wir uns für die Dauer mit einer so federscheinigen Naturbeschreibung werden begnügen müssen).

Zu beachten ist, dass die (exakten) Prognosen, zu welchen ich (je nach freier Wahl der Messungsart an A) für das System B gelangen kann, sich zu einander sehr wohl wie Impulsmessung und Ortsmessung verhalten können. Man kann also nicht wohl um die Auffassung herumkommen, dass das System B thatsächlich einen bestimmten Impuls und eine bestimmte Koordinate hat. Denn was ich nach freier Wahl prophezeihen kann, das muss auch in der Wirklichkeit existieren. —

Meiner Meinung nach ist die gegenwärtige prinzipiell statistische Beschreibung nur ein Durchgangsstadium. — Ich möchte nochmals sagen, dass ich Ihre Behauptung, dass aus einer deterministischen Theorie keine statistischen Sätze gefolgert werden können, nicht für richtig halte. Denken Sie nur an die klassische statistische Mechanik (Gastheorie, Theorie der Brown'schen Bewegung). Beispiel: ein materieller Punkt läuft gleichförmig auf einer geschlossenen Kreisbahn; sie kann die Wahrscheinlichkeit rechnerisch bestimmen, ihn zu einer bestimmten Zeit auf einem bestimmten Teil der Peripherie anzutreffen. Wesentlich ist nur, dass sie den Anfangszustand nicht oder nicht genau kenne!

Freundlich grüsst Sie Ihr
A. Einstein

INDICES

compiled by J. Agassi

Index of Names

(The letter 'q' stands for 'quoted')

Index of Subjects

Italicized page numbers indicate that the reference is of special importance. A page number followed by 't' indicates a page on which a term is discussed.

ABSOLUTE, the, 111*n. See also* Uniqueness

Abstract, abstractness, 423, 425. *See also* Generalization

Acceptability, 53–54, 106–108, 123, 142, 145, 267–268, 394–395, 414–415, 418–419. *See also* Appraisal; Belief; Corroboration; Decisions, concerning acceptance of a theory

Accidental discovery, 54

Adequacy, (38–39,) 55, 418. *See also* Appraisal

Addition theorem, 228, 290

Ad hoc hypothesis, *see* Hypothesis

Aesthetics, 109, 137, 145, 429, 430

After-effect, 163*n*, 175*n*, 179, 367; freedom from, 162*nt*, 165–166, 207; in finite sequences, *section* 55, 160–162*t*, 184, 291, 292&*n*, 293&*n*; in infinite sequences, *section* 57, 165–166, 171, 186*n*; absolute, 171*t*, 173, 175, 176*t*, 177, 180, 182, 183, 185, 186, 292, 293*n*, 294; invariant to certain transformations, 172. *See also* Randomness; Selections, insensitivity to; Sequences, random

Agreement, concerning the outcome of a test, 104, 107. *See also* Decisions, concerning the outcome of tests

Aim of Science, *see* Science; Decisions; Fruitfulness

All-and-some statements, 193&*nt*

All-or-nothing principle, 366*t*

Alternative, 152*t*, 159*t*, 160, 164, 185, 187, 192; random, 163&*n*, 361; *see also* Sequence

Analytic statements, *see* Tautologies

Appraisal, of a theory's adequacy, 263*t*&*n*–264, 265–266, 268, 269, 275, 276

Approximation, 163*n*, 184, 191, 199, 252, 268, 276, 277, 364, 374. *See also* Modification.

Apriorism, 29, 30, 45*n*, 208, 254, 264, 312, 315, 316, 368*n*, 370, 371. *See also* Transcendental argument.

Argument, *see* Discussion; Criticism

Association theorem, 327, 328&*n*, 329, 351, 352

Asymmetry between verification (or confirmation) and falsification, 41–42, 70&*n*, 262, 265, 268, 313, 314, 424*n*.

Atomic statements, 35, 36, 128*t*&*nt*, 313, 378–379; relatively, 128*t*, 130, 285&*n*–286, 379, 380, 381, 405. *See also* Field of application

Atomism (metaphysical), 19, *38, 278*, 443

Authority of experience, non-existence of, 51–52&*n*, 106–108. *See also* Basic statements, uncertainty of

Auxiliary hypothesis, *see* Hypothesis

Axioms, Axiomatization, Axiomatized systems, 71–75, 79, 92, 171, 327; 'organic', 334*n*; independence of, *see* Independence, logical; interpretation of, *section* 17, 72–75, 83–84, 327. *See also* Formalization ; Probability, formal theory of

Axiomatics, 320

BASIC STATEMENTS OR TEST STATEMENTS, 35*nt*, 43, 46–48, 70*n*, 78, 84, 85, 90–93, *sections* 28&29, 100–103*t*, 105, 106, 111, 261, 262, 264, 266–268, 272, 273, 275, 314; *see also* Potential falsifier; degree of composition of, 114–115&*nt*, 127, 128&*n*, 129, 140*n*–141*n*, 285; falsifiability of, 84, *102–103*, 110–111, 424*n*; formal and material requirements for, 101–103; homotypic, 90*t*, 112, 119; permitted, 86&*n*, 113, 124, 129, 267*n*, 374*n*; prohibited, 41, 86, 89, *90*, 91, 114, 124, 258; in probability, *see* decidability; their negation are instantial statements, 86*n*, 91, 101*n*, 102, 141*n*, 252, 258&*n*, 266, 267*n*; relativity of, *section* 29, *104–105*, 111&*n*, 128&*n*; rule for accepting, 86, 87*n*, *104–105*,

473

statistical, statistical estimate of frequency, or statistical extrapolation, *section* 57, 165*n*, 167–169&*n*, 179, 181*n*, 184, 186, *192–194*, 204–207, *208*, 209, 210, 211, 247, 261, 295, 367, 385*n*, 409, 412, 415; *see also* Distribution; Equidistribution; decidability of, *see* Decidability; universal-existential, 193*t*&*n*

IDEALIZATION, its critical use 444, 447
Idempotence, 328, 350–351
Imaginary experiments, *appendix*＊ xi, *442, 443, 444*; author's, 216, 232, *section* 77, 236–246, 301–302 *appendix* vii, 303–305, 452; invalidity of, 216*n*, 232*n*, 236*n*, 239*n*, 242*n*, 299*n*, 444*n*, 452, *457–458*; replaceable by that of Einstein, Podolski, and Rosen, 236*n*, 244*n*; Bohm's, 448–450; Bohr's, 242, *appendix* v, 296–298, 445–446, 447, 454–456; Carnot's 443, 444; Einstein's, 443, 447; of Einstein and Pauli, 448, 449–450; of Einstein, Podolski, and Rosen, 221*n*, 244*n*–245*n*, 444, 445, 446, 447–448, 450, *appendix* ＊xii, *458–459*; Galileo's, 442&*n*–443; Heisenberg's, 229–230, 242, 443, 450–451, 452, 454
Implication or conditional, 62&*n*, 67&*n*, 68&*n*, 120*n*, 122, 123&*n*, 438; counterfactual, so-called, 434*t*&*nt*, 438, 440; material, 76*n*, 91*n*, 439; modal or strict, 434, 439, 440; necessary, or subjunctive, or nomic, 434*t*&*n*, 435, 436, 439, 440, 441; *see also* Necessity
Independence, logical, of an axiom or of a part of a system, 71*t*, 76&*n*, 107; of probability axioms, *see* Probability, formal; probabilistic, 157*t*, 158&*n*, 170, 173, 366, 367*t*, *368*, 369, 371, 375*n*, 396–397; *see also* irrelevance; logical and probabilistic, compared, 184, 405, 406
Indeterminism, metaphysical, 206&*n*, 212 &*n*, 216, *section* 78, 247&*n*–250
Indifference, principle of, 168*n*. *See also* Equidistribution.
Induction, 27, 33, 34, 35, *43*, 53, 87&*n*, 107, 138&*n*, 168–169, *279*&*n*, 286 appendix＊ I, 312–317, *420*, 432, 437, 439; eliminative, *279n*, *419*; problem of, *section* I, 28*t*, *42*, 63, 66, 68, 93, 94, 106, 107, 263&*n*–266, *312*, 369–370, 371*n*, 422; solved, 42, 313, 314, *418*; principle of, *28*, 29, 52, 138*n*, 253–254, 264–265, 370; *see also* Apriorism; Infinite regress; Transcendental argument; falsification of, *254*&*n*; superfluity of, 315.

Induction, mathematical, 40*n*, 157*n*, 290–291
Inductive direction, deductive movement in, quasi-induction, 41*t*, 77, *section* 85, 276–278, 314. *See also* Universality, levels of
Inductive inference, *see* Method; inductivist view of; Probability logic.
Inference, *see* Deduction; inductive and probable, *see* Method, inductivist view of; Probability logic.
Infinite regress, 29, 30, 47, 87*n*, 94, 105, 254, 264, 315, 369
Information, *see* Content
Information theory, 404
Initial conditions, 59*t*, 60&*n*, 85, 86&*n*, 100–101&*n*, 119, 127, 133, 161, 205, 206, 208, 209, 229, 230, 430, 433–436&*n*. *See also* Experimental conditions
Insensitivity, *see* Selection
Instantiation, 86*n*, *91*, 101*n*, 141*n*, 252, 258&*n*, 267*n*, 272*n*, *374nt*.
Instrumentalism, 36*n*, 59&*n*, 61*n*, 100&*n*, 373, *423t*, 425, 426. *See also* Operationalism; Pragmatism
Interference by measurement, *see* Heisenberg's formula, orthodox interpretation of; Imaginary experiment, Bohr's and Heisenberg's
Interpretation, of axioms, 73–75; of Bernoulli's theorem, *see* there; of Heisenberg's formula, *see* there; of observations in the light of theories, 59&*n*, 75,80, *106*, 107&*n*, 131, 279, 280, *412–413*, 423; *see also* Theory, and experiment; of probability statements, *see* Probability, formal, interpretations of; of quantum theory, *see* there; of science, 262, *278–281*
Intersensuality of scientific experience, 103
Intersubjectivity of scientific experience, *44*&*n*–45&*n*, 46–47, 56, 84, *87n*, 98, *102–103*, 104, *111n*
Intuition, creative, 31, 32, 76*n*
Invariance, *see* Transformations
Irrelevance, probabilistic, 158*t*&*n*, 161&*nt*. *See also* Independence, probabilistic
Iterations, *see* Blocks

JUDGEMENT, concerning facts, 99–100, 109–111
Justification, 43, 44, 109–111, 315, 369, 420, *436–438*
Justification versus objectivity, 92–100. *See also* Fruitfulness

128–135; the two measures compared, 130; increases with content, *section 35*, 119–121, 124, 140&*n*, 141, 142&*n*, 270*n*, 272&*n*, 273, 373–374, 385, 399, 400; increases with improbability, 118–119, 126, 212, 267*n*, 269, 270, 273, 385; increases with simplicity, *section 43*, 140–142, 267*n*, 270*n*, 273; increases with universality and predision, *section 36*, 121–123, 124–126, 141, 269, 273, 411*n*, 413*n*, 425

Terminology, 274*n*, 394, 418

Theory, Theoretical systems, 6, 27, 28, 32, 34, 40, 50, CHAPTER III, 59&*n*, 61&*n*, *section 16*, 71–72, 75, 77, 81, 83, 86 88, 92, 102, *106–111*, 113, 119, 124, 126, 127, 136, 276, 277, 285–286, 314, 373; *see also* Laws; Universal statements; and experiment, *section 30*, 106–108, 268, 422–426&*n*, 441; *see also* Interpretation; origin of, 31–32, 169, 315, 451

Thermodynamics, 196–197, 198*n*, 200, 203, 207–208&*n*, 209, 443, 460

'Tolerance', Carnap's principle of, 53*n*

Transcendental argument, 368*t*&*nt*, 370*n*, 383&*n*

Transformations, mathematical, invariance to, 143, 144, 404, 405. *See also* Co-ordinates

Transformations, probabilistic, *see* Probability, theory of

Translation, from realistic to formal mode of speech, 88–89

True, Truth, 61*n*, 71, 73, 74, *88–89*, 94, 109, 139, 248, 251, 256, 258, 262, 263, 264, *265*, *section 84*, 274&*n*–267&*n*, 278, 316, 317, 415, 419, 424*n*, 428, 437, 438

Truth-frequency, 256*t*&*nt*–260, 316

Truth-function, 128*n*, 286, 313

Truth-value, 275

Two-slit experiment, *appendix v*, 296&*n*–298, 454–456

UNCERTAINTY, *see* Hypothesis

Uncertainty principle, *see* Heisenberg's formula

Uniformity of nature, principle of, 90*n*, 252–253, 369, 437&*n*, 438. *See also* Metaphysical faith in regularities

Uniqueness, 46

Universal statements, 27, 28, 37, 41, 45, 62, 70*n*, 90*n*, 209, 258, 286, 373–377; *see also* Laws; Names, universal; strict or non-existence, 62*t*–63, 67*n*, 101&*n*, 102, 194, 426, 431, 438; strict versus numerical, *section 13*, 62–64, existential, 193; zero probability of, *see* there

Universality, levels of, 47, 75, *section 18*, 75–77, 84, *section 36*, 121–123*t*&*nt*, 124, 268, 269, 271, 273, 276, 277, *278*, 425, 438

Universality, accidental and necessary, 427–428, *432–438*

Universals, problem of, 66, 68, 74, 94–95, 441. *See also* Names, universal

Use or usage of words, 15, 65, 66, 68, 84&*n*, 276

VALIDITY, Bolzano's concept of, 119*n*

Value of judgements concerning science, necessary, 38, 50, 55. *See also* Decisions

Verdict, 108–110

Verification, or confirmation in the sense of weak verification, 33, 37*n*, 38, 50, 53, 80, 252*n*, 258&*n*, 261–262, 267*n*, 272, 278, 312, 316; of a basic statement, impossible, 91, 93–94, 423–424&*n*; of an existential statement, possible, 33, 69–70&*n*, 90; of probability statements, impossible, 191&*n*, 192, 193&*n*, 196&*n*; of universal statements, impossible in one way and too easy in another, 40, 41, 42, 63, 70, 78*n*, 91, 101*n*, 107*n*, 149*n*, 169, 206, 232, *section 79*, 252–254, 257, 264– 265, 315, 371, 422–425; *see also* Instantiation; Zero probability

Vienna circle, 51*n*, 59*n*, 251*n*, 265*n*, 311, 312

WAVE-PACKET, 222*t*, 224, 234, 301; reduction of, 235&*nt*–236, 445*n*, 448

ZERO PROBABILITY, of a universal statement, 40*n*, 257, *appendix *vii*, 363–373, 375*n*, 382, 383, 388, 405, 412

Zero probability of the second argument, 330–331, 335, 357*n*, 388, 389

CPSIA information can be obtained
at www.ICGtesting.com
Printed in the USA
LVHW091535110820
662922LV00014B/1787

9 781614 277439